SCIENCE AND TECHNOLOGY OF AROMA, FLAVOR, AND FRAGRANCE IN RICE

SCIENCE AND TECHNOLOGY OF AROMA, FLAVOR, AND FRAGRANCE IN RICE

Edited By

Deepak Kumar Verma
Prem Prakash Srivastav, PhD

APPLE ACADEMIC PRESS

Apple Academic Press Inc.
3333 Mistwell Crescent
Oakville, ON L6L 0A2
Canada

Apple Academic Press Inc.
9 Spinnaker Way
Waretown, NJ 08758
USA

Exclusive worldwide distribution by CRC Press, a member of Taylor & Francis Group
No claim to original U.S. Government works

ISBN 13: 978-1-77463-534-6 (pbk)
ISBN 13: 978-1-77188-660-4 (hbk)

Library and Archives Canada Cataloguing in Publication

Science and technology of aroma, flavor, and fragrance in rice / edited by Deepak Kumar Verma, Prem Prakash Srivastav, PhD.

Includes bibliographical references and index.
Issued in print and electronic formats.
ISBN 978-1-77188-660-4 (hardcover).--ISBN 978-0-203-71145-3 (PDF)

1. Rice--Sensory evaluation. I. Verma, Deepak Kumar, 1986-, editor
II. Srivastav, Prem Prakash, editor

TX558.R5S35 2018	664'.725	C2018-903436-X	C2018-903437-8

CIP data on file with US Library of Congress

Apple Academic Press also publishes its books in a variety of electronic formats. Some content that appears in print may not be available in electronic format. For information about Apple Academic Press products, visit our website at **www.appleacademicpress.com** and the CRC Press website at **www.crcpress.com**

CONTENTS

ABOUT THE EDITORS

Deepak Kumar Verma
Research Scholar, Agricultural and Food Engineering Department, Indian Institute of Technology Kharagpur (WB), India

Deepak Kumar Verma is an agricultural science professional and is currently a PhD Research Scholar in the specialization of food processing engineering in the Agricultural and Food Engineering Department, Indian Institute of Technology, Kharagpur (WB), India. In 2012, he received a DST-INSPIRE Fellowship for PhD study by the Department of Science & Technology (DST), Ministry of Science and Technology, Government of India. Mr. Verma is currently working on the research project "Isolation and Characterization of Aroma Volatile and Flavoring Compounds from Aromatic and Non-Aromatic Rice Cultivars of India." His previous research work included "Physico-Chemical and Cooking Characteristics of Azad Basmati (CSAR 839-3): A Newly Evolved Variety of Basmati Rice (*Oryza sativa* L.)". He earned his BSc degree in agricultural science from the Faculty of Agriculture at Gorakhpur University, Gorakhpur, and his MSc (Agriculture) in Agricultural Biochemistry in 2011. He also received an award from the Department of Agricultural Biochemistry, Chandra Shekhar Azad University of Agricultural and Technology, Kanpur, India. Apart from his area of specialization in plant biochemistry, he has also built a sound background in plant physiology, microbiology, plant pathology, genetics and plant breeding, plant biotechnology and genetic engineering, seed science and technology, food science and technology etc. In addition, he is a member of different professional bodies, and his activities and accomplishments include conferences, seminars, workshops, training, and also the publication of research articles, books, and book chapters.

Prem Prakash Srivastav, PhD

Associate Professor of Food Science and Technology, Agricultural and Food Engineering Department, Indian Institute of Technology, Kharagpur (WB), India

Prem Prakash Srivastav, PhD, is Associate Professor of Food Science and Technology in the Agricultural and Food Engineering Department at the Indian Institute of Technology, Kharagpur (WB), India, where he teaches various undergraduate-, postgraduate-, and PhD-level courses and guides research projects. His research interests include the development of specially designed convenience, functional, and therapeutic foods; the extraction of nutraceuticals; and the development of various low-cost food processing machineries. He has organized many sponsored short-term courses and completed sponsored research projects and consultancies. He has published various research papers in peer-reviewed international and national journals and proceedings and many technical bulletins and monographs as well. Other publications include books and book chapters along with many patents. He has attended, chaired, and presented various papers at international and national conferences and delivered many invited lectures at various summer/winter schools. Dr. Srivastav has received several best poster paper awards for his presentations. He is a member of various professional bodies, including the International Society for Technology in Education (ISTE), the Association of Food Scientists and Technologists (India), the Indian Dairy Association (IDA), the Association of Microbiologists of India (AMI), and the American Society of Agricultural and Biological Engineers, and the Institute of Food Technologists (USA).

LIST OF CONTRIBUTORS

Paramita Bhattacharjee
Reader, Department of Food Technology & Biochemical Engineering Jadavpur University, Kolkata–700 032, India, Mobile: +91-9874704488, E-mail: pb@ftbe.jdvu.ac.in, yellowdaffodils07@gmail.com

Sudhanshi Billoria
Research Scholar, Agricultural and Food Engineering Department, Indian Institute of Technology, Kharagpur–721302, West Bengal, India, Mobile: +91-8768126479, E-mail: sudharihant@gmail.com

Jyoti P. Dhakane
Research Scholar, Indian Agricultural Research Institute, Pusa campus, New Delhi–110012, India, Mobile: +00-91-8745000441, E-mail: jyotip.dhakane@gmail.com

Arpit Gaur
Research Scholar, Division of Genetics and Plant Breeding, Sher-e-Kashmir University of Agricultural Sciences and Technology of Kashmir, Shalimar, Srinagar–190025, Jammu and Kashmir, India, E-mail: arpitgaur@skuastkashmir.ac.in

Vidya Hinge
Women Scientist (WOS-A), Department of Botany, Savitribai Phule Pune University, Pune – 411007, India, Mobile: +00-91-7588611028, Tel.: +91-20-25601439, Fax: +91-20-25601439, E-mail: vidyahinge17@gmail.com

Kiran Khandagale
Research Scholar, Department of Botany, Savitribai Phule Pune University, Pune–411 007, India, Mobile: +00-91-9860898859, Tel.: +91-20-25601439, Fax: +91-20-25601439, E-mail: kirankhandagale253@gmail.com

Subramanian Radhesh Krishnan
Research Scholar, Department of Biotechnology, Alagappa University, Karaikudi 630003, India, Tel.: +91-4565 225215, Mobile: +91-9566422094, Fax: +91-4565 225202, E-mail: radheshkrishnan.s@gmail.com

Dipendra Kumar Mahato
Senior Research Fellow, Indian Agricultural Research Institute, Pusa campus, New Delhi–110012, India, Mobile: +00-91-9911891494, +00-91-9958921936, E-mail: kumar.dipendra2@gmail.com

Chakravarthi Mohan
Postdoctoral Fellow, Molecular Biology Laboratory (LBM), Department of Genetics and Evolution (DGE), Federal University of Sao Carlos (UFSCar), SP, Brazil – 13565905, Tel. +551633518378, Mobile: +5516996196465, E-mail: chakra3558@gmail.com

Pandiyan Muthuramalingam
Research Scholar, Department of Biotechnology, Alagappa University, Karaikudi 630 003, India, Tel.: +91-4565 225215, Mobile: +91-9597771342, Fax: +91-4565 225202, E-mail: pandianmuthuramalingam@gmail.com

Altafhusain Nadaf
Associate Professor, Department of Botany, Savitribai Phule Pune University, Pune – 411007, India,
Mobile: +00-91-7588269987, Tel.: +91-20-25601439, Fax: +91-20-25601439,
E-mail: abnadaf@unipune.ac.in

Manikandan Ramesh
Associate Professor, Department of Biotechnology, Alagappa University, Karaikudi 630 003, India,
Tel.: +91-4565 225215, Mobile: +91-9442318200, Fax: +91-4565 225202,
E-mail: mrbiotech.alu@gmail.com

Deo Rashmi
Research Scholar, Department of Botany, Savitribai Phule Pune University, Pune – 411007, India,
Mobile: +00-91-9960844746, Tel.: +91-20-25601439, Fax: +91-20-25601439,
E-mail: deo.rashmi9@gmail.com

Saroj Kumar Sah
Research Scholar, Department of Biochemistry, Molecular Biology, Entomology and Plant
Pathology, Mississippi State University, MS 39762, USA, E-mail: saroj1021@gmail.com

Asif B. Shikari
Associate Professor-cum-Scientist, Division of Genetics and Plant Breeding, Sher-e-Kashmir
University of Agricultural Sciences and Technology of Kashmir, Shalimar, Srinagar–190025,
Jammu and Kashmir, India, E-mail: asifshikari@gmail.com

Bhagavathi Sundaram Sivamaruthi
Postdoctoral Fellow, Faculty of Pharmacy, Department of Pharmaceutical sciences,
Chiang Mai University, Chiang Mai, Thailand, Tel. +66-5394-4341; Fax: +66-5389-4163,
Mobile: +66-61-3216651, E-mail: sivasgene@gmail.com

Prem Prakash Srivastav
Associate Professor, Agricultural and Food Engineering Department, Indian Institute of Technology,
Kharagpur–721302, West Bengal, India, Mobile: +91-9434043426, Tel.: +91-3222–283134,
Fax: +91-3222–282224, E-mail: pps@agfe.iitkgp.ernet.in

Deepak Kumar Verma
Research Scholar, Agricultural and Food Engineering Department, Indian Institute of Technology,
Kharagpur–721302, West Bengal, India, Mobile: +91-7407170260, +91-9335993005,
Tel.: +91-3222-281673, Fax: +91-3222-282224, E-mail: deepak.verma@agfe.iitkgp.ernet.in;
rajadkv@rediffmail.com

Shabir H. Wani
Assistant Professor-cum-Scientist, Division of Genetics and Plant Breeding, Sher-e-Kashmir
University of Agricultural Sciences and Technology of Kashmir, Shalimar, Srinagar–190025,
Jammu and Kashmir, India, E-mail: shabirhussainwani@gmail.com

Rahul Zanan
Young Scientist (SERB), Department of Botany, Savitribai Phule Pune University, Pune – 411007,
India, Mobile: +00-91-9689934273, Tel.: +91-20-25601439, Fax: +91-20-25601439,
E-mail: rahulzanan@gmail.com

LIST OF ABBREVIATIONS

2-AP	2-acetyl-1-pyrroline
2-MF	2-methyl-3-furanthiol
2,6-DMP	2,6-dimethylpyridine
AACs	aroma active compounds
AATs	alcohol acyltransferases
AB-ald	4-aminobutyraldehyde
AC	amylose content
ADH	alcohol dehydrogenase
AEDA	aroma extract dilution analyses
AMADH	amino aldehyde dehydrogenase
AOAC	Association of Analytical Communities
ATV	aroma threshold value
BAC	bacterial artificial chromosome
BADH	betaine aldehyde dehydrogenase
BPRs	back pressure regulators
BRRI	Bangladesh Rice Research Institute
CAR/PDMS	carboxen/polydimethylsiloxane
CCDs	carotenoid cleavage dioxygenases
CCF	cooled coated fiber device
cDNA	complementary DNA
CF–SPME	cold-fiber solid phase micro extraction
CO_2	carbon dioxide
CRISPR	clustered regularly interspaced short palindromic repeats
CW/DVB	carbowax/divinylbenzene
CW/TPR	carbowax/templated resin
DI-SPME	direct immersion SPME
DMAPP	dimethylallyl diphosphate
DNA	deoxyribonucleic acid
DOM	degree of milling
DVB/CAR/PDMS	divinylbenzene/carboxen/polydimethylsiloxane

EHH	extended haplotype homozygocity
EST	expressed sequence tag
FAO	Food and Agriculture Organization
FBS	fiber bakeout station
FD	factor flavor dilution factor
Fgr	fragrance
fgr	gene fragrant related gene
FT	flavor threshold
GABA	γ-amino butyric acid
GABald	$\Delta 1$-pyrroline/γ-aminobutyraldehyde
GC	gel consistency
GC-FID	gas chromatography–flame ionization detector
GC-MS	gas chromatography–mass spectrometry
GC-O	gas chromatography–olfactometry
GC-PCI-IT-MS	gas chromatography–positive chemical ionization-ion trap-tandem mass spectrometry
GE	genetic engineering
GMF	genetically modified food
GT	gelatinization temperature
Hairpin RNA	hairpin ribonucleic acid
HMMF	4-hydroxy-2-methylene-5-methyl-3(2H)-furanone
Hp-RNA	Hairpin loop RNA
HPLC	high performance liquid chromatography
HPLC-MS	high performance liquid chromatography–mass spectrometry
HS	headspace
HS-SPME	head pace solid phase micro-extraction
ICABGRRD	Indonesian Centre for Agricultural Biotechnology and Genetic Resources Research and Development
ICRR	Indonesian Centre for Rice Research
In-SPME	in-needle SPME
IPP	isopentenyl diphosphate
ISFET	ion-selective field effect transistor
IT-SPME	*in-tube* SPME
IUPAC	International Union of Pure and Applied Chemistry
KDML	Khao Dawk Mali

KEaC	kernel elongation after cooking
KEbC	kernel elongation before cooking
KEGG	Kyoto Encyclopedia of Genes and Genomes
KO	KEGG orthology
KOH	potassium hydroxide
LC	liquid chromatography
LC-MS	liquid chromatography–mass spectrometry
LOD	limit of detection
LOQ	limit of quantification
LOX	lipoxygenase
MAS	marker assisted selection
MDC	minimum detectable concentration
MEP	methylerythritol phosphate
MEPS	micro extraction by packed syringe
MIPs	molecular imprinted polymers
MPE	membrane protection extraction
MS	mass spectrometry
MVA	mevalonic acid
NAD	nicotinamide adenine dinucleotide
NADH	nicotinamide adenine dinucleotide dehydrogenase
NGS	next generation sequencing
NMD	nonsense-mediated decay
NT device	needle trap device
NTCD	Nang Thom Cho Dao
OAC	odor active compounds
OAV	odor active value
OI	odor intensity
Os2-AP	oryza sativa 2-acetyl-1- pyrroline
OTV	odor threshold value
P&T	purge and trap
P5CS	pyrroline-5 carboxylate synthetase
PA	polyacrylate
PAR	phenylacetaldehyde reductases
PCA	principle component analysis
PDMS	polydimethylsiloxane
PDMS/DVB	polydimethylsiloxane/divinylbenzene

PID controllers	proportional–integral–derivative controllers
ppb	parts-per-billion (10^{-9})
PPI	protein–protein interactions
ppm	parts-per-million (10^{-6})
QTLs	quantitative trait locus
RAM	restricted access material
RAP	rice annotation project
RAPD	random amplified polymorphic DNA
RDL	reliable detection limit
RFLP	restriction fragment length polymorphism
RGAP	rice genome annotation project
RNA	ribonucleic acid
RNAi	RNA interference
RRI	Rice Research Institute
RRII	Rice Research Institute of Iran
S/N Ratio	*signal-to-noise* ratio
SBSE	stir-bar sorptive extraction
SBSE-GC	stir-bar sorptive extraction gas chromatography
SC-CO$_2$	supercritical carbon dioxued
SCFE	supercritical fluid extraction
SCFs	supercritical fluids
SD	steam distillation
SDE	simultaneous distillation extraction
SFE	supercritical fluid extraction
SFR	stir-fried rice
SIDA	stable isotope dilution assay
SIM	selected ion monitoring mode
SIMCA	soft independent modeling of class analogy
SMM	single magnet mixer
SNIF	surface of nasal impact frequency
SNP	single nucleotide polymorphism
SPE	solid-phase extraction
SPME	solid phase micro extraction
SR	steamed rice
SSIIa	starch synthase IIa
SSR	simple sequence repeats

STRING	Search Tool for the Retrieval of Interacting Genes
STS	sequence tagged sites
TALEN	transcription activator-like effector nucleases
TFME	thin-film micro extraction
TMP	2,4,6-trimethylpyridine
TPI	triose phosphate isomerase
TPP	thiamine pyrophosphate
USA	United State of America
USD	United State Dollars
V-SDE	vacuum-simultaneous steam distillation/extraction
VOCs	volatile organic compounds
VTEs	vertical tube evaporators
WAH	weeks after heading

PREFACE

Aroma is one of the diagnostic aspects of rice quality that can determine acceptance or rejection of rice before it is tested. It is also considered as an important property of rice that indicates its preferable high quality and price in the market. An assessment of all known data reveals that more than 450 chemical compounds have been documented in various aromatic and non-aromatic rice cultivars. The primary goals were to identify the compounds responsible for the characteristics and rice aroma. Many attempts were made to search for key compounds for rice aroma, but no single compound or group of compounds could be reported that are fully responsible for rice aroma.

The first volatile aroma compound, 2-acetyl-1-pyrroline (2-AP), was reported in 1982 by Buttery et al., and again discovered by them in 1983 from aromatic rice due to its impact volatile character. 2-AP is the most important flavoring compound of cooked rice since its discovery. Rice consists of balanced complicated mixtures of volatile aroma compounds that impart its characteristic flavor. There are no single analytical techniques that can be used for investigation of volatile aroma compounds in a rice sample. Currently many technologies are available for extraction of rice volatile aroma compounds, and these technologies have been also modified from time to time according to need, and many of them are still in process and may emerge into a new form, particularly in the distillation, extraction and quantification concept.

Early research for this book, *Science and Technology of Aroma, Flavor, and Fragrance in Rice,* can be traced 30 years back. Scientific progress has been achieved in discovery, development, and application in broad areas, including: 1) Traditional and Modern Techniques for Extraction and Isolation; 2) Recent Research, Advances and Innovations in Extraction and Isolation Techniques; 3) Traditional and Advanced Modern Analytical Techniques for Identification, Characterization and Quantification of Responsible Chemical Compounds; 4) Biochemical and Molecular Biology Based Evaluation; 5) Biotechnological Developments and Genetic

Engineering in Aroma, Flavor, and Fragrance; 6) Biochemistry and Genetics Based Research, etc.

No previous book on the market throughout the world has been devoted and focused to aroma, flavor, and fragrance of rice. *Science and Technology of Aroma, Flavor, and Fragrance in Rice* is specially designed to collect all the information on the progress and developments of recent research, advances, and innovations on aroma, flavor and fragrance of rice. The broad area of research presented here is from a broad range of leading researchers. The main objective of this book is to provide a powerful collection of knowledge to help advance progress and development, and to help researchers enrich their sound knowledge on science and technology of rice aroma, flavor, and fragrance.

Finally, we wish to extend our sincere thanks to all the authors who have made contributions with dedication, persistence, and cooperation in completing their chapters in a timely manner and whose cooperation has made our task as editors a pleasure. Unquestionably, the information they have provided in this book, as well as in past publications, will contribute greatly to the advancement of flavor and fragrance research.

—Deepak Kumar Verma
Prem Prakash Srivastav

PART I

PROSPECTIVE AND TECHNOLOGICAL DEVELOPMENTS IN AROMA, FLAVOR, AND FRAGRANCE OF RICE

CHAPTER 1

INTRODUCTION TO RICE AROMA, FLAVOR, AND FRAGRANCE

DEEPAK KUMAR VERMA[1] and PREM PRAKASH SRIVASTAV[2]

[1]Research Scholar, Agricultural and Food Engineering Department, Indian Institute of Technology, Kharagpur–721302, West Bengal, India, Mobile: +91-7407170260, +91-9335993005, Tel.: +91-3222-281673, Fax: +91-3222-282224, E-mail: deepak.verma@agfe.iitkgp.ernet.in; rajadkv@rediffmail.com

[2]Associate Professor, Agricultural and Food Engineering Department, Indian Institute of Technology, Kharagpur–721302, West Bengal, India, Mobile: +91-9434043426, Tel.: +91-3222-283134, Fax: +91-3222-282224, E-mail: pps@agfe.iitkgp.ernet.in

CONTENTS

1.1 INTRODUCTION

Aroma is one of the diagnostic aspects of rice quality that determines the acceptance or rejection of rice before it is tested (Verma et al., 2012, 2013, 2015). It is also considered as an important property of rice that indicates its preferable high quality and price in the market (Paule and Powers, 1989; Ishitani and Fushimi, 1994). Early research on rice aroma can be traced 30 years back (Buttery et al., 1982, 1983), but more scientific progress in modern analytical techniques has been achieved with the discovery, development, and application of gas chromatography. Gas chromatography received tremendous advances and innovations when combined with mass spectrometry to form the technique known as gas chromatography-mass spectrometry (GC-MS). This technique made identification and quantification of volatile compounds much easier in mixtures of sample matrix and greatly enhanced interest in chemistry of rice aroma. An assessment of all known data reveals that more than 500 compounds have been documented in various aromatic and non-aromatic rice cultivars (Table 1.1) (Widjaja et al., 1996a; Verma and Srivastav, 2016). The primary goals were to identify the compounds responsible for the characteristic rice aroma (Buttery et al., 1982). Many attempts are made to date to search the key compounds for rice aroma (Buttery et al., 1982, 1983), but any single compound or group of compounds that are fully responsible for rice aroma have not yet been reported. The first volatile aroma compound, 2-acetyl-1-pyrroline (2-AP), was reported in 1982 by Buttery et al. (1982) and discovered by Buttery et al. (1983) in 1983 from aromatic rice due to its strong volatile characteristic. 2-AP is the most important flavoring compound of cooked rice since its discovery (Buttery et al., 1982, 1983). Rice consists of balanced complicated mixtures of volatile aroma compounds that impart their characteristic flavor. There are no single analytical techniques that can be used for investigation of volatile aroma compounds in rice sample. Although, currently, many technologies are available for the extraction of rice volatile aroma compounds and these technologies have been also modified time-to-time according to the need, many of them are still under process to emerge into a new form, particularly in the distillation and extraction concept.

TABLE 1.1 List of Rice Aroma Compounds Reported in Rice Cultivars (Verma and Srivastav, 2016)

Aroma Compounds	Aroma Compounds	Aroma Compounds
1. (1S, Z)-calamenene	2. 2-methyl-decane	3. Heptadecane
4. (2-aziridinylethyl) amine	5. 2-methyl-dodecane	6. Heptadecyl-cyclohexane
7. (2E)-dodecenal	8. 2-methyl-furan	9. Heptanal
10. (E)-3-Octene-2-one	11. 2-methyl-hexadecanal	12. Heptane
13. (E) 6, 10-dimethyl-5,9 undecadien-2-one	14. 2-Methyl-I-propenyl benzene	15. Heptanol
16. (E)-14-hexadecenal	17. 2-methyl-naphthalene	18. Hepten-2-ol<6-methyl-5->
19. (E)-2, (E)-4-decadienal	20. 2-methyl-tridecane	21. Heptene-dione
22. (E)-2, (E)-4-heptadienal	23. 2-methyl-undecanal	24. Heptylcyclohexane
25. (E)-2, (E)-C heptadienal	26. 2-methyl-undecanol	27. Hexacosane
28. (E)-2, (E)-C nonadienal	29. 2-n-butyl-furan	30. Hexadecanal
31. (E)-2-decenal	32. 2-nonanone	33. Hexadecane
34. (E)-2-heptanal	35. 2-nonenal	36. Hexadecane 2,6-bis-(t-butyl)-2,5
37. (E)-2-heptenal	38. 2-ocetenal (E)	39. Hexadecanoic acid
40. (E)-2-hexenal	41. 2-octanone	42. Hexadecanol
43. (E)-2-nonen-1-ol	44. 2-octenal	45. Hexadecyl ester, 2,6-difluro-3-methyl benzoic acid
46. (E)-2-nonenal	47. 2-pentadecanone	48. Glycerin
49. (E)-2-octen-1-ol	50. 2-pentylfuran	51. Hexane
52. (E)-2-octenal	53. 2-phenyl-2-methyl-aziridine	54. Hexanoic acid
55. (E)-2-undecenal	56. 2-phenylethanol	57. Hexanoic acid dodecate
58. (E)-3-nonen-2-one	59. 2-propanol	60. Hexanol
61. (E)-4-nonenale	62. 2-tetradecanone	63. Hexatriacontane
64. (E)-hept-2-enal	65. 2-tridecanone	66. Hexyl furan
67. δ-cadinol	68. 2-undecanone	69. Hexyl hexanoate
70. (E,E)-2,4-nonadienal	71. Hexylpentadecyl ester-sulphurous acid	

TABLE 1.1 (Continued)

Aroma Compounds	Aroma Compounds	Aroma Compounds
72. (*E,E*)-2,4-octadienal	73. 3-(*t*-Butyl)-phenol	74. Indane
75. (*E,E*)-farnesyl acetone	76. 3,5,23-trimethyl-tetracosane	77. Indene, 2,3-dihydro-1,1,3 trimethyl-3-phenyl
78. (*E,Z*)-2,4-decadienal	79. 3,5,24-trimethyl-tetracontane	80. Indole
81. (*S*)-2-methyl-1-dodecanol	82. 3,5-dimethyl-1-hexene	83. I-S, *cis*-calamene lonol 2
84. (*Z*)-2-octen-1-ol	85. 3,5-dimethyl-octane	86. Isobutyl hexadecyl ester oxalic acid
87. \(*Z*)-3-Hexenal	88. 3,5-di-tert-butyl-4-hydroxybenzaldehyde	89. Isobutyl nonyl ester oxalic acid
90. [(1-methylethyl)thio]cyclohexane	91. 3,5-heptadiene-2-one	92. Isobutyl salicylate
93. <6-methyl->Heptan-2-ol	94. 3,5-octadien-2-ol	95. Isolongifolene
96. <*cis*-2-*tert*-butyl->Cyclohexanol acetate	97. 3,7,11,15-tetramethyl-2-hexadecimalen-1-ol	98. Isopropylmyristate
99. 1-(1H-pyrrol-2-yl)-ethanone	100. 3,7,11-trimethyl-1-dodecanol	101. Limonene
102. 1, 2–Bis, cyclobutane	103. 3,7-dimethyl-hexanoic acid	104. Lirnonene
105. 1,1-dimethyl-2-octyl-cyclobutane	106. 3-dimethyl-2-(4-chlorphenyl)-thioacrylamide	107. l-octen-3-o1
108. 1,2,3,4-Tetramethyl benzene	109. 3-ethyl-2-methyl-heptane	110. 2-methyl-butanal
111. 1,2,3-trimethyl-benzene	112. 3-Hexanone	113. Methoxy-phenyl-oxime
114. 1,2,4-trimethyl-benzene	115. 3-hydroxy-2,4,4-trimethylpentyl *iso*-butanoate	116. Methyl (*E*)-9-octadecenoate
117. 1,2-dichloro-benzene	118. 3-methyl-1-butanol	119. Methyl (*Z,Z*)-11,14-eicosadienoate
120. 1,2-dimethyl-benzene	121. 3-methyl-2-butyraldehyde	122. Methyl (*Z,Z*)-9,12-octadecadienoate
123. 1,3,5-trimethyl-benzene	124. 3-Methyl-2-heptyl acetate	125. Methyl butanoate
126. 1,3-diethyl-benzene	127. 3-methyl-butanal	128. Methyl decanoate

TABLE 1.1 (Continued)

Aroma Compounds	Aroma Compounds	Aroma Compounds
129. 1,3-dimethoxy-benzene	130. 3-methyl-hexadecane	131. Methyl dodecanoate
132. 1,3-dimethyl-benzene	133. 3-methyl-pentadec-ane	134. Methyl ester trideca-noic acid
135. 1,3-dimethyl-naph-thalene	136. 3-methyl-pentane	137. Methyl heptanoate
138. 1,4-dimethyl-benzene	139. 3-methyl-tetradecane	140. Methyl hexadecano-ate
141. 1,4-dimethyl-naph-thalene	142. 3-nonene-2-one	143. Methyl isocyanide
144. 10-heptadecene-1-ol	145. 3-octdiene-2-one	146. Methyl isoeugenol
147. 10-pentadecene-1-ol	148. 3-octene-2-one	149. Methyl linolate
150. 17-hexadecyl-tetratri-acontane	151. 3-tridecene	152. Methyl naphthalene
153. 17-pentatricontene	154. 4,6-dimethyl-dodec-ane	155. Methyl octanoate
156. 1-chloro-3,5-bis(1,1-dimethylethyl)2-(2-propenyloxy)benzene	157. 4-[2-(methylamino)ethyl]-phenol	158. Methyl oleate
159. 1-chloro-3-methyl butane	160. 4-acetoxypentadec-ane	161. Methyl pentanoate
162. 1-chloro-nonadecane	163. 4-cyclohexyl-dodec-ane	164. Methyl tetradecano-ate
165. 1-decanol	166. 4-ethyl-3,4-dimethyl-cyclohexanone	167. Myrcene
168. 1-docosene	169. 4-ethylbenzaldehyde	170. *N,N*-dimethyl chloestan-7-amine
171. 1-dodecanol	172. 4-hydroxy-4-methyl-2-pentanone	173. *N,N*-dinonyl-2-phen-ylthio ethylamine
174. 1-ethyl-2-methyl-benzene	175. 4-methyl benzene formaldehyde	176. Naphthalene
177. 1-ethyl-4-methyl-benzene	178. 4-methyl-2-pentyne	179. *n*-decanal
180. 1-ethyl-naphthalene	181. 4-vinyl-guaiacol	182. *n*-decane
183. 1-heptanol	184. 4-vinylphenol	185. *n*-dodecanol
186. 1-hexacosene	187. 5-amino-3,6-dihydro-3-imino-1(2*H*) pyr-azine acetonitrile	188. *n*-eicosanol

TABLE 1.1 (Continued)

Aroma Compounds	Aroma Compounds	Aroma Compounds
189. 1-hexadecanol	190. 5-butyldihydro-2(3*H*)-furanone	191. *n*-heneicosane
192. 1-hexadecene	193. 5-ethyl-4-methyl-2-phenyl-1,3-dioxane	194. *n*-heptadecanol
195. 1-hexanol	196. 5-ethyl-6-methyl-(*E*)3-hepten-2-one	197. *n*-heptadecylcyclo-hexane
198. 1-methoxy-naphtha-lene	199. 5-ethyl-6-methyl-2-heptanone	200. *n*-heptanal
201. 1-methyl-2-(1-methylethyl)-benzene	202. 5-ethyldihydro-2(3*H*)-furanone	203. *n*-heptanol
204. 1-methyl-4-(1-methylethylidene)-cyclohexene	205. 5-*iso*-Propyl-5*H*-furan-2-one	206. *n*-hexadecane
207. 1-methylene-1*H*-indene	208. 5-methy-3-hepten-2-one	209. *n*-hexadecanoic acid
210. 1-nitro-hexane	211. 5-methyl-2-hexanone	212. *n*-hexanal
213. 1-nonanol	214. 5-methyl-pentadec-ane	215. *n*-hexanol
216. 1-octadecene	217. 5-methyl-tridecane	218. Nitro-ethane
219. 1-octanol	220. 5-pentyldihydro-2(3*H*)-furanone	221. *n*-nonadecane
222. 2-methyl-5-isopro-pyl-furan	223. 6,10,13-trimethyl-tetradecanol	224. *n*-nonadecanol
225. 1-pentanol	226. 6,10,14-trimethyl-2-pentadecanone	227. *n*-nonanal
228. 1-tetradecyne	229. 6,10,14-trimethyl-pentadecanone	230. *n*-nonanol
231. 1-tridecene	232. 6,10-dimethyl-2-un-decanone	
233. 1-undecanol	234. 6,10-dimethyl-5,9 undecadien-2-one	235. *n*-octanal
236. 2,2,4-trimethyl-3-carboxyisopropyl, isobutyl ester penta-noic acid	237. 6,10-dimethyl-5,9-undecandione	238. *n*-octane
239. 2,2,4-trimethyl-heptane	240. 6-dodecanone	241. n-octanol

TABLE 1.1 (Continued)

Aroma Compounds	Aroma Compounds	Aroma Compounds
242. 2,2,4-trimethyl-pentane	243. 6-methyl-2-hepta-none	244. Heptacosane
245. 2,2-dihydroxy-1-phe-nyl-ethanone	246. 6-methyl-3,5-heptadi-ene-2-one	247. *n*-octyl-cyclohexane
248. 2,3,5-trimethyl-naph-thalene	249. 6-methyl-5-ene-2-heptanone	250. Nonadecane
251. 2,3,6-trimethyl-naph-thalene	252. 6-methyl-5-hepta-none	253. Nonanal
254. 2,3,6-trimethyl-pyridine	255. 6-methyl-5-hepten-2-one	256. Nonane
257. 2,3,7-trimethyl-octanal	258. 6-methyl-octadecane	259. Nonanoic acid
260. 2,3-butandiol	261. 7,9-di-tert-butyl-1-oxaspiro-(4,5) deca-6,9-diene-2,8-dione	262. Nonene
263. 2,3-dihydrobenzo-furan	264. 7-chloro-4-hydroxy-quinoline	265. Nonnenal
266. 2,3-dihydroxy-suc-cinic acid	267. 7-methyl-2-decene	268. Nonyl-cyclohexane
269. 2,3-octanedione	270. 7-tetradecene	271. n-pentanol
272. 2,4,4-trimethylpen-tan-1,3-diol di-*iso*-butanoate	273. 9-methyl-nonadecane	274. *n*-tetradecanol
275. 2,4,6-trimethyl-decane	276. Acetic acid	277. *n*-tricosane
278. 2,4,6-trimethyl-pyridine (TMP) or Collidine	279. Acetic acid tetra-decate	280. n-undecanal
281. 2,4,7,9-tetramethyl-5-decyn-4,7-diol	282. Acetone	283. *n*-undecanol
284. 2,4-bis(1,1-dimethylethyl)-phenol	285. Acetonitrile	286. Oct-1-en-3-ol
287. 2,4-di(*tert*butyl) phenol	288. Acetophenone	289. Octacosane
290. 2,4-diene dodecanal	291. Alk-2-enals	292. Octadecane
293. 2,4-hexadienal	294. Alka-2,4-dienals	295. Octadecyne

TABLE 1.1 (Continued)

Aroma Compounds	Aroma Compounds	Aroma Compounds
296. 2,4-hexadiene aldehyde	297. Alkanals	298. Octanal
299. 2,4-nonadienal	300. Alkyl-cyclopentane	301. Octanoic acid
302. 2,4-pentadiene aldehyde	303. Anizole	304. Octanol
305. 2,4-pentandione	306. Azulene	307. Octyl formate
308. 2,5,10,14-tetramethyl-pentadecane	309. Benzaldehyde·	310. O-decyl hydroxamine
311. 2,5,10-trimethyl-pentadecane	312. Benzene	313. Oxalic acid-cyclohexyl methyl nonate
314. 2,5-dimethyl-undecane	315. Benzene acetaldehyde	316. Palmitate
317. 2,6,10,14-tetramethyl-heptadecane	318. Benzene formaldehyde	319. Pentacontonal
320. 2,6,10,14-tetramethyl-hexadecane	321. Benzothiazole	322. Pentacosane
323. 2,6,10,14-tetramethyl-pentadecane	324. Benzyl alcohol	325. Pentadecanal
326. 2,6,10-trimethyl-dodecane	327. Bezaldehyde	328. Pentadecane
329. 2,6,10-trimethyl-pentadecane	330. Bicyclo[4,2,0]octa-1,3,5-triene	331. Pentadecanoic
332. 2,6,10-trimethyl-tetradecane	333. Bis-(I-methylethyl) hexadecanoate	334. Pentadecyl-cyclohexane
335. 2,6-bis(t-*butyl*)-2,5 cyclohexadien-I-one	336. Butan-2-one-3-Me	337. Pentanal
338. 2,6-bis(t-*butyl*)-2,5-cyclohexadien-1, 4-dione	339. Butanal	340. Pentanoic acid
341. 2,6-di(*tert*-butyl)-4-methylphenole	342. Butandiol	343. Pentyl hexanoate
344. 2,6-diisopropyl-naphthalene	345. Butanoic acid	346. Phenol
347. 2,6-dimethoxy-phenol	348. Butylated hydroxy toluene	349. Phenyl acetic acid-4-tridecate
350. 2,6-dimethyl-aniline	351. Citral	352. Phenylacetaldehyde
353. 2,6-dimethyl-decane	354. Cubenol	355. Phthalic acid

TABLE 1.1 (Continued)

Aroma Compounds	Aroma Compounds	Aroma Compounds
356. 2,6-dimethyl-heptadecane	357. Cyclodecanol	358. Phytol
359. 2,6-dimethyl-naphthalene	360. Cyclosativene	361. Propane
362. 2,7-dimethyl-octanol	363. *d*-dimonene	364. Propiolonitrile
365. 2-acetyl-1-pyrroline (2-AP)	366. Decanal	367. Propyl acid
368. 2-acetyl-naphthalene	369. Decane	370. *p*-xylene
371. 2-butyl-1,2-azaborolidine	372. Decanoic acid	373. Pyridine
374. 2-butyl-1-octanol	375. Decanol	376. Pyrolo[3,2-d]pyrimidin-2,4(1*H*,3*H*)-dione
377. 2-butyl-2-octenal	378. Decyl aldehyde	379. Styrene
380. 2-butylfuran	381. Decyl benzene	382. Tetracosane
383. 2-butyl-octanol	384. Diacetyl	385. Tetradecanal
386. 2-chloro-3-methyl-1-phenyl-1-butanone	387. Dichloro benzene	388. Tetradecane
389. 2-chloroethyl hexyl ester isophthalic	390. Diethyl phthalate	391. Tetradecanoic acid
392. 2-decanone	393. Diethyl phthalate	394. Tetradecanol
395. 2-decen-1-ol	396. Di-*iso*-butyl adipate	397. Tetradec-I-ene
398. 2-decenal	399. Dimethyl disulphide	400. Tetrahydro-2,2,4,4-tetramethyl furan
401. 2-dodecanone	402. Dimethyl sulphide	403. Toluene
404. 2-ethyl-1-decanol	405. Dimethyl trisulphide	406. *trans*-2-octenal
407. 2-ethyl-1-dodecanol	408. *d*-limonene	409. *trans*-Caryophylene
410. 2-ethyl-1-hexanol	411. Docosane	412. Triacontane
413. 2-ethyl-2-hexenal	414. Dodecanal	415. Tricosane
416. 2-ethyl-decanol	417. Dodecane	418. Tridecanal
419. 2-heptadecanone	420. Dotriacontane	421. Tridecane
422. 2-heptanone	423. Eicosane	424. Trimethylheptane
425. 2-heptenal	426. Eicosanol	427. Tritetracontane
428. 2-heptene aldehyde	429. Ethanol	430. Turmerone
431. 2-heptylfurane	432. Ethenyl cyclohexane	433. Undecanal
434. 2-hexadecanol	435. Ethyl (*E*)-9-octadecenoate	436. Undecane

TABLE 1.1 (Continued)

Aroma Compounds	Aroma Compounds	Aroma Compounds
437. 2-hexanone	438. Ethyl benzene	439. Undecanol
440. 2-hexyl-1-decanol	441. Ethyl decanoate	442. Undecyl-cyclohexane
443. 2-hexyl-1-octanol	444. Ethyl dodecanoate	445. Vanillin
446. 2-hexyl-decanol	447. Ethyl hexadecanoate	448. Z-10-pentadecen-1-ol
449. 2-hexylfurane	450. Ethyl hexanoate	451. α-cadinene
452. 2-methoxy-4-vinyl-phenol	453. Ethyl linoleate	454. α-cadinol
455. 2-methoxy-phenol	456. Ethyl nonanoate	457. α-terpineol
458. 2-methyl-butanol	459. Ethyl octanoate	460. β-bisabolene
461. 2-methyl-1,3-pentanediol	462. Ethyl oleate	463. β-terpineol
464. 2-methyl-1-hexadecanol	465. Ethyl tetradecanoate	466. γ-nonalacton
467. 2-methyl-2,4-diphenylpentane	468. Ethylbenzene	469. γ-nonalactone
470. 2-methyl-3-furanthiol (2-MF)	471. Farnesol	472. γ-terpineol
473. 2-methyl-3-octanone	474. Formic acid hexate	475. δ-cadinene
476. 2-methyl-5-decanone	477. Geranyl acetone	

(*Source*: Adapted from Verma, D. K. and Srivastav, P. P. (2016). Extraction Technology for Rice Volatile Aroma Compounds. *In*: Food Engineering: Emerging Issues, Modeling, and Applications (eds. Meghwal, M., and Goyal, M. R.). Used with permission from Apple Academic Press.)

1.2 AROMA, FLAVOR, AND FRAGRANCE AT A GLANCE

The terms "aroma," "flavor," and "fragrance" are synonyms of each other and do not differ much from each other. Although difficult, an effort is made here to define the terms.

Aroma is a very complex sensation that may be defined as typically pleasant smell arising from plants, (such as leaves, stem, root, fruits, or grains), cooking of plant's parts, etc., and should be agreeable with odor, flavor, and fragrance. Very limited stimuli are available that create sensations of the taste. However, more than 500 volatile compounds (Table 1.1) have been identified (Verma and Srivastav, 2016). Each of these may potentially contribute to the perception of rice aroma, depending on their concentrations and sensory thresholds.In the case of rice, aroma is a result of a large number of compounds that are present in specific proportion in the field at the time of flowering. It also includes something pleasant smell that we feel from cooked or uncooked rice with distinctive odor, flavor, and fragrance. Efferson (1985) suggested that rice without distinctive aroma is like food without salt for those consumers who prefer rice with strong aroma.

A fragrance is defined as a pleasant or sweet smell and has no more difference with "aroma," and "flavor" because of the synonyms of each other. All the characteristics of the material that produces the sensation of taste, smell, and texture when the material is taken in the mouth are termed as flavor and are also perceived by the general pain, tactile, and temperature receptors in the mouth. Flavor is one of the most important parts of the three main sensory properties that are decisive in the selection, acceptance, and ingestion of a food. Flavor is often the main reason and considered by the consumer as an important quality attribute on the basis of which rice is accepted or rejected.

1.3 IMPORTANCE OF AROMA, FLAVOR, AND FRAGRANCE IN RICE RESEARCH

Aroma in rice grains is contributed by several volatile aroma compounds that are synthesized due to distinct biochemical pathways. More than 200 rice cultivars are recognized to contain numerous volatile aroma compounds (Nijssen et al., 1996). These volatile aroma compounds include a series of compounds like aldehydes, ketones, organic acids, alcohols, esters, hydrocarbons, phenols, pyrazines, pyridines, and other compounds. Apart from these, some other compounds are also present, but they are extracted by the breakdown of chemicals like fatty acids present in samples (Bergman et al., 2000). The volatile aroma compounds released after cooking are different from those released in the field at the time of flowering.

Aroma is considered world's 3rd highest desired trait in rice as well as in India, followed by taste and elongation after cooking (Bhattacharjee et al., 2002; Verma et al., 2012, 2013, 2015). Several types of rice aroma, flavor, and fragrance are reported by many researchers in various type of rice and rice's products depicted in Table 1.2, for instance, bland-like, bran-like, brown rice, burned-like, burnt-like, buttery-like, cold-steam-bread-like, corn-like, corn leaf-like, cracker-like, dusty-like, earthy-like, fermented sour-like, floral-like, gasoline aroma-like, grainy-like, hot-steam-bread-like, musty-like, nut-like, paint-like, pandan-like, pear-barley-like, plastic-like, popcorn-like, potato-like, rancid-like, raw dough-like, rice milk-like, smoky-like, spicy-like, sulfur-like, tortilla-like, vegetable-like, and white glue-like (Withycombe et al., 1978; Paule and Powers, 1989; Chastril, 1990; Yau and Liu, 1999). Popcorn aroma, which sometimes described

TABLE 1.2 Different Types of Aroma, Flavor, and Fragrance Reported in Rice (Cooked, Uncooked, and Its Product)

Types of Aroma, flavor and fragrance		
✓ Acetone like odor	✓ Fatty type flavor	✓ Oligosulfide odor
✓ Acidic, fruity	✓ Fatty-waxy	✓ Orange-like
✓ Aldehydic type odor	✓ Fermented type odor	✓ Paraffinic
✓ Almond-like	✓ Fishy odor	✓ Peach-like
✓ Aromatic odor	✓ Floral citrus like odor	✓ Peanut type odor
✓ Aromatic, gasoline	✓ Flower-like	✓ Peppermint-like odor
✓ Balsamic odor	✓ Foul odor of rotten eggs	✓ Phenolic-like
✓ Banana-like	✓ Fresh sweet-like	✓ Popcorn-like
✓ Beany-like	✓ Fresh woody-like	✓ Pungent, aldehyde odor
✓ Berry; blossom	✓ Fruity and floral-like	✓ Pungent, suffocating-like odor
✓ Bitter-like	✓ Fruity-like	✓ Pungent-like
✓ Bland type odor	✓ Fungal type odor	✓ Rancid-like
✓ Burnt-like	✓ Fusel oil-like	✓ Raspberry-like
✓ Buttery flavor	✓ Gasoline type odor	✓ Raw mushroom-like
✓ Buttery, creamy	✓ Grape-like	✓ Repulsive pungent odor
✓ Cabbage type	✓ Grassy alcoholic	✓ Roasted nuts-like
✓ Camphor-like odor	✓ Green and beany-like	✓ Roasted-like
✓ Caramel-like	✓ Green and grass-like	✓ Rose-like
✓ Cheesy type	✓ Green and slightly fruity-like	✓ Rotten cabbage or eggs

TABLE 1.2 (Continued)

Types of Aroma, flavor and fragrance		
✓ Cherry-like scent	✓ Green leaf like	✓ Rubbery, pungent, acid type
✓ Chicken-like	✓ Green tomato-like	✓ Seasoning-like
✓ Chocolate type	✓ Green waxy oily fatty	✓ Slight aromatic
✓ Citrus-like	✓ Green-like	✓ Slight phenolic
✓ Citrus-woody profile	✓ Hay-like	✓ Smoky-like
✓ Clove, phenolic type	✓ Herbaceous-like	✓ Sour type
✓ Cocoa type	✓ Herbal-like	✓ Soapy-like
✓ Coconut-like	✓ Honey-like	✓ Spicy and clove-like
✓ Coffee type	✓ Hydrocarbon-like odor	✓ Spicy-like
✓ Cooked jasmine rice	✓ Indole-like	✓ Sulfury-like
✓ Cooked meat	✓ Lemon-like odor	✓ Sweet (dilute)-like
✓ Cooked potato	✓ Licorice type odor	✓ Sweet and cheesy-like
✓ Cooked rice-like	✓ Magnolia-like	✓ Sweet and coconut-like
✓ Cooling, minty	✓ Meaty type	✓ Sweet and floral-like
✓ Cucumber-like	✓ Medicine-like	✓ Sweet and pleasant-like
✓ Dairy type odor	✓ Metallic	✓ Sweet-like
✓ Detergent-like odor	✓ Mild alcohol-like	✓ Tallow-like
✓ Distinct odor	✓ Mild chlorine odor	✓ Terpenic type odor
✓ Earthy type odor	✓ Mild hydrocarbon	✓ Toasted
✓ Estery type odor	✓ Mild pleasant odor	✓ Unpleasant and Phenolic-like
✓ Ethereal type odor	✓ Mild sweet odor	✓ Vanilla-like
✓ Ethereal, and pungent	✓ Mild, orange blossom odor	✓ Waxy-like
✓ Ethereal-like	✓ Minty type odor	
✓ Ether-like odor	✓ Moderately strong odor	
✓ Naphthalene-like	✓ Moldy aroma	
✓ Naphthyl type odor	✓ Mothball-like	
✓ Nutty and roasted-like	✓ Mushroom-like	
✓ Nutty-like	✓ Musty aroma	

as pandan (*Pandanus amaryllifolius*)-like aroma (Laksanalamai and Ilangantileke, 1993) is due to the presence of 2-AP in rice. 2-AP is the most important and prominent aroma chemical since it was used to identify in rice cultivars (Petrov et al., 1996). Yajima et al. (1979) identified α-pyrrolidone and indole. The former is found as a key odorant during the study of volatile flavor components in cooked Kaorimai (scented rice, *O. sativa japonica*). Buttery et al. (1988) identified seven compounds, viz. (*E*)-2-decenal, (*E*)-2-nonenal, (*E, E*)-2,4-decadienal, 2-AP, decanal, nonanal, and octanal with low odor thresholds from listed 64 aroma volatile chemical compounds from rice. Later, the chemical compound 2-AP was considered as the major odorant contributor of popcorn-like rice aroma (Buttery et al., 1982; Petrov et al., 1996). This odorant contributor in rice is found at different concentration range from a minimum of 1–10 ppb to a maximum of 2 ppm level (Buttery et al., 1988). Widjaja et al. (1996b) made comparative study of the products of lipid oxidation and aroma volatile compounds, viz. (*E*)-2-decenal, (*E, E*)-2, 4-nonadienal, and (*E, E*)-2,4-decadienal from fragrant and non-fragrant rice and reported that these aroma chemical compounds are responsible for aroma in waxy rice. These three aroma compounds are also reported by Grimm et al. (2000) for contributing its distinctive aroma in glutinous or waxy rice.

1.4   CLASSIFICATION OF RICE AROMA CHEMICALS

The majority of rice volatile aroma chemicals have popcorn-like smell due to 2-AP content. Aroma similar to taste and elongation after cooking of rice is considered as desirable quality for consumer acceptability (Bergman et al., 2000; Bhattacharjee et al., 2002). Generally, rice aroma may be classified according to the flavor, fragrance, and smell as well as on the basis of their chemical structure as described in Sections 1.4.1 and 1.4.2.

1.4.1   CLASSIFICATION OF RICE AROMAS BASED ON THEIR FLAVOR, FRAGRANCE, AND SMELL

Rice aromas were classified into five major groups as shown in Table 1.3. Although any single chemical compound is not responsible for any

TABLE 1.3 Classification of Rice Volatile Aromas Based on Their Flavor, Fragrance, and Smell

S No.	Type of Rice Aroma	Responsible Chemical Compounds
1.	Green	Aldehydes, Some alcohol and Ketones, (E)-2,(E)-4-hexadienal, (E)-2-hexenal, (E)-2-octenal, (E)-3-octen-2-onea, 2-heptanone, 2-methyl-2-pentanol, Benzaldehyde, Decanal, Geranyl acetone, Methyl heptanoate and n-hexanal
2.	Fruity/Floral	(E)-2,(E)-4-hexadienal, (E)-3-octen-2-onea, 2-hexanone, 2-nonanone, 2-undecanone, 6-methyl-5-hepten-2-one, Heptanone, Ketones, Methyl heptanoate, n-heptanal, n-nonanal and n-octanol
3.	Roasty	2, 3-octanedione
4.	Nutty	(E,E)-2,4-nonadienal, 2-pentyfuran, 4-vinyl guaiacol and Benzaldehyde
5.	Bitter	Benzaldehyde and Pyridine

(Source: Adapted from Verma, D. K. and Srivastav, P. P. (2016). Extraction Technology for Rice Volatile Aroma Compounds In: Food Engineering: Emerging Issues, Modeling, and Applications (eds. Meghwal, M. and Goyal, M. R.). Used with permission from Apple Academic Press.)

specific aroma in cooked rice, a mixture of compounds in fixed proportion is responsible for specific aromas in cooked rice. This is shown very clearly in Table 1.3. For instance, fresh green or woody smell was contributed mainly by aldehydes, ketones, and some alcohols; fruity and floral smell comes due to the presence of heptanone, ketones, and 6-methyl-5-hepten-2-one, while benzaldehyde and 2-pentyfuran provided nutty aromas (Widjaja et al., 1996a).

1.4.2 BASED ON THEIR CHEMICAL STRUCTURE

Rice aroma chemicals can be classified according to their chemical structures in their families (Table 1.4). These alternative classifications could be established as a function of the chemical family of the precursor used for their production by chemical conversion.

TABLE 1.4 Classification of Rice Volatile Aromas Based on Their Chemical Structure

S No.	Family of Aroma	Aroma Chemicals
1.	Alcohols	(2Z)-2-Octen-1-ol; (3Z)-3-Hexen-1-ol; (E)-2-Nonen-1-ol; (E)-2-hexenol; (E)-2-Octen-1-ol; (E)-2-Pentenol; (S)-2-Methyl-1-dodecanol; 1,2-Propanediol; 10-Heptadec en-1-ol; 10-Pentadecene-1-ol; 1-Butanol; 1-Heptanol; 1-Hepten-3-ol; 1-Hexadecanol; 1-Hexanol; 1-Hexanol, 2-ethyl; 1-Nonanol; 1-Octanol; 1-Octen-3-ol; 1-Pentanol; 1-Propanol; 2-(2-Propoxyethoxy)ethanol; 2,3-Butanediol; 2,4-Hexadien-1-ol; 2-Butanol; 2-Butoxyethanol; 2-Butyl-1-octanol; 2-Decen-1-ol; 2-Ethyl-1-decanol; 2-Ethyl-3-buten-1-ol; 2-Ethyl-4-methyl-1-pentanol; 2-Ethylhexanol; 2-Heptanol; 2-Hexadecanol; 2-Hexyl-1-decanol; 2-Hexyl-1-octanol; 2-Methyl-1-hexadecanol; 2-Methyl-undecanol; 2-Penten-1-ol; 2-Propanol; 3,4-Dimethyl-1-cyclo-hexanol; 3,7,11,15-Tetramethyl-2-hexadecimalen-1-ol; 3,7,11-Trimethyl-1-Dodecanol; 3,7,11-Trimethyl-1-octanol; 3-Methyl-1-butanol; 6,10,13-Trimethyl-tetradecanol; 7-Octen-4-ol; Anizole; Carveol; cis-2-Tert-butyl-cyclohexanolacetate; cis-Linalool oxide; Cubenol; Cyclodecanol; Dodecanol; Ethanol; Hexanol; Isobutanal; Isopentanol; Linalool; L-α-Terpineol; Menthol; Methanol; n-Dodecanol; n-Eicosanol; n-Heptadecanol; n-Heptanol; n-Hexadecane; n-Hexanol; n-Nonadecanol; n-Nonanol; n-Octanol; Pentacosene; Pentanol; Phytol; Tetradecanol; trans-Linalool oxide; Undecanol; Z-10-pentadecen-1-ol; α-Terpineol; δ-Cadinol
2.	Aldehydes	(4E)-4-Nonenal; (E)-2-Decenal; (E)-2-Hexenal; (E)-2-Nonenal; (E)-2-Octenal; (E)-2-Undecenal; (E,E)-2,4-Decadienal; (E,E)-2,4-Nonadienal; (E,Z)-2,6-Nonadienal; (E,Z)-Deca-2,4-dienal; (Z)-Dec-2-enal; (Z)-Hex-3-enal; (Z)-Non-2-enal; (Z,E)-2,4-Heptadienal; 2,3,7-Trimethyl-octanal; 2,4-Diene dodecanal; 2,4-Heptadienal; 2,4-Hexadienal; 2,4-Hexadiene aldehyde; 2,4-Octadienal; 2,4-Pentadiene aldehyde; 2-Butyl-1-octenal; 2-Butyl-2-octenal; 2-Dodecenal; 2-Ethyl-2-hexenal; 2-Ethylbenzaldehyde; 4-Ethylbenzaldehyde; 2-Ethylhexanal; 2-Heptenal; 2-Heptene aldehyde; 2-Hexenal; 2-Hydroxybenzaldehyde; 2-Methyl-2-pentenal; 2-Methylbutanal; 2-Methyl-hexadecanal; 2-Methylpentanal; 2-Methylpropanal; 2-Ocetenal (E); 2-Octenal; 2-Pentenal; 2-Undecenal; 3,5-Di-tert-butyl-4-hydroxybenzaldehyde; 3-Methyl-2-butyraldehyde; 4,5-Epoxy-(E)-dec-2-enal; 4-Methyl benzene formaldehyde; Benzaldehyde; Benzene acetaldehyde; Benzene formaldehyde; Butanal; Cinnamaldehyde; Citral; Decanal; Decyl aldehyde; Dodecanal; Dodecenal; Ethanal; Formaldehyde; Heptanal; Hexadecanal; Hexanal; Isovaleraldehyde; m-Tolualdehyde; Myrtenal; Nonanal; Nonenal; Octanal; para-Methyl-benzaldehyde; Pentacontonal; Pentadecanal; Pentanal; Phenylacetaldehyde; p-Methylbenzaldehyde; Propanal; Safranal; Tetradecanal; Tridecanal; Undecanal; Undecane; α-Hexylcinnamic aldehyde; β-Cyclocitral; β-Homocyclocitral

S No.	Family of Aroma	Aroma Chemicals
3.	Ketones	(3E)-3-Octen-2-one; (3E)-6-Methyl-3,5-heptadien-2-one; (E)-3-Nonen-2-one; 1-(1H-pyrrol-2-yl)-ethanone; 1-(2-Methylphenyl)-ethanone; 1-Hydroxy-2-propanone; 1-Octen-3-one; 2,2,6-Trimethylcyclohexanone; 2,2-Dihydroxy-1-phenylethanone; 2,3-Octanedione; 2,6,6-Trimethyl-2-cyclohexene-1,4-dione; 2,6-Bis(1,1-dimethylethyl)-2,5-cyclohexadiene-1,4-dione; 2-Butanone; 2-Decanone; 2-Dodecanone; 2-Heptadeca-none; 2-Heptanone; 2-Hexanone; 2-Methyl-3-octanone; 2-Methyl-5-decanone; 2-Nonanone; 2-Octanone; 2-Pentadecanone; 2-Tetradecanone; 2-Tridécanone ; 2-Undecanone; 3,5,5-Trimethyl-2-cyclopenten-1-one; 3,5-Heptadiene-2-one; 3,5-Octadien-2-one; 3-Hydroxy-2-butanone; 3-Methyl-3-buten-2-one; 3-Nonanone; 3-Nonen-2-one; 3-Octanone; 3-Octene-2-one; 3-Penten-2-one; 4-Cyclopentylidene-2-butanone; 4-Hydroxy-4-methyl-2-pentanone; 5-Ethyl-6-methyl-2-heptanone; 5-Ethyl-6-methyl-3-hepten-2-one; 5-Methy-3-hepten-2-one; 5-Methyl-2-hexanone; 6,10,14-Trimethyl pentadecanone; 6,10,14-Trimethyl-2-pentadecanone; 6,10-Di-methyl-2-undecanone; 6,10-Dimethyl-5,9-undecadien-2-one; 6,10-Dimethyl-5,9-undecandione; 6-Dodecanone; 6-Methyl-2-heptanone; 6-Methyl-3,5-heptadiene-2-one; 6-Methyl-5-ene-2-heptanone; 6-Methyl-5-heptanone; 6-Methyl-5-hepten-2-one; 7,9-Di-tert-butyl-1-oxaspiro-(4,5)deca-6,9-diene,2,8-dione; Acetone; β-Damascenone; Butan-2,3-dione; Carvone; Cyclohexanone; Diphenylketone; Farnesyl acetone; Geranyl acetone; Heptene-dione; Isophorone; Methylheptenone; Neryl acetone; Non-1-en-3-one; Nonadecanone; Nonalactone; Pentan-2-one; p-Menthan-3-one; Undecalactone; α/β-Ionone; β-Damascone; γ-Octalactone
4.	Esters and Fatty acids	**Esters:** (Z)-9-Octadecenoic acid ethyl ester; 2,4,4-Trimethylpentan-1,3-diodli-iso-butanoate; 2-Hydroxy, 2-methyl-benzoic acid; 3-Hydroxy-2,4,4-trimethylpentyliso-butanoate; Acetic acid tetradecate; Benzenebutyr-ate; Benzyl acetate; Dibutyl phthalate; Diethyl carbonate; Diethyl phthalate; Diisobutyladipate; Ethyl-benzoate; Ethyl-hexadecanoate; Ethyl-hexanoate; Ethyl-linoleate; Ethyl-oleate; Ethyl-palmitate; Formic acid hexate; Hexadecanoic acid, methyl ester; Hexanoic acid dodecate; Isobutyl-salicylate; Isopropyl-myristate; Isopropyl-dodecanoate; Methyl-2-furoate; Methyl-benzoate; Methyl-linoleate; Methyl-Oleate; Methyl-palmitate; Methyl-stearate; Methylsalicylate; Octadecanoic acid, ethyl ester; Octyl-formate; Oxalic acid-cyclohexyl methyl nonate; Phenyl acetic acid-4-tridecate

TABLE 1.4 (Continued)

S No.	Family of Aroma	Aroma Chemicals
		Fatty acids: (2E)-2-Hexadecenoic acid; 3-Methylbutanoic acid; Acetic acid; Caprylic acid; Decanoic acid; Eicosanoic acid; Hexadecanoic acid; Lauric acid; Linoleic acid; Myristic acid; Nonadecylic acid; Nonanoic acid; Oleic acid; Palmitate; Pentadecanoic acid; Stearic acid; Tetradecanoic acid; Tridecanoic acid
5.	Heterocyclic Compounds	2-Methyl-furan; 2,3-Dihydrobenzofuran; 2-Methyl-5-isopropyl-furan; 2-Pentyl-furan; Indole; Methoxy-phenyl-oxime; Methyl-naphthalene
6.	Volatile Hydrocarbons	**Alkane Hydrocarbons:** 17-Hexadecyl-tetratriacontane; 3,5,24-Trimethyltetratriacontane; 4-Acetoxy-pentadecane; 6-Methylo-ctadecane; 2-Methyl decane; 2,6,10,14-Tetramethyl-hexadecane; [(1-Methylethyl)thio]cyclohexane; 2,6,10-Trimethyl-pentadecane; 2-Methyl-decane; 2-Methylheptane; 2-Methyl-tridecane; 3,5,23-Trimethyl-tetracosane; 3,5,24-Trimethyl-tetracontane; 3-Methyl-pentadecane; 3-Methyl-tetradecane; 4-Methyldecane; 5-Methyl-pentadecane; 5-Methyl-tridecane; 6-Methyl-octadecane; 9-Methyl-nonadecane; Butyl acetate; Decane; Docosane; Dodecane; Dotriacontane; Eicosane; Ethyl acetate; Geranyl acetate; Heneicosane; Heptadecane; Hexadecane; Hexatriacontane; n-Eicosane; Nonadecane; Nonane; Octadecane; Pentacosane; Pentadecane; Tetracosane; Tetradecane; Triacontane ; Tricosane; Tridecane; Tritetracontane **Alkene Hydrocarbons:** Propane; Undecane; (E)-5-Methyl-4-decene; (Z)-3-Undecene; 17-Pentatricontene; 1-Butene; 1-Docosene; 1-Hexacosene; 1-Tetradecene; 2-Tetradecene; 3,5-Dimethyl-1-hexene; 3-Ethyl-1,5-octadiene; 3-Tridecene; 5-Octadecene; 7-Methyl-2-decene; 7-Tetradecene; Butyl cyclopropane; Dodecene; Ethylene; Nonene; Thujopsene; α-Cadinene; δ-Cadinene **Alkyne Hydrocarbons:** 1-Tetradecyne; 4-Methyl-2-pentyne; Octadecyne

S No.	Family of Aroma	Aroma Chemicals
		Aromatic Hydrocarbons: (1S,Z)-Calamenene; 1,2,3,4-Tetramethyl-benzene; 1,2,3,5-Tetramethyl-benzene; 1,2,3-Trimethyl-benzene; 1,2,4,5-Tetramethyl-benzene; 1,2,4-Trimethyl-benzene; 1,2-Dimethyl-3-ethyl-benzene; 1,3,5-Trimethyl-benzene; 1,3-Dimethoxy-benzene; 1,3-Dimethyl-4-ethylbenzene; 1,3-Dimethyl-5-ethylbenzene; 1,4-Diethyl-benzene; 1,4-Dimethyl-2-ethylbenzene; 1,4-Dimethylnaphthalene; 1-Chloro-3,5-bis(1,1-dimethylethyl)2-(2-propenyloxy) benzene; 1-Ethyl-3-methyl-benzene; 1-Ethyl-naphthalene; 1-Methoxy-naphthalene; 1-Methyl-naphthalene; 2,3,5-Trimethylnaphthalene; 2,3,6-Trimethyl-naphthalene; 2,6-Bis(1,1-dimethylethyl)-2,5-cyclohexadiene-1,4-dione; 2,6-Dimethylnaphthalene; 2-Acetyl-naphthalene; 2-Ethyl-naphthalene; 2-Methylnaphthalene; 4-Ethyltoluene; Benzene; Carene; Decyl-benzene; Ethyl-benzene; Indene; m-Diethyl-benzene; m-Xylene; Naphthalene; n-Propyl-benzene; O-Diethyl-benzene; O-Xylene; P-Diethyl-benzene; p-Xylene; Styrene; β-Bisabolene
		Non-aromatic Cyclic Hydrocarbon: 4-Cyclohexyl-dodecane; Azulene; Cyclosativene; Germacrene-D; Heptyl-cyclohexane; n-Heptadecylcyclohexane; n-Octyl-Cyclohexane
		Other Volatile Hydrocarbons: 2,6-Dimethyl-decane; 2,4,6-Trimethyl-decane; 3,5-Dimethyl-octane; 3-Ethyl-2-methyl-heptane; 2,5-Dimethyl-undecane; 2-Methyl-dodecane; 2,6,10-Trimethyl-dodecane; 2,6,10-Trimethyl-tetradecane; 3-Methyl-hexadecane; 2,6-Dimethyl-heptadecane; 2,6,10,14-Tetramethyl-heptadecane; Hexacosane; Heptacosane; Octacosane; 1,1-dimethyl-2-octyl-cyclobutane; Alkyl-cyclopentane; Nonyl-cyclohexane; Undecyl-cyclohexane; Pentadecyl-cyclohexane; Heptadecyl-cyclohexane; 1-ethyl-2-methyl-benzene
7.	Organic acids	1,2-Benzenedicarboxylic acid; 1-Hexanoic acid; 2,2,4-Trimethyl-3-carboxyisopropyl, isobutyl ester pentanoic acid; 2,3-Dihydroxy-succinic acid; 2-Ethylcaproic acid; 2-Methylbutanoic acid; 3,7-Dimethyl-hexanoic acid; Benzoic acid; Butanoic acid; Ethyl myristate; Furoic acid; Hexadecyl ester, 2,6-difluro-3-methyl benzoic acid; Hexylpentadecyl ester-sulphurous acid; Isobutyl hexadecyl ester oxalic acid; Isobutyl nonyl ester oxalic acid; n-Heptanoic acid; Pentanoic acid; Phenylacetic acid; Propyl acid; Succinic acid
8.	Chlorine-containing	1-Chloro-3-methyl butane;1-Chloro-nonadecane; 2-Chloro-3-methyl-1-phenyl-1-butanone

TABLE 1.4 (Continued)

S No.	Family of Aroma	Aroma Chemicals
9.	Nitrogen-containing	1*H*-Indole; 1-Nitrohexane; 2,3,5-Trimethylpyrazine; 2,3-dimethyl-5-ethylpyrazine; 2,3-Dimethylpyrazine; 2,5-Dimethylpyrazine; 2,6-Dimethylpyrazine; 2-Acetopyridine; 2-Acetyl-1-pyrroline; 2-Acetyl-2-thiazoline; 2-Acetylpyrrole; 2-Amino acetophenone; 2-Butyl-1,2-azaborolidine; 2-Ethyl-3,5-dimethylpyrazine; 2-Ethyl-3-methylpyrazine; 2-Ethyl-5-methylpyrazine; 2-Ethyl-6-methylpyrazine; 2-Isoamyl-6-methylpyrazine; 2-Isobutyl-3-methoxypyrazine; 2-Methoxy-3,5-dimethylpyrazine; 2-Methylpyrazine; 2-Methylpyridine; 3-Ethyl-2,5-dimethylpyrazine; 3-Methylindole; 3-Methylpyridine; 3-Vinylpyridine; 4-Methylpyridine; 5-Amino-3,6-dihydro-3-imino-1(2*H*) pyrazine acetonitrile; 7-Chloro-4-hydroxyquinoline; Benzonitril; Benzothiazole; Ethenylpyrazine; Ethylpyrazine; Isocyanato-methylbenzene; *N,N*-dimethyl chloestan-7-amine; *N,N*-dinonyl-2-phenylthioethylamine; *N*-FurfurylPyrrole; Nicotine; N-methoxymethanamine; O-Decylhydroxamine; Propiolonitrile; Pyrazine; Pyridine; Pyrolo[3,2-d]pyrimidin-2,4(1H,3H)-dione; Pyrrole; α-Pyrrolidinone; β-Quinoline
10.	Phenol-containing	2,2-Dihydroxy-1-phenyl-ethanone; 2,6-Dimethoxy-phenol; 2,6-Dimethylaniline; 2-Methylphenol; 2-Phenoxy-ethanol; 2-Phenylethanol; 4-Vinylguaicol; 4-Vinylphenol; 5-Ethyl-4-methyl-2-phenyl-1,3-dioxane; Acetophenone; Benzyl alcohol; Biphenyl; Butylatedhydroxytoluene; Guaiacol; Isoeugenol; Isocugenol; *m*-Cresol; *p*-Cresol; Phenol; Toluene; Vanillin
11.	Sulphur-containing	[(1-Methylethyl)thio] cyclohexane; 1-Propene-1-thiol; 2-Methyl-3-furanthiol; 3-Dimethyl-2-(4-chlorphenyl)-thioacrylamide; 3-Methyl-2-butene-1-thiol; Butanethiol; Dimethyl disulfide; Dimethyl sulphide; Dimethyl trisulfide; Hydrogen sulfide; Methanethiol;Methional; Prenylthiol
12.	Terpene	**Sesquiterpenes:** Camphene; *d*-limonene; Eucalyptol; L-Limonene; Menthone; *p*-Cymene; α-Phellandrene; α-Pinene **Monoterpene:** (E,E)-Farnesol; Aromadendrene; Isolongifolene; Longifolene; Turmerone; Valencen; α-Farnesene; β-Caryophyllene; β-Elemene

1.5 FACTORS AFFECTING RICE AROMA, FLAVOR, AND FRAGRANCE

Many factors such as genetic factor, stickiness/glutinousness, pre-harvesting, harvesting time, moisture content at harvesting, storage conditions, degree of milling, storage temperature and time, washing, cooking method, and serving temperature of cooked rice affect the development of aroma, flavor, and fragrance in rice at different stages of growth and development. Some of them are described as follows.

1.5.1 GENETIC FACTOR

Genetic factor is the main important source of variability for the development of aroma, flavor, and fragrance in rice through crop improvement. Lorieux et al. (1996), Jin et al. (2003), and Chen et al. (2006) have been reported that this is due to an eight-base pair deletion in exon 7 of a gene. This base pair deletion occurs at chromosome 8. Bradbury et al. (2005b) reported that the putative gene for the development of fragrance genotyping in rice is coded by a putative enzyme betaine aldehyde dehydrogenase 2 (*BAD2*). After the deletion, the encoded enzyme loses its function and, subsequently, the accumulation of 2-AP takes place in aromatic rice cultivars. In 2008, Fitzgerald et al. (2008) studied 464 samples collected from the Centre of Genetic Resource of International Rice Research Institute (IRRI) Manila, Philippines, in the search of the second fragrance gene in rice. In the report of Fitzgerald et al. (2008), the majority of rice cultivars belonged to South, and rice cultivars from Southeast Asia had 2-AP, but did not carry 8-bp deletion, whereas Champagne et al. (2008) suggested that the 8-bp deletion in the fragrance allele is also a cause of mutation that drives the accumulation of 2-AP in addition to contributing the aroma in rice cultivars.

1.5.2 STICKINESS/GLUTINOUSNESS

Stickiness/Glutinousness is an important physical characteristic in addition to other characteristics, based on which consumers prefer overall acceptability of rice grain quality (Bhonsle and Krishnan, 2010). Among the glutinous rice cultivars, quality and environmental adaptations are considered to vary widely; some of them are aromatic, from white to purple and black in color, and with diverse grain size. These glutinous rice cultivars are the staple food for the peoples of Lao PDR and northeast Thailand since ancient times (Almanac, 2012). These rice cultivars have very unique characteristics such as being opaque (somewhat translucent) when raw; most nonglutinous rice varieties (Chaudhary, 2003) have 2.6–4.8% amylose content, which is lower than that (10–30%) found in nonglutinous rice varieties (KBIRRI, 2016). Glutinous rice cultivars have high amylopectin content, which is responsible for glue-like stickiness but does not contain dietary gluten (i.e., it does not contain glutenin and gliadin), and due to this reason, such rice cultivars are considered as safe for gluten-free diets (Chaudhary, 2003).

1.5.3 PRE-HARVESTING

Quality traits of rice cultivars, like aroma, amylose and protein content, etc., are affected by cultural practices and environmental conditions. Rice samples with low protein content are found to have more aroma than those with high protein content (Juliano et al., 1965). Champagne et al. (2007) grew five diverse rice cultivars with aroma and flavor conventionally with typical use of nitrogen at 50% and 100%, and chicken litter was used as organic management for comparison. Rice samples were found to have lower protein content of 7.7% and 7.5% with organic management and 50% N rate, respectively. The result concluded that rice samples showed no difference in aroma or flavor from those with higher protein content (mean 9.2% with 100% N rate). Similarly, Terao et al. (2005) reported that in the rice cultivar Akitakomachi grown under elevated CO_2 concentration, the protein content will decrease, but it did not change the sensory properties to a level that could be detected by taste panel evaluation. The concentration of 2-AP varies with environmental conditions. Itani et al. (2004) reported higher concentration of 2-AP in brown rice ripened at low temperature (day 25°C; night 20°C) than that ripened at high temperature (35°C in day; 30°C at night).

1.5.4 HARVESTING TIME

For harvesting at the proper stage, some physiological parameters (such as maturity of crop, moisture content, and meteorological conditions) must be considered that permit farmers to foster conditions for high recovery of head rice yield. Champagne et al. (2005) studied the effects of drain and harvest dates on rice sensory and reported that a predominant variety, M-202 produced in California, was found to be the representative of stable flavor with 14-day span timing of field draining and harvesting of 32–48 days after flowering. Tamaki et al. (1989) reported that rice flavor declines with maturity. In immature rice, flavor was deliberated to be rich but was found to be poor in over-ripened rice. In an early-heading cultivar of brown rice, the concentration of 2-AP during grain development stage was increased at four or five weeks after heading (WAH) and thereafter observed decreased rapidly with 20% at the maximum of seven or eight WAH. However, it was found increase at four WAH and then steadily decreased with 40% at the maximum of eight WAH (Itani et al., 2004).

1.5.5 STORAGE TEMPERATURE AND TIME

Storage temperature and time are also very important factor that affects the aroma quality of the milled rice. Lipase is a pancreatic enzyme that catalyzes the breakdown of fats to fatty acids and glycerol or other alcohols. This pancreatic enzyme in the residual bran is present on the surface of milled rice which form free fatty acid during the period of rice storage. On hydrolysis, the surface lipids form free fatty acids that are susceptible to oxidation (Yasumatsu and Moritaka, 1964). There are a number of secondary oxidation products (viz., acids, alcohols, aldehydes, furanones, ketones, and lactones), and hydrocarbons are formed by oxidation of unsaturated fatty acids; particularly, linoleic and linolenic acids were reported for the development of off-flavors and odors (Yamamatsu et al., 1966; Grosch, 1987). Champagne et al. (2005) observed based on aroma values (AV) that hexanal (grassy flavor) and 2-pentylfuran (beany) probably contributed more to flavor change in milled rice early in storage rather than at later period. 2-Nonenal (rancid flavor) and octanal (fatty flavor) contributed more to the overall flavor of milled rice during long-term storage.

1.5.6 SOAKING

Many countries over the globe, such as Japan, Korea, and other Asian countries, pre-soak rice for >30 min as a traditional practice. This pre-soaking traditional practice helps in shortening of gelatinization time and facilitates uniform cooking. Whereas the chemical changes have altered the rice flavor and aroma in the grain by soaking.

1.5.7 COOKING METHOD

The method of cooking also affects the flavor in rice, i.e., a consumer panel found rice cooked by the Pilaf method had more acceptable flavor than the excess cooking method (Crowhurst and Creed, 2001). Possible flavor compounds were lost during draining following cooking in the excess cooking method.

1.5.8 SERVING TEMPERATURE OF COOKED RICE

Serving temperature affects the contents of certain compounds of individual cultivars differently, which affect the aroma of rice. For example, Liu et al. (1996) reported that aroma of cooked rice samples rated higher in sweetness at 18°C, whereas samples evaluated at 60°C scored higher in burnt rice and rice with earthy, moldy, rancid, and sulfur aromas.

1.6 CONCLUSION

Rice is an important food crop that sustains to the daily food requirement of the people throughout the world, including Asia, South and North America, Africa, and parts of Europe. The rice-the food of daily life is preferred on the basis of aroma, flavor, and fragrance after cooking that is also considered as an important quality trait preferred by most people. The goal of this chapter is to offer a brief overview of what is known about aroma compounds in rice, addressing and discussing especially on the importance of rice aroma, flavor, and fragrance, and what are the factors that affect the aroma quality of rice.

ACKNOWLEDGMENTS

Deepak Kumar Verma and Prem Prakash Srivastav are indebted to the Department of Science and Technology, Ministry of Science and Technology, Government of India, for an individual research fellowship (INSPIRE Fellowship Code No.: IF120725; Sanction Order No. DST/INSPIRE Fellowship/2012/686; Date: 25/02/2013).

KEYWORDS

- 2-acetyl-1-pyrroline
- aroma chemicals
- aroma compounds
- *Pandanus amaryllifolius*
- rice-rice aroma
- rice quality
- volatile aroma

REFERENCES

Aggarwal, R. K., Brar, D. S., Nandi, S., Huang, N., & Khush, G. (2002). SPhylogenetic relationships among Oryza species revealed by AFLP markers. *Theor. Appl. Genet., 98,* 1320–1328.

Ahn, S. N., Bollich, C. N., & Tanksley, S. (1992). DRFLP tagging of a gene for aroma in rice. *Theor. Appl. Genet., 84,* 825–828.

Ahuja, S. C., Panwar, D. V. S., Ahuja, U., & Gupta, K. R. (1995). *Basmati Rice – The Scented Pearl.* CCS Haryana Agricultural University, Hisar, pp. 63.

Ahuja, U., Ahuja, S. C., Thakrar, R., & Rani, S. N. (2008). Scented rices of India. *Asian Agri-Hist., 12*(4), 267–283.

Amarawathi, Y., Singh, R., Singh, A. K., Singh, V. P., Mohapatra, T., & Sharma, T. R. (2008). Mapping of quantitative trait loci for basmati quality traits in rice (*Oryza sativa* L.). *Mol. Breed., 21,* 49–65.

Anacleto, R., Cuevas, R. P., Jimenez, R., Llorente, C., Nissila, E., Henry, R., & Sreenivasulu, N. (2015). Prospects of breeding high quality rice using post-genomic tools. *Theor. Appl. Genet., 128,* 1449–1466.

Anonymous, (2003–2004). *Economic Survey*, Government of India, Ministry of Finance, pp. 16–18.

Arai, E., & Itani, T. (2000). Effects of early harvesting of grains on taste characteristics of cooked rice. *Food Sci. Technol. Res., 6*, 252–256.

Bandillo, N., Raghavan, C., Muyco, P. A., Sevilla, M. A., Lobina, I. T., Dilla-Ermita, C. J., Tung, C. W., McCouch, S., Thomson, M., Mauleon, R., Singh, R. K., Gregorio, G., Redona, E., & Leung, H. (2013). Multi-parent advanced generation inter-cross (MAGIC) populations in rice: progress and potential for genetics research and breeding. *Rice, 6*, 11.

Bhonsle, S. J., & Sellappan, K. (2010). Grain quality evaluation of traditional cultivated rice varieties of Goa (India). *Recent Research in Science and Technology, 2*(6), 88–97.

Bligh, H. F. J. (2000). Detection of adulteration of Basmati rice with non-premium long-grain rice. *Int. J. Food Sci &Tech., 35*, 257–265.

Bourgis, F., Guyot, R., Gherbi, H., Tailliez, E., Amabile, I., Salse, J., Lorieux, M., Delseny, M., & Ghesquiere, A. (2008). Characterization of the major fragrance gene from an aromatic japonica rice and analysis of its diversity in Asian cultivated rice. *Theor. Appl. Genet., 117*, 353–368.

Bradbury, L. M. T., Fitzgerald, T. L., Henry, R. J., Jin, Q., & Waters, D. L. E. (2005a). The gene for fragrance genotyping in rice. *Plant Biotech. J., 3*, 363–370.

Bradbury, L. M. T., Gillies, S. A., Brushett, D. J., Waters, D. L. E., & Henry, R. J. (2008). Inactivation of an aminoaldehyde dehydrogenase is responsible for fragrance in rice. *Plant Mol. Biol., 68*, 439–449.

Bradbury, L. M. T., Henry, R. J., Jin, Q., Reinke, F., & Waters, D. L. E. (2005b). A perfect marker for fragrance genotyping in rice. *Molecular Breed., 16*, 279–283.

Burr, B., & Burr, F. A. (1991). Recombinant inbred lines for molecular mapping in maize. *Theor. Appl. Genet., 85*, 55–60.

Buttery, R. G., Ling, L. C., Juliano, B. O., & Turnbough, J. G. (1983). Cooked rice aroma and 2-acetyl, 1-pyrrolline. *J. Agri. Food Chemistry, 31*, 823–826.

Champagne, E. T., Bett, K. L., Vinyard, B. T., Webb, B. D., McClung, A. M., Barton, II, F. E., Lyon, B. G., Moldenhauer, K., Linscombe, S., & Kohywey, D. (1997). Effects of drying conditions, final moisture content, and degree of milling on rice flavor. *Cereal Chem., 74*, 566–570.

Champagne, E. T., Bett-Garber, K. L., Grimm, C. C., & McClung, A. M. (2007). Effects of organic fertility management on physicochemical properties and sensory quality of diverse rice cultivars. *Cereal Chem., 84*, 320–327.

Champagne, E. T., Bett-Garber, K. L., Thompson, J., Mutters, R., Grimm, C. C., & McClung, A. M. (2005). Effects of drain and harvest dates on rice sensory and physico-chemical properties. *Cereal Chem., 82*, 369–374.

Champagne, E. T., Bett-Garber, K. L., Thomson, J. L., Shih, F. F., Lea, J., & Daigle, K. (2008). Impact of presoaking on the flavor of cooked rice. *Cereal Chem., 85*, 706–710.

Chang, T. T. (1976). The origin, evolution, cultivation, dissemination and diversification of Asian and African rices. *Euphytica, 25*, 425–441.

Chaudhary, R. C. (2003). 'Specialty rices of the world: Effect of WTO and IPR on its production trend and marketing'. *Food Agriculture & Environment, 1*(2), 34–41.

Chaudhary, R. C., Shrinivasan, K., & Tran, D. V. (2001). Speciality rices of the world –a prologue. In: *Speciality Rices of the World: Breeding, Production and Marketing*

(Choudhary, R. C., & Tran, D. V., eds.), FAO, Rome, Italy, and Oxford, IBH Publishers, India, pp. 14.

Chen, S. H., Wu, J., Yang, Y., Shi, W. W., & Xu, M. L. (2006). The FGR gene responsible for rice fragrance was restricted within 69 kb. *Plant Sci., 171*, 505–514.

Cordeiro, G. M., Christopher, M. J., Henry, R. J., & Reinke, R. F. (2002). Identification of microsatellite markers for fragrance in rice by analysis of the rice genome sequence. *Mol. Breed., 9*(4), 245–250.

Crowhurst, D. G., & Creed, P. G. (2001). Effect of cooking method and variety on the sensory quality of rice. *Food Serv. Technol., 1*, 133–140.

Darell, K., Johnson, P., Zeeman, S., & Smith, A. M. (2000). Association between physiochemical characteristics and cooking qualities in high yielding rice varieties. *IRRN, 28*, 28–29.

Dhulappanavar, C. V. (1976). Inheritance of scent in rice. *Euphytica, 25*, 659–662.

Edwards, D., Henry, R. J., & Edwards, K. J. (2012). Advances in DNA sequencing accelerating plant biotechnology. *Plant Biotechnol. J., 10*, 621–622.

Fitzgerald, M. A., Sackville, H. N. R., Calingacion, M. N., Verhoeven, H. A., & Butardo, V. M. (2008). Is there a second fragrance gene in rice? *Plant Biotech. J., 6*, 416–423.

Fukai, Y., & Tukada, K. (2006). Influence of pre-washing on quality of cooked rice maintained at a constant temperature. 1. Influence of cooking conditions on quality of cooked rice. *Journal of the Agricultural Chemical Society of Japan, 53*, 587–591.

Garland, S., Lewin, S., Blakeney, A., Reinke, R., & Henry, R. (2000). PCR based molecular markers for the fragrance gene in rice (*Oryza sativa* L.). *Theor. Appl. Genet., 101*, 364–371.

Glaszmann, J. C. (1987). Isozymes and classification of Asian rice varieties. *Theore. Appl. Genet., 74*, 21–30.

Goff, S. A., Ricke, D., Lan, T. H., Presting, G., Wang, R. L., Dunn, M., Glazebrook, J., Sessions, A., Oeller, P., Varma, H., Hadley, D., Hutchinson, D., Martin, C., Katagiri, F., Lange, B. M., Moughamer, T., Xia, Y., Budworth, P., Zhong, J. P., Miquel, T., Paszkowski, U., Zhang, S. P., Colbert, M., Sun, W. L., Chen, L. L., Cooper, B., Park, S., Wood, T. C., Mao, L., Quail, P., Wingh, R., Dean, R., Yu, Y. S., Zharkikh, A., Shen, R., Sahasrabudhe, S., Thomas, A., Cannings, R., Gutin, A., Pruss, D., Reid, J., Tavtigian, S., Mitchell, J., Eldredge, G., Scholl, T., Miller, R. M., Bhatnagar, S., Adey, N., Rubano, T., Tusneem, N., Robinson, R., Feldhaus, J., Macalma, T., Oliphant, A., & Briggs, S. (2002). A draft sequence of the rice genome (*Oryza sativa* L. ssp *japonica*). *Science, 296*, 92–100.

Grosch, W., & Schieberle, P. (1997). Flavor of cereal products – A review. *Cereal Chem., 74*, 91–97.

Grosch, W. (1987). Reactions of hydroperoxides – Products of low molecular weight. In: *Autoxidation of Unsaturated Lipids*. 1st (ed.). Chan, H. W. S. (ed.). Academic Press, London, pp. 95–140.

Hashemi, F. S. G., Rafii, M. Y., Ismail, M. R., Mahmud, T. M. M., Rahim, H. A., Asfaliza, R., Malek, M. A., & Latif, M. A. (2013). Biochemical, genetic and molecular advances of fragrance characteristics in rice. *Crit. Rev. Plant Sci., 32*, 445–457.

Heffner, E. L., Sorrells, M. E., & Jannink, J. L. (2009). Genomic selection for crop improvement. *Crop Sci., 49*, 1–12.

Huang, Q., He, Y., Jin, R., Huang, H., & Zhu, Y. (1999). Tagging of restorer gene for rice H. L type CMS using microsatellite markers. *Rice Genet. Newsletter, 16*, 75–77.

Itani, T., Tamaki, M., Hayata, T., & Fushimi, K. H. (2004). Variation of 2-acetyl-1-pyrroline concentration in aromatic rice grains collected in the same region in Japan and factors affecting its concentration. *Plant Prod. Science, 7*, 178–183.

Jain, N., Jain, S., Saini, N., & Jain, R. K. (2006). SSR analysis of chromosome 8 regions associated with aroma and cooked kernel elongation in Basmati rice. *Euphytica, 152*(2), 259–273.

Jain, S., Jain, R. K., & McCouch, S. R. (2004). Genetic analysis of Indian aromatic and quality rice (*Oryza sativa* L.) germplasm using panels of fluorescently labeled microsatellite markers. *Theor. Appl. Genet., 109*, 965–977.

Jannink, J. L., Lorenz, A. J., & Iwata, H. (2010). Genomic selection in plant breeding: from theory to practice. *Brief Funct. Genomics, 9*, 166–177.

Jezussek, M., Juliano, B. O., & Schieberle, P. (2002). Comparison of key aroma compounds in cooked brown rice varieties based on aroma extract dilution analyses. *J. Agri. Food Chem., 50*, 1101–1105.

Jin, Q., Waters, D., Cordeiro, G. M., Henry, R. J., & Reinke, R. F. (2003). A single nucleotide polymorphism (SNP) marker linked to the fragrance gene in rice (*Oryza sativa* L.). *Plant Science, 165*, 359–364.

Juliano, B. O., & Duff, B. (1991). Rice grain quality as an emerging priority in national rice breeding programmes. In: *Rice Grain Marketing and Quality Issues*. Selected papers from International Rice Research Conference held in Seoul, the Korea Republic, from 27–31 August 1990. International Rice Research Institute, Los Banos, Philippines, pp. 55–64.

Juliano, B. O., Onate, L. U., & Del Mundo, A. M. (1965). Relation of starch composition, protein content, and gelatinization temperature to cooking and eating qualities of milled rice. *Food Technol., 19*, 1006–1011.

Katherine, A. S., Ogden, R., McEwing, R., Briggs, H., & Gorham, J. (2008). InDel markers distinguish Basmatis from other fragrant rice varieties. *Field Crops Res., 105*, 81–87.

KBIRRI (Knowledge Bank, International Rice Research Institute). (2016). URL: knowledgebank.irri.org. Accessed on 02–03–2016.

Khush, G. S., & Dela Cruz, N. (1998). Developing Basmati rices with high yield potential. *Cahiers Options Mediterraneennes 24, Rice Quality*. A pluridisciplinary approach. (CD-ROM computer file) CIHEAM, Paris.

Khush, G. S., & Juliano, B. O. (1985). Breeding for high-yielding rices of excellent cooking and eating qualities. In: *Proceedings of Rice Grain and Quality Marketing*. International Rice Research Institute, Los Banos, Laguna, pp. 61–69.

Khush, G. S. (2000). Taxonomy and origin of rice. In: Singh, R. K., Singh, U. S., Khush, G. S., (ed.). *Aromatic Rices*. Oxford and IBH Publishing Co. Pvt. Ltd., New Delhi, pp. 47–69.

Khush, G. S., Paul, C. M., & Cruz, N. M. (1979). Rice grain quality evaluation and improvement at IRRI, In: *Chemical Aspects of Rice Grain Quality*, IRRI, Los Banos, Philippines, pp. 21–31.

Kibria, K., Islam, M. M., & Begum, S. N. (2008). Screening of aromatic rice lines by phenotypic and molecular markers. *Bangladesh J. Bot., 37*, 141–147.

Kovach, M. J., Calingacion, M. N., Fitzgerald, M. A., & McCouch, S. R. (2009). The origin and evolution of fragrance in rice (*O. sativa* L.). *Proc. Nat. Acad. Sci., 106*(34), 14444–14449.

Kumari, P., Ahuja, U., Chawla, V., & Jain, R. K. (2011). Phenotyping of agronomic and grain quality traits of recombinant inbred lines (RIL) derived from the cross CSR 10 x HBC 19. *Oryza, 48*(2), 97–102.

Kumari, P., Ahuja, U., Jain, S., & Jain, R. K. (2012). Molecular and biochemical analysis of aroma in CSR10 x Taraori basmati derived Recombinant Inbred Lines. *Elec. J. Plant Breeding, 3*(1), 782–787.

Kuo, S. M., Chou, S. Y., Wang, A. Z., Tseng, T. H., Chueh, F. S., & Yen, H. E. (2005). The betaine aldehyde dehydrogenase (BAD2) gene is not responsible for the aroma trait of SA0420 rice mutant derived by sodium azide mutagenesis. 5th International Rice Genetics Symposium, Philippines, *IRRI*, pp. 166–67.

Li, J. H., Wang, F., Liu, W. G., Jin, S. J., & Liu, Y. B. (2006). Genetic analysis and mapping by SSR marker for fragrance gene in rice Yuefeng, B., *Mol. Plant Breed., 4,* 54–58.

Li, J. Y., Wang, J., & Zeigler, R. S. (2014). The 3000 rice genomes project: New opportunities and challenges for future rice research. *GigaScience, 3,* 8.

Li, J., Xiao, J., Grandillo, S., Jiang, L., Wan, Y., Deng, Q., Yuan, L., & McCouch, S. R. (2004). QTL detection for rice grain quality traits using an interspecific back-cross population derived from cultivated Asian (*O. sativa* L.) and African (*O. glaberrima* S.) rice. *Genome, 47,* 697–704.

Liu, T. T., Wu, C. M., & Yau, N. J. N. (1996). Descriptive analysis of odor of cooked rice maintained at 18°C and 60°C. *Food Sci., 23,* 761–766.

Lorieux, M., Petrov, M., Huang, N., Guiderdoni, E., & Ghesquiere, A. (1996). Aroma in rice-genetic analysis of a quantitative trait. *Theor. Appl. Genet., 93,* 1145–1151.

Mahdav, M. S., Pandey, M. K., Kumar, P. R., Sundaram, R. M., Prasad, G. S. V., Sudarshan, L., & Rani, N. S. (2009). Identification and mapping of tightly linked SSR marker for aroma trait for use in marker assisted selection in rice. *Rice Genet. Newsl., 25,* 38–39.

Mahindra, S. N. (1995). *Manual of Basmati Rice.* Metropolitan Book Co. Pvt. Ltd., New Delhi, India, pp. 307.

Melissa, A., Fitzgerald, T., Susan, R., McCouch, S. R., & Hall, R. D. (2009). Not just a grain of rice: the quest of quality. *Trends in Plant Sci., 14*(3), 133–139.

Meullenet, J. F., Marks, B. P., Griffin, K., & Daniels, M. J. (1999). Effects of rough rice drying and storage conditions on sensory profiles of cooked rice. *Cereal Chem., 76,* 438–486.

Meullenet, J. F., Marks, B. P., Hankins, J. A., Griffin, V. K., & Daniels, M. J. (2000). Sensory quality of cooked long-grain rice as affected by rough rice moisture content, storage temperature, and storage duration. *Cereal Chem., 77,* 259–263.

Monsoor, M. A., & Proctor, A. (2002). Effect of water washing on the reduction of surface total lipids and FFA on milled rice. *J. AOCS, 79,* 867–870.

Morishima, H. (1984). Wild plants and domestication. In: Tsunoda, S., & Takahashi, N., (ed.). *Biology of Rice.* Elsevier, Amsterdam, pp. 3–30.

Myint, K. M., Courtois, B., Risterucci, A. M., Frouin, J., Soe, K., Thet, K. M., Vanavichit, A., & Glaszmann, J. C. (2012). Specific patterns of genetic diversity among aromatic rice varieties in Myanmar. *Rice, 5,* 20.

Nagaraju, J., Kathirvel, M., Ramesh, K. R., Siddiq, E. A., & Hasnain, S. E. (2002). Genetic analysis of traditional and evolved basmati and non-basmati rice varieties by using fluorescence based ISSR-PCR., & SSR markers. *Proc. Nat. Acad. Sci.*, USA *99*(1), 5836–5841.

Navarro, M., Butardo, V., Bounphanousay, C., Reano, R., Hamilton, R. S., & Verhoeven, H. (2007). The good, the BAD and the fragrant – Understanding fragrance in rice. International network on quality rices – clearing old hurdles with new science: improving rice grain quality. Philippines, *IRRI*, pp. 16, Abstract, April 17–19.

Park, J. K., Kim, S. S., & Kim, K. O. (2001). Effect of milling ratio on sensory properties of cooked rice and on physicochemical properties of milled and cooked rice. *Cereal Chem., 78,* 151–156.

Petrov, M., Danzart, M., Giampaoli, P., Faure, J., & Richard, H. (1996). Rice aroma analysis: Discrimination between a scented and a nonscented rice. *Sci., 16,* 347–360.

Piggott, J. R., Morrison, W. R., & Clyne, J. (1991). Changes in lipids and in sensory attributes on storage of rice milled to different degrees. *Int. J. Food Sci. Technol., 26,* 615–628.

Pinson, S. R. M. (1994). Inheritance of aroma in six rice cultivars. *Crop Science, 34,* 1151–1157.

Pummy, K., Ahuja, U., Chawla, V., & Jain, R. K. (2009). Genetics of apiculus, hull and pericarp color and awnedness in CSR10 x Taraori Basmati derived recombinant inbred lines. *Oryza, 46*(2), 57–69.

Rai,V. P, Singh, A. K., Jaiswal, H. K., Singh, S. P., Singh, R. P., & Waza, S. A. (2015). Evaluation of molecular markers linked to fragrance and genetic diversity in Indian aromatic rice. *Turk. J. Bot., 39,* 209–217.

Rani, S. N. (1992). Paper presented at the Special Seminar for the awards, IRRC, *IRRI*, Los Banos, Philippines.

Rao, M. B., Rani, S. N., Saradhi, U. V. R., Prasad, G. S. V., Pandey, M. K., & Virmani, M. (2006). Quantification of principle aroma compounds and possibility of allelic/genic diversity in indigenous aromatic rices (*Oryza sativa* L.) In: *Proceedings of 2nd International Rice Congress*, New Delhi, India, pp. 316–317.

Reddy, P. R., & Satyanarayana, K. (1980). Inheritance of aroma in rice. *Indian J. Genet., 40,* 327–329.

Rice Association, (2007). Rice Market Briefing 65.A.001 (www.riceassociation.org.uk/).

Richharia, R. H., & Misro, B. (1965). Studies in world genetic stock of rice. IV-Distribution of Scented Rices. *Oryza, 2,* 57–59.

Saini, N., Jain, N., Jain, S., & Jain, R. K. (2004). Assessment of genetic diversity within and among Basmati and non-Basmati rice varieties using AFLP, ISSR and SSR markers. *Euphytica, 140,* 133–146.

Sakthivel, J., Shobha, R. N., Pandey, M. K., Sivaranjani, A. K. P., Neeraja, C. N., & Balchandran, S. M. (2009). Development of a simple functional marker for fragrance in rice and its validation in Indian Basmati and non-Basmati fragrant rice varieties. *Mol. Breed., 24,* 185–190.

Shao, G., Tang, A., Tang, S., Luo, J., Jiao, G., Wu, J., & Hu, P. (2011). A new deletion mutation of fragrant gene and the development of three molecular markers for fragrance in rice. *Plant Breed., 130,* 172–176.

Shi, W., Yang, Y., Chen, S., & Xu, M. (2008). Discovery of a new fragrance allele and the development of functional markers for the breeding of fragrant rice varieties. *Mol. Breed., 22,* 185–192.

Shull, G. H. (1908). The composition of a field of maize. *Ann. Breed. Assoc., 4,* 296–299.

Sidhu, G. S., Javier, E. L., Hel, P. K., Hak, L., & Puthea, S. (2001). Evaluation of Basmati rices in Cambodia. *Crop Improv., 28*(1), 40–49.

Singh, A. K., Gopalakrishnan, S., Singh, V. P., Prabhu, K. V., Mohapatra, T., Singh, N. K., Sharma, T. R., Nagarajan, M., Vinod, K. K., Singh, D., Singh, U. D., Chander, S., Atwal, S. S., Seth, R., Singh, V. K., Ellur, R. K., Singh, A., Anand, D., Khanna, A., Yadav, S., Goel, N., Singh, A., Shikari, A. B., Singh, A., & Marathi, B. (2011). Marker assisted selection: a paradigm shift in Basmati breeding. *Indian J. Genet., 71*(2), 1–9.

Singh, R. K., Mani, S. C., Singh, N., Singh, G., Singh, H. N., Rohilla, R., & Singh, U. S. (2003). Aromatic rices of Uttar Pradesh and Uttaranchal. In: *A Treatise on the Scented Rices of India* (Singh, R. K., & Singh, U. S., eds.). Kalyani Publishers, New Delhi, India, pp. 403–420.

Singh, R. K., Singh, U. S., Khush, G. S., & Rohilla, R. (2000). Genetics and biotechnology quality traits in aromatic rices. In: *Aromatic Rices* (ed. Singh, R. K., Singh, U. S., Khush, G. S.), Oxford and IBH Publishing Co. Pvt. Ltd., New Delhi, pp. 47–69.

Siwach, P., Jain, S., Saini, N., Chowdhury, V. K., & Jain, R. K. (2004). Allelic diversity among Basmati and non-Basmati long grain *indica* rice varieties using microsatellite markers. *J. Plant Biochem. Biotech., 13,* 25–32.

Sood, B. C., & Siddq, E. A. (1978). A rapid technique for scent determination in rice. *Indian J. Genet.Plant Breed., 38,* 268–271.

Sood, B. C. (1978). Studies in cooking and nutritive quantity of cultivated rice *Oryza sativa* with specific reference to genetics of kernel elongation, aroma and protein content. PhD. *Thesis, IARI,* New Delhi.

Spindel, J., Begum, H., Akdemir, D., Virk, P., Collard, B., Redoña, E., Atlin, G., Jannink, J. L., & McCouch, S. R. (2015). Genomic selection and association mapping in rice (*Oryza sativa*): effect of trait genetic architecture, training population composition, marker number and statistical model on accuracy of rice genomic selection in elite tropical rice breeding lines. *PLoS Genet., 11,* e1004982.

Sun, S. H., Gao, F. Y., Lu, X. J., Wu, X. J., Wang, X. D., Ren, G. J., & Luo, H. (2008). Genetic and gene fine mapping of aroma in rice (*Oryza sativa* L. Cyperales, Poaceae). *Genet. Mol. Biol., 31,* 532–538.

Tamaki, M., Ebata, M., Tashiro, T., & Ishikawa, M. (1989). Physico-ecological studies on quality formation of rice kernel. II. Effects of ripening stage and some ripening conditions on free amino acids in milled rice kernel and in the exterior of cooked rice. *Jpn. J. Crop Sci., 58,* 695–703.

Terao, T., Miura, S., Yanagihara, T., Hirose, T., Nagata, K., Tabuchi, H., Kim, H. Y., Lieffering, M., Okada, M., & Kobayashi, K. (2005). Influence of free-air CO_2 enrichment (FACE) on the eating quality of rice. *J. Sci. Food Agri., 85,* 1861–1868.

Tragoonrung, S., Shen, J. Q., & Vanavichit, A. (1996). Tagging an aromatic gene in low land rice using bulk segregant analysis. International Rice Research Institute 1996. Rice Genetics III. *Proc. 3rd Int. Rice Genet Symp.* 16–20 Oct. 1995. Manila, 613–618.

Tsugita, T., Kurata, T., & Kato, H. (1980). Volatile components after cooking rice milled to different degrees. *Agri. Biol. Chem., 44,* 835–840.

Varshney, R. K., Terauchi, R., & McCouch, S. R. (2014). Harvesting the promising fruits of genomics: applying genome sequencing technologies to crop breeding. *PLoS Biol., 12*, e1001883.

Verma, D. K., & Srivastav, P. P. (2016). Extraction technology for rice volatile aroma compounds In: *Food Engineering Emerging Issues, Modeling, and Applications* (eds. Meghwal, M., & Goyal, M. R.). In book series on Innovations in Agricultural and Biological Engineering, Apple Academic Press, USA, pp. 245–291.

Verma, D. K., Mohan, M., & Asthir, B. (2013). Physicochemical and cooking characteristics of some promising basmati genotypes. *Asian Journal of Food Agro-Industry, 6*(2), 94–99.

Verma, D. K., Mohan, M., Prabhakar, P. K., & Srivastav, P. P. (2015). Physico-chemical and cooking characteristics of Azad basmati. *International Food Research Journal, 22*(4), 1380–1389.

Verma, D. K., Mohan, M., Yadav, V. K., Asthir, B., & Soni, S. K. (2012). Inquisition of some physico-chemical characteristics of newly evolved basmati rice. *Environment and Ecology, 30*(1), 114–117.

Wan, X. Y., Wan, J. M., Jiang, L., Wang, J. K., Zhai, H. Q., Weng, J. F., Wang, H. L., Lei, C. L., Wang, J. L., Zhang, X., Cheng, Z. J., & Guo, X. P. (2006). QTL analysis for rice grain length and fine mapping of an identified QTL with stable and major effects. *Theor. Appl. Genet., 112*, 1258–1270.

Wanchana, S., Kamolsukyunyong, W., Ruengphayak, S., Toojinda, T., Tragoonrung, S., & Vanavichit, A. (2005). A rapid construction of a physical contig across a 4.5 cM region for rice grain aroma facilitates marker enrichment for positional cloning. *Sci. Asia, 31*, 299–306.

Whitworth, M. B., Greenwell, P., & Fearn, T. (1996). *Physical Techniques for Establishing the Authenticity of Rice.* Presented at 10th International Cereal and Bread Congress, Chalkidiki, Greece, June 9–12.

Widjaja, R., Craske, J. D., & Wootton, M. (1996). Comparative studies on volatile components of non-fragrant and fragrant rices. *J. Science Food Agri., 70*, 151–161.

Xing, Y. Z., Tan, Y. F., Xu, C. G., Hua, J. P., & Sun, X. L. (2001). Mapping quantitative trait loci for grain appearance traits of rice using a recombinant inbred line population. *Acta. Bot. Sin., 43*, 840–845.

Yamamatsu, K., Moritaka, S., & Wada, S. (1966). Stale flavor of stored rice. *Agri. Biol. Chem., 30*, 483–486.

Yano, M., & Sasaki, T. (1997). Genetic and molecular dissection of quantitative traits in rice. *Plant Mol. Biol., 35*, 145–153.

Yasumatsu, K., Moritaka, S., & Wada, S. (1966). Studies on cereals. V. Stale flavor of stored rice. *Agri. Biol. Chem., 30*, 483–486.

Yoshihashi, T. Huong, N. T. T., Surojanametakul, V., Tungtrakul, P., & Varanyanond, W. (2005). Effect of storage conditions on 2- acetyl-lpyrroline content in aromatic rice variety. Khao Dawk Mali 105. *J. Food Sci., 70*, 34–37.

Yoshihasthi, T., Nguyen, T. T. H., & Kabaki, N. (2004). Area dependency of 2-acetyl-1-pyrroline content in an aromatic rice vaiety. Khao Dawk Mali 105. *Jarq-Japan Agric. Res. Q., 38*, 105–109.

CHAPTER 2

AROMATIC RICE FROM DIFFERENT COUNTRIES: AN OVERVIEW

DEEPAK KUMAR VERMA,[1] PREM PRAKASH SRIVASTAV,[2] and ALTAFHUSAIN NADAF[3]

[1]Research Scholar, Agricultural and Food Engineering Department, Indian Institute of Technology, Kharagpur–721302, West Bengal, India, Mobile: +91-7407170260, +91-9335993005 Tel.: +91-3222-281673, Fax: +91-3222-282224 E-mail: deepak.verma@agfe.iitkgp.ernet.in; rajadkv@rediffmail.com

[2]Associate Professor, Agricultural and Food Engineering Department, Indian Institute of Technology, Kharagpur–721302, West Bengal, India, Mobile: +91-9434043426, Tel.: +91-3222-283134 Fax: +91-3222-282224, E-mail: pps@agfe.iitkgp.ernet.in

[3]Associate Professor, Department of Botany, Savitribai Phule Pune University, Pune–411007, India Mobile: +91-7588269987 Tel.: +91-20-25601439 Fax: +91-20-25601439, E-mail: abnadaf@unipune.ac.in

CONTENTS

2.1 INTRODUCTION

Aroma is a major characteristic of rice quality that increases the potential value of rice in the international trade (Nayak et al., 2002; Verma and Srivastav, 2016). Over the past half-century, the demand of aromatic rice for special purpose has dramatically increased. The leading aromatic rice in the world trade is Basmati, which produces a distinctive aroma in the field at the time of plant growth; grain development; during harvesting and storage; and during milling, cooking and eating, and it has long grain and slender shape with less chalkiness and medium texture on cooking (Kamath et al., 2008). This variety is produced in the Punjab area along both sides of the Indus River in Pakistan and India. Many countries such as India, Thailand, Vietnam, USA, China, etc., are involved in developing special aromatic rice cultivars (Table 2.1) and earning of millions USD dollars by the annual export in addition to being the main grower and consumer of this aromatic rice (Singh et al., 2000e).

Although aromatic rice is very popular in the world market due to its aroma, but in the majority of the Indian, indigenous aromatic cultivars of rice are small and medium grained and not like long grain of Basmati type (Singh et al., 2000a). This rice is an important commodity in the international market among the various types of rice grown in South Asian countries. Non-aromatic rice constitutes the bulk of rice trade, but most of the market business in aromatic rice is from India, Pakistan, and Thailand.

TABLE 2.1   List of Some Prominent Aromatic Rice from Different Country

Country	Aromatic Rice	References
India	Basmati-370, Basmati-386, Dehradun or Amritsari Basmati, Haryana Basmati-1, Improved Sabarmati, Kalanamak, Karnal local, Kasturi, Mahi Sugandha, Punjab Basmati-l, Pusa Basmati-1 Pusa-33, Ranbir Basmati, Sabarmati, Taraori Basmati, Type-3, Dubraj, Ambemohar, Chinoor	Itani, (1993); Kiani, et al. (2012); Kunlu, & Fuming, (1995)
Thailand	Hawm Klong Luang, Hawm Supanburi, Jasmine, KDML-105, Khao Hawm (Thai Fragrant Rice), RD-15, RD-6, Siamati, Thai Hom Mali Rice	Anonymous, (1996); Anonymous, (1997b); Itani, et al. (2004)
Vietnam	Can Duoc, Di Huong, Hai Hau, Jasmine 85, Lam Thao, Lua Tam, Nam Dinh, Nang Thom, Nang Thom Cho Dao, Nep Bac, Nep Hoa Vang, Nep Rong, OM3536, Phu Xuyen, Tam Canh, Tam On, Tam Xoan, VD20	Anonymous, (1997c); Bhasin, (2000)
USA	A-201, Della, Dellamont, Jasmine-85, Kasmati, Texamati	Kato, et al. (2006b); Khush, (2000)
China	Congjiangxixiangmi, De Chang Xiang Mi, Ganwan-xian 22, Huanglongxiangmi, Jiangyongxiangdao, Jingcixiangdao, Jingxixiangdao, Minghou-tsiutsao, Qingbuxiangjingmi, Qufuxiangmi, Shuang-Zhu-Zhan, SR 5041, Tainung Sen 20, Ta-pei-mang, Xiang Keng 3, Xiang Keng 3, Xiang Nuo 4, Xiangyou 63, Yongshunxiangdao, Zhao Xing 17, Zhe 9248	Anonymous, (1994); Baqui, et al. (1997); Buu, (2000); Chang, (1976); Chaudhary, (2003); Gay, et al. (2004)
Indonesia	Batang Gadis, Bengawan Solo, Gunung Perak, Mentik Wangi, Pandanwangi, Pare Kembang, Pulu mandoti, Rojo Lele, Rojolele, Sintanur, Situpatenggang	Islam, et al. (1996); Kamath, et al. (2008)
Japan	Haginokaori, Hieri, Izayoi, Jakou, Jakouine, Kabashiko, Miyakaori, Oitakoutou, Sari Queen, Sawakaori, Sasaminori	Arashi, (1975); Bisne, et al. (2008); Boriboon, (1996); Chuanyuang, et al. (1994); Efferson, (1985); Fushimi, et al. (1994); Haryanto, et al. (2008); Hien, et al. (2007)
	Basmati 50021-1, Basmati Pak, Basmati-185, Basmati-370, Basmati-385, Basmati-6129, PK 4048-3, PK 50005-3, PK-50010, Super Basmati	Kato et al. (2006a); Khush, (2000).

TABLE 2.1   (Continued)

Country	Aromatic Rice	References
Pakistan	Basmati 50021-1, Basmati Pak, Basmati-185, Basmati-370, Basmati-385, Basmati-6129, PK 4048-3, PK 50005-3, PK-50010, Super Basmati	Kato, et al. (2006a); Khush, (2000)
Iran	Ahalami tarom, Anbar-boo, Champa, Domsiah, Doroudzan, Fajr, Gerdeh, Hasan, Hassan Saraie, Hassani, Kodus, Mehr, Mir tarom, Mirza, Mosa Tarom, Neda, Nemat, Poya, Qaserdashti, Sadri, Salari, Sang tarom, Shiroudi, Tarom amiri, Tarom deylamani, Tarom mahalli,Tarom,	CBS (Central Bureau of Statistic) (2004); Das, et al. (2000); Fukuoka, et al. (2006); Kunlu, (1995)
Bangla-desh	Badshabhog, Banshful, Basmati-D, Bau-pagal, BR-16 (Shahibalam), BR-26, BR-34, BR-36, BR-38, BR-39, BR-6, BRRI Dhan-34 (locally known as Khashkhani), Chinigura, Dadkhani, Dolhabhog (BR-5), Jotalo, Kalazira, Kataktara, Kataribagh, Kataribhog, Nizershail, Sakkorkhora, Radhuni Pagal, Tulshimala, Ukunamodhu	Anik, et al. (2002); Anonymous, (1997a); BBS (Bangladesh Bureau of Statistics), (1994); Kondo, (1987)
Afghani-stan	Bahra, Bala, Germa Bala, Lawangi, Luke Qasan, Monda Mashruqi, Pashadi, Pashadi Konar, Permel, Sarda Bala, Sela Doshi, Sela Takhar, Sherkati, Surkha-Bala, Surkha-Daraz-Baghlan, Surkhamabain, Torishi,	Itani, (2002) ; Kato, et al. (2006a)

2.2   AROMATIC RICE IN INDIA

India has immense wealth of aromatic rice, and this sub-continent is known as home for aromatic rice diversity (Bisne and Sarawgi, 2008). India is the fourth major exporter of rice after Thailand, Vietnam, and USA and emerges as one of the major rice exporters. The exported rice from these countries is characterized by certain unique features, which may be graded as export quality rice with normal nutritional quality in international markets (Verma and Srivastav, 2017). But many varieties have already been lost as an outcome of the green revolution, where foremost importance was given to yield capacity rather than to quality (Singh and Singh, 1998). A large number of rice races are grown and adapted to the specific agro-ecological conditions of the different rice growing regions of the country (Table 2.2), and some of them are mostly for home consumption; in contrast, Basmati is found in the Tarai region of Uttar Pradesh, Bihar, and

TABLE 2.2   List of Some Indigenous Grain Aromatic Rice from India

Place	Aromatic Rice Cultivars
Andhra Pradesh	Amritsari, Jeeragasambha, Kaki Rekhalu, Sukhdas
Assam	[4]Badshahbhog, Bhaboli-Joha, Bhugui, Boga Tulsi, Boga-Joha, Boga-manikimadhuri, Bogi-Joha, Bokul-Joha, Borjoha, Borsal, Cheniguti, Chufon, Goalporia-Joha-1, Goalporia-Joha-2, Govindbhog, Joha-Bora, Kaljeera, Kamini-Joha, Kataribhog, Khorika-Joha, Kola-Joha, Koli-Jo-ha, [4]Kon-Joha-1, [4]Kon-Joha-2, [4]Krishna-Joha, Kunkuni-Joha, Malbhog, Manikimadhuri-Joha, Prasadbhog, [4]Raja-Joha, Ramphal-Joha, Ranga-Joha
Bihar	Abdul, [4]Badshahbhog, [*]Badshapasand, Bahami, [3]Baikani Jeera, Bhilahi Basmati, Champaran Basmati (Kali), Champaran Basmati (Lal), Champaran Brahmabhusi, Chenaur, Deobhog, Dewta Bhog, [*,4]Kalanamak, Kamina, Kanak Jeera, Karia Kamod, Katami, [4]Katarani, Kesar, Mircha, Mohanbhog, Pasand, Ram Tulsi, Ramjain, Sataria, Shyam Gopalbhog, Sona Lari, [4]Sonachur, Tulsi Basmati (Bhuri), Tulsi Manjari,
Gujarat	Kamod-118, Kolhapur Scented, Pankhali-203, Zeerasal
Haryana	Basmati 370[3], Khalsa-7[3], Pakistani Basmati[3], Taraori Basmati[3],
Himachal Pradesh	Achhu, [3]Baldhar Basmati, Begmi, [3]Chimbal Basmati, [3,4]Madhumalti, [3,4]Mushkan, Panarsa local, [3]Seond Basmati
Jammu & Kashmir	Muskh Budgi
Karnataka	Kagasali, Sindigi Local, Kumud, Kali kumud
Kerala	Gandhkasala , Jeerakasala
Madhya Pradesh	Badshabhog, Baspatri, Chatri, Chattri, Chinni gauri, Chinni sagar, Chinoor, Chinore, [4]Dhubraj-354, [4]Dhubraj, Jauphool, Kali Kamod, Kalu Mooch, Laloo[3], Madhuri, Tulsi Manjari, Vishnu Parag, [4]Vishnubhog,
Maharashtra	[4]Ambemohor (Ambemohar-102, Ambemohar-157, Ambemohar-159), Chinore, Kagasali, Prabhavati, Sakoli-7, [1]Kala bhat, [2,4]Champakali, [2,4]Ghansal, Girga, [1]Kothmirsal, [1]Chimansal, [2,4]Kamod, Kalsal, Kamavatya, Khadkya, [2,4]Raibhog, Tamsal
Manipur	[4]Chakhao, Chahao Amubi (black scented rice), Chahao Angangbi (pink/red scented rice)
Orissa	[4]Badshahbhog, Barangomati, Karpurkali, Sagartara, T-412, T-812, Takurabhoga
Punjab	[3]Basmati-370, [3]Basmati-385, [3]Pakistani Basmati
Rajasthan	[3]Basmati (local), [3]Basmati-370
Tamilnadu	Jeeraga Samba

TABLE 2.2 (Continued)

Place	Aromatic Rice Cultivars
Utter Pradesh	[2]Adamchini, [*,1,2]Badshahpasand, Bahantapul, [1]Basmati safeda, [3]Basmati-370, Bengal Juhi, Benibhog, Bhanta Phool, [4]Bindli, Chhoti Karmuhi, Chinnawar, Dehraduni Basmati[3], [4]Dhania, [3,4]Dhubraj, [3]Duniapat (T-9), [2,3]Hansraj, Jeerabattis, [3]Kala Sukhdas, [*,1,2,4]Kalanamak, Kanakjeera, Kesar, [*,1,3]Lalmati, Laungchoor, Moongphali, Motachinaeum, [3]Nagina-12, Parsam, Phool Chameli, Ram Jawain, Rambhog, Ramjawain, [3]Ramjinwain (T-1), Safeda[3], [4]Sakarchini, Selhi; Dulahniya, Shakkarchini, [4]Sonachur, [3]T-9, [3]Tapovan Basmati, Thakurbhog, [*,1,4]Tilak chandan, Tinsukhia, Tulsi Manjari, Tulsi Prasad, [3]Type-3, [3]Vishnuparag, Yuvraj
West Bengal	Badshabhog, [*]Badshahpasand, Chinisakkar, Danaguri, Gandheshwari, Kalo Kanakchur, Katanbhog, Kataribhog, Nunia, Radhuni Pagal, [4]Randhunipagal, Sitabhog, Tulai Panji, Tulsibhog

Sources: Singh, (2000), Singh, et al. (2000d), Nadaf, et al. (2016); *Popular land races; [1]Popular but grown in small areas only; [2]Still popular and grown in large areas; [3]Non-basmati aromatic rice; [4]Potential candidates for export.

West Bengal, indicating that this area is probably the origin of aromatic rice which have either been lost or are at the threshold of extinction.

India owns a large number of different types of aromatic rice cultivars with great diversity. Some of them are globally known for important Indian aromatic cultivars as depicted in Table 2.1. Non-basmati aromatic rice (Table 2.2) are comparatively much superior to Basmati types (Table 2.3) depending on the several quality traits, such as excellent aroma, a likely kernel elongation before cooking (KEbC), kernel elongation after cooking (KEaC), fluffiness, taste, texture, etc., and therefore, are in great demand (Singh et al., 1997; Verma et al., 2012, 2013, 2015). This type of aromatic rice, if exploited properly, can gratify the taste and aromatic preference of individuals from diverse regions of the world. Unfortunately, the available diversity of this aromatic rice is not exploited, except for Basmati types. Various varieties of non-basmati type aromatic rice can exceed Basmati with respect to traits (Table 2.3). Besides, many of them are acclimated to cultivate under local conditions and in regions where Basmati rice has been found to be incapable to grow. A few such potential candidates for exports (Table 2.2) could be identified and promoted for cultivation; these candidates are perceived by the farmers to be as superior as Basmati in aroma, grain elongation, taste, and digestibility (Singh et al.,

TABLE 2.3 Traditional Indian Basmati Rice Considered Premium Owing to Unique Features and Quality Traits

Unique Feature	Remarks
Color	The color of a Basmati is translucent, creamy white. Brown Basmati rice is also available but the most commonly used is white Basmati.
Grain	Long grain. The grain is long (6.61–7.5 mm) or very long (more than 7.50 mm and 2 mm breadth).
Shape	Shape or length-to-width ratio is another criteria to identify basmati rice. This needs to be over 3.0 in order to qualify as Basmati.
Texture	Dry, firm, separate grains. Upon cooking, the texture is firm and tender without splitting, and it is non-sticky. This quality is derived from the amylose content in the rice. If this value is 20-22%, the cooked rice does not stick.
Elonga-tion	The rice elongates almost twice upon cooking but does not fatten much. When cooked, the grains elongate (70 to 120%) over the pre-cooked grain more than other varieties.
Flavor	Distinctive fragrance. The most important characteristic of them all is the aroma. Incidentally, the aroma in Basmati arises from a cocktail of 100 compounds- hydrocarbons, alcohols, aldehydes and esters. A particular molecule of note is 2-acetyl-1-pyrroline.

Source: Adapted from Verma, D. K. (2011). Physico-chemical and cooking characteristics of Azad Basmati (CSAR 839-3) – A Newly Evolved Variety of Basmati Rice (*Oryza sativa* L.). M.Sc. (Agri.) Thesis, Chandra Shekhar Azad University of Agriculture and Technology, Kanpur (UP), India.

1997; Verma et al., 2012, 2015). Most of the traditional rice cultivars have low yield capacity; however, less progress in the development of geno-types of aromatic rice has been seen to date despite their high value and demand (Singh et al., 1997).

2.3 AROMATIC RICE IN THAILAND

In the international rice trade, Thailand is a major player to produce and export aromatic rice of fine grain. Several aromatic rice varieties are pro-duced in Thailand, which has economic and social importance among Thai consumers. Popular aromatic rice cultivars are shown in Table 2.1. Jasmine rice with a unique jasmine aroma is the most renowned rice in

Thailand and has special markets in Hongkong and Singapore at relatively high prices, whereas Khao Hawm (also called as Thai fragrant rice) is known for the national pride of the Thai people and mostly cultivated by every household mainly for their own consumption. The diversity of Thai aromatic rice cultivars is very wide, and only a few cultivars are in high demand in domestic and international markets. Khao Dawk Mali (KDML) is one of them, which is grown by farmers in their fields for both domestic consumption and export. KDML-105 variety differs in cooking quality and other traits and is also considered to be one of the topmost earners of foreign trade among the agricultural products (Anonymous, 1996; Boriboon, 1996; Anonymous, 1997a, 1997b, 1997c; Sarkarung et al., 2000; Singh et al., 2000e).

2.4   AROMATIC RICE IN VIETNAM

Many aromatic cultivars are available in Vietnam (Table 2.1), and some of these prominent cultivars of aromatic rice (Table 2.4) belong to the three distinguished classes as traditional, improved, and exotic (Nghia et al., 2001b). Among these classes, the aromatic varieties of traditional class have not been cultivated from a long time due to their long-growing period and low-yielding capacity, which prohibited intensive production. However, the Vietnamese people highly appreciated the grain quality of this rice (Nayak et al., 2002) and paid more attention for in situ conservation or on-farm conservation due to their specified targeted areas to express the aroma. For instances, Lua Tam and Nang Thom, Nang Thom Cho Dao (NTCD), and Tam Xoan are aromatic rice cultivars in the North and South, in the Mekong Delta, and in the Red River Delta area of Vietnam, respectively. Currently, due to their small market, Vietnam has limited cultivation of traditional varieties in some areas. Jasmine 85, VD20, and OM3536 have shorter growing period and higher potential yielding capacity and are known as the most popularly improved and modern varieties of aromatic rice. Hai Hau is another important aromatic rice that is slightly white, soft, and fragrant.

Since the mid-1990s, Vietnam has been more concerned about aromatic rice production as it surveyed, described, and studied to some degree.

TABLE 2.4   List of Aromatic Rice Cultivars from Vietnam

Vietnam aromatic rice		
Bang Thom	Nang Huong Ran	Phu Xuyen
Can Duoc	Nang Huong	Re Thom Ha Dong
Di Huong	Nang Tet 2	Tam Canh
Gie Vang 2	Nang Thom Cho Dao-1	Tam Cao Vinh Phuc
Hai Hau	Nang Thom Cho Dao-2	Tam Den Bac Ninh
Heo Thom	Nang Thom Cho Dao-3	Tam Den Hai Phong
Hoa Lai	Nang Thom Cho Dao-4	Tam Den
Jasmine 85	Nang Thom Duc	Tam On
Lam Thao	Nang Thom Muon	Tam Thom Ngoc Khe
Lang Thom Som	Nang Thom Som	Tam Thom Trang Liet
Lua Ngu	Nang Thom	Tam Xoan
Lua Tam	Nep Bac	Tau Huong Lo
Me Huong2	Nep Hoa Vang	Tau Huong Muon
MTL250−1	Nep Huong	Tau Huong Som
MTL250−2	Nep Rong	Tau Huong-2
Nam Dinh	Nep Thom	Thanh Tra
Nang Huong 2	OM3536	VD20

Sources: Buu, (2000); Gay, et al. (2004); Hien, et al. (2007).

Aromatic rice in the Northern Province of Vietnam is of economic importance, and only this area has the maximum diversity of rice (Chang, 1976). The existing maximum areas of rice diversity in Vietnam are known as lowland rice productive area. Okuno et al. (1998) suggested that a great need to conserve this area because the conservation of genetic diversity in the lowland area of rice is related to the welfare of the Vietnamese local community (Fukuoka et al., 2006).

Aromatic rice production in Vietnam are on both domestic and international markets which may participate to support competitively like other countries and also comparable to the ordinary rice. This rice yielded higher price to the income of Vietnamese rice farmers' (Gay et al., 2004; VTPAEPC, 2008). The market of South Vietnam received about 2.0–2.5 time's higher price from aromatic rice than that for normal rice. But, aromatic rice is usually adulterated with other varieties, which make it very difficult to control the purity (Fukuoka et al., 2006; VTPAEPC, 2008). In Vietnam, many studies have focused on aromatic rice with an aim to understand breeding, production, and future prospects (Nghia et al., 2001a, 2001b).

2.5   AROMATIC RICE IN THE UNITED STATES OF AMERICA

The United States of America (USA) cultivate aromatic rice on a limited scale. Long grain aromatic rice is most common and very familiar in the USA; on cooking, the rice grains separate easily, and kernels are 4 times longer than its breadth (Rohilla et al., 2000). The USA has at least 3 main types of the aromatic rice cultivars (Table 2.5). Della is the predominant and most liked rice cultivars developed by the Louisiana Rice Experiment Station 15 years ago. This cultivar is marketed under several trade names, including Texmati, Pecan Rice, and Della; Jasmine-85, A-201, and Dellamont are other known prominent varieties of aromatic rice in the USA (Table 2.1) (Singh et al., 2000b, 2000e).

2.6   AROMATIC RICE IN CHINA

China is the largest producer of rice in the world and has played a fundamental role in changing the overall world's rice scenario (Bhasin, 2000). The cultivars of aromatic rice in China are less known outside the country due to the strict regional characterization and low yield (Yang et al., 2012). China also has rich famous traditional regional aromatic rice cultivars such as Congjiangxixiangmi, Huanglongxiangmi, Jiangyongxiangdao, Jingcixiangdao, Jingxixiangdao, Qingbuxiangjingmi, Qufuxiangmi and

TABLE 2.5   Main aromatic rice cultivars in the USA

S. N.	Types of Aromatic Rice	Examples of USA Rice Cultivars
1.	Basmati type	Calmati-201 and Calmati-202
2.	American long grain type	Sierra, Della, A201, Delrose, Delmont
3.	Jasmine Type	Jasmine-85, Jazzman, JES and a number of private company varieties

Yongshunxiangdao for many centuries (Yang et al., 2012) and has developed its own specific aromatic rice cultivars (Table 2.1). The residents of China prefer to eat the semi-aromatic rice as compared to pure aromatic rice (Singh et al., 2000c).

In 1985, China started a breeding program with the objective to develop semi-dwarf, high-quality scented rice (Singh et al., 2000c). Yang et al. (1988) developed an aromatic rice variety "Tainung Sen 20." In 1994, Gan-wan-xian 22, an aromatic rice cultivar, was released with many promising characteristics such as intermediate gelatinization temperature (GT), high amylose content (AC), soft gel consistency (GC), and agronomical parameters (Singh et al., 2000c). Shuang-Zhu-Zhan is also an important high-quality rice variety that is widely planted in south China (Chuanyuang and Shuzhan, 1994). A team of Crop Breeding and Cultivation Institute, Zhejiang Academy of Agricultural Sciences, China, under the leadership of Professor Qiu Baigin developed a new scented rice "Zhe 9248," which has excellent eating quality like glossy, smooth, and soft grains on cooking (Singh et al., 2000c). China also received a credit in 1995 to develop the first aromatic rice hybrid (Kunlu, 1995; Kunlu and Fuming, 1995).

2.7 AROMATIC RICE IN INDONESIA

In Indonesia, aromatic rice was already cultivated from a long time ago. Many important quality characteristics, viz. a perfumed, nutty flavor, aroma and a light, fluffy texture after cooking were noted in Indonesian aromatic cultivars as well as in cultivars of other countries such as India (Ramiah and Rao, 1953; Singh et al., 1997; Singh and Singh, 1998; Singh, 2000; Verma et al., 2012, 2013, 2015), Thailand (Anonymous, 1997c; Sarkarung et al., 2000; Singh et al., 2000e; Nayak et al., 2002), Vietnam (Buu, 2000; Nghia et al., 2001a, 2001b; Nayak et al., 2002), USA (Rohilla et al., 2000; Singh et al., 2000b, 2000e), Pakistan (Singh et al., 2000c, 2000e), Bangladesh (Efferson et al., 1985; Islam et al., 1996; Baqui et al., 1997; Das and Baqui, 2000), Iran (Nematzadeh et al., 2000; Moumeni et al., 2003; Nematzadeh and Kiani, 2007; Vazirzanjani et al., 2011), and Afghanistan (Singh et al., 2000b; Sarhadi et al., 2006, 2008, 2009, 2015). Usually, farmers of Indonesia prefer irrigated lands such as Pandanwangi and Rojolele to grow local aromatic rice cultivars (Haryanto et al., 2008).

Indonesian Centre for Agricultural Biotechnology and Genetic Resources Research and Development (ICABGRRD) and Indonesian Centre for Rice Research (ICRR) are the major bodies involved in developing new cultivars of rice. In addition to Indonesian aromatic rice cultivars (Table 2.1), several new cultivars of aromatic rice such as Bengawan Solo, Sintanur, and Batang Gadis have already been developed in 1993, 2001, and 2002, respectively; these cultivars are known for high yield, early maturing, and tolerance against diseases and pests (Seno et al., 2013). Bengawan Solo is a short duration aromatic rice from Indonesia with strong aroma and good grain quality. The aroma of Bengawan Solo is maintained across the regions, and thus, it could be grown widely in Indonesia (Partoatmodjo et al., 1994; Singh et al., 2000c).

In Indonesia, around 11.6 million ha upland area is not used for crop production, according to the Central Bureau of Statistic report (CBS, 2004). Researchers from Indonesia are focusing to develop upland aromatic rice (Lafitte and Courtois, 2002; Totok, 2004; Totok and Utari, 2005; Totok et al., 2005; Kato et al., 2006a, 2006b) to improve upland rice quality and upland productivity, i.e., only 1.2 million ha upland area contributes to rice production, which is 2.6 million tons per year with productivity of 2.27 t ha^{-1} (CBS, 2004). Studies on developed upland aromatic rice have been reported by Lafitte and Courtois (2002) and Kato et al. (2006a, 2006b). Situpatenggang is an aromatic rice cultivar released in 2002 for upland aromatic rice. However, information about the cultivation of aromatic rice specifically in the upland is still limited.

2.8   AROMATIC RICE IN JAPAN

In Japan, aromatic rice has been cultivated on a small scale, particularly in hilly areas for daily use, special guests, festivals, or religious celebrations in many regions for over a thousand years (Arashi, 1975; Kondo, 1987; Itani, 2002; Itani et al., 2004; Okoshi et al., 2016). Japanese aromatic rice cultivars are now cultivated only for domestic consumption in the inner hilly areas (Okoshi et al., 2016). A large number of aromatic rice landraces and important cultivars from Japan are distributed in hilly regions and

TABLE 2.6 List of Aromatic Rice Cultivars from Japan

Japanese aromatic rice	Kaisen	Sasaminori
Haginokaori	Kamari	Nioi-akako
Hendoyori	Kaori-ine	Nioi-kichi
Hieri	Kaori-wase	Nioi-mai
Hieri-kochi	Karasu-mochi	Nioi-mochi
Hokkaido-40nichi-wase	Keitoku	Nioi-wase
Iwaka	Koutou	Nioi-yoshi
Izayoi	Kumamoto	Oitakoutou
Jakou	Kumamoto-zairai	Saigoku-kawachi
Jakou-ibaraki	Mangoku	Sari Queen
Jakouine	Meragome	Sawakaori
Jakou-ine	Miyakaori	Sasaminori
Jakou-mochi	Narukosan-koutou	Shiro-wase
Jakou-niigata	Nezumi-gome	Wase-hieri
Jakou-wase		

Sources: Ootsuka, et al. (2014), Okoshi, et al. (2016).

other parts of Japan (listed in Tables 2.1 and 2.6, respectively). The first fragrant (aroma) landraces were recorded during the period of 17th century (Ootsuka et al., 2014). Japanese aromatic rice cultivars are generally differentiated by their region-based geographical distribution and characteristics (Okoshi et al., 2016). For instance, the eastern and western parts of Japan were represented by Jakou and Kabashiko aromatic rice cultivars, respectively, whereas Hieri or Sawakaori aromatic rice are cultivated in Kochi Prefecture (Nakamura et al., 1996). These aromatic rice cultivars differ in their morphology and agronomical characteristics from each other (Miyagawa and Nakamura, 1984; Itani, 2002). In Japan, the most active producer of aromatic rice is Kubokawa area in Kochi Prefecture (Itani, 1993; Fushimi and Ishitani, 1994). *Hieri* is local, late-heading, japonica-type, and leading cultivar of aromatic rice. Miyakaori, released in 1983 as the first improved cultivar of aromatic rice, has early to medium heading, semi-dwarf with moderate lodging resistance characteristics (Oikawa et al., 1991). Two other varieties of aromatic rice released in Japan are Haginokaori (traditional improved) and Sari Queen, which is a Japanese/

Indian-Basmati-related, late-heading cultivar of aromatic rice (Itani, 1993; Chaudhary, 2003; Itani et al., 2004).

2.9 AROMATIC RICE IN BANGLADESH

Aromatic rice varieties are the immense gift of Mother Nature to Bangladesh, which has major market potentials in other countries. There are many varieties of aromatic rice varying from small grain to long grain and slender types are cultivated in Bangladesh (Table 2.1), though they are not classified by any international standard grade. All aromatic rice varieties in Bangladesh are grown without chemical fertilizer (Baqui et al., 1997; Das and Baqui, 2000; Anik and Talukder, 2002; Tama et al., 2015).

All the Bangladeshi aromatic rice cultivars belong to special aromatic, specialty, or gourmet class of rice. All these cultivars are of traditional type, photoperiod-sensitive, and are normally transplanted during Aman season in the lowland rainfed ecosystem. Most of them are the main cultivars of a particular region of Bangladesh, which are popularly grown in a specific location (Table 2.7). This is believed to be due to the variations in agro-ecological conditions (Efferson et al., 1985; Shahidullah et al., 2010). Bangladeshi aromatic rice could find significant markets in both domestic and international trade, which fetch premium prices and thus deserve special treatment.

In the 19th century, Bangladesh Rice Research Institute (BRRI) developed quite a good number of aromatic rice varieties. BR-16 (Shahibalam), BR-36, BR-38, BR-39, and BR-6 are the fine quality aromatic rice with slender grains and have attractive qualities in the export trade. Some of

TABLE 2.7   Principal Cultivars of Aromatic Rice Cultivated in a Particular Area of Bangladesh

Aromatic Rice Cultivars	Area of Cultivation
Badshabhog	Dinajpur and Kustia
Banshful	Naogaon
Chinigura	Mymensingh and Dhaka
Dadkhani	Natore and Rajshahi
Kalajira	Sherpur and Mymensingh
Kataribhog	Dinajpur
Tulshimala	Chittagong

them have slightly pleasant aroma, while BR-5 (Dolhabhog), BR-34, and BR-38 aromatic rice are considered excellent among cultivars with small to medium grain size, have fine aroma, and generally used by consumers for polao and biryani type preparations. BRRI Dhan-34 is locally known as Khashkhani. BR-26 rice is also an important aromatic rice type with long slender grain that has soft and slightly sticky consistency qualities that should have demand in the East Asian markets (viz., Japan and Korea) (BBS, 1994; Baqui et al., 1997; Das and Baqui, 2000; Anik and Talukder, 2002).

In Bangladesh, Kalijira rice (very small grain type) is considered the "prince of rice." Kalijira and Kataribhog ((long and slender type) are important aromatic cultivars predominant in Mymensingh and Dinajpur, respectively. Chinigura, very small grain aromatic rice, is another important aromatic cultivar predominant among the Bangladeshi aromatic rice culti-vars and is grown in more than 70% area in the northern districts of Naogaon and Dinajpur. For regular consumption, local cultivars such as Banshful, Bau-pagal, Kataktara, Kataribhog, and Nizershail are among those that are currently grown. The cultivars Kalajira and Badshabhog of northern Ban-gladesh are known for more scented varieties. On cocking, these cultivars emits very strong aroma. In addition, different organs such root, stem, leaf, and flowers of such cultivars also emit fragrance in their cultivated field. Kataribhog, a long and slender grain aromatic rice cultivar, is also scented but to a lesser degree (Efferson et al., 1985; BBS, 1994; Islam et al., 1996; Baqui et al., 1997; Das and Baqui, 2000; Shahidullah et al., 2010).

The international trade of Bangladeshi aromatic rice is growing very rapidly. High demand of such aromatic rice in both local and foreign trade has increased to a great extent for domestic consumption and for-eign export in recent years, which fetches premium prices especially in the markets of the EU and USA at a great comparative advantage.

2.10   AROMATIC RICE IN PAKISTAN

Pakistan is an important country among the most aromatic rice produc-ing countries in the world. Aromatic rice is cultivated most in the parts of Punjab, Sindh, North-West Frontier Province (NWFP, now may refer as Khyber Pakhtunkhwa), and province of Baluchistan, which contributed 44%, 44%, 3%, and 9%, respectively, of total production. Punjab has 80%

contribution of total area for cultivation with respect to Basmati-type rice among all the other areas. One-third part of the total production of Pakistani aromatic rice is exported, whereas two-thirds are consumed by local residents. The Rice Research Institute (RRI), Kala Shah Kaku (Punjab), has a long history to devote to basmati rice research. RRI has developed may rice cultivars that have excellent grain quality traits and aroma. Some of them are Basmati-370 (released in 1933), Basmati Pak (in 1968), Basmati-185, Basmati-385, Super Basmati (in 1996), etc., as listed in Table 2.1. In Pakistan, Super Basmati is most popular and regarded as a premium variety (Singh et al., 2000c, 2000e).

2.11 AROMATIC RICE IN AFGHANISTAN

In Afghanistan, rice is the 2nd most important crop. Aromatic rice in Afghanistan and its surrounding areas is in very high demand, and this demand is interestingly increasing among the consumers because of its aroma, good taste, and soft texture after cooking (Sarhadi et al., 2006, 2009, 2015). Many aromatic rice cultivars are credited to this country, as shown in Table 2.1. Bahra is one of most important aromatic rice cultivars from Afghanistan (Singh et al., 2000b), and other cultivars such as Bala, Lawangi, and Pashadi are also introduced from Afghanistan as desirable aromatic rice cultivars (Sarhadi et al., 2008).

2.12 AROMATIC RICE IN IRAN

In Iran, rice is the 2nd main staple food crop after wheat among the cereals that has an important role in Iranian food habits (Moumeni et al., 2003; Nematzadeh and Kiani, 2007; Vazirzanjani et al., 2011; Kiani et al., 2012). Rice cultivation is over 600,000 hectares with a yield of 4–5 ton/ha in Iran. Indica type of rice cultivars are mostly grown in Iran under irrigation condition. Mazandaran and Gilanare are the two states of Iran that contribute more than 80% of Iran's rice production. Consumers in Iran generally prefer two important quality factors such as grain shape and aroma (Moumeni et al., 2003). The aromatic rice cultivars of Iran are listed in Table 2.1. Some of them such as Anbar-boo, Domsiah, Hassan, Hassani, Mehr, Mirza, Mosa Tarom, Sadri, Salari, Saraie, and Tarom are the most important local cultivars that are highly aromatic and considered as high-quality rice cultivars (Nematzadeh et al., 2000; Moumeni et al., 2003). Local Iranian varieties that had good quality became rare (Nematzadeh and Kiani, 2007; Vazirzanjani et al., 2011). The Rice Research Institute of Iran (RRII), Rasht in Gilan Province, is making a concerted effort in the development of rice varieties (Nematzadeh et al., 2000).

2.13 CONCLUSION

Aromatic rice varieties are playing a vital role in global rice trading. In aromatic rice, flavor is considered as one of the most significant factors in market business, which distinguishes it from ordinary rice. Rice grain of aromatic cultivars is characterized on the basis of certain unique features on which they may be graded as export quality rice with good unique nutritional values. Consumers have become more quality conscious about the rice cultivars they consume. When farmers became conscious of their rice quality, they were driven to produce better quality rice. The present chapters address an overview on aromatic rice grown in different countries, which has good quality attributes along with production. Therefore, there is a strong need to focus and explore the quality of rice such as aromatic rice along with production.

ACKNOWLEDGMENT

Deepak Kumar Verma and Prem Prakash Srivastav are indebted to Department of Science and Technology, Ministry of Science and Technology, Govt. of India for an individual research fellowship (INSPIRE Fellowship Code No.: IF120725; Sanction Order No. DST/INSPIRE Fellowship/2012/686 & Date: 25/02/2013).

KEYWORDS

- **aroma**
- **aromatic rice**

- **basmati**
- **cultivars**
- **genotypes**
- **Khao Dawk Mali (KDML)**
- **rice diversity**
- **rice quality**

REFERENCES

Anik, A. R., & Talukder, R. K. (2002). Economic and financial profitability of aromatic and fine rice production in Bangladesh. *Bangladesh Journal of Agriculture Economics, 25*(2), 103–113.

Anonymous, (1994). A report on Zhe 9248-new scented rice line came out in Zhejiang. *Chinese Rice Research Newslt., 2,* 11.

Anonymous, (1996). National yield contest of KDML 105. Department of Internal Trade. *Ministry of Commerce*, Thailand.

Anonymous, (1997a). *Khao Hawm Klong Luang; Varietal Description*. Rice Research Institute. Department of Agriculture, Thailand.

Anonymous, (1997b). *Khao Hawm Supanburi: Varietal Description*. Rice Research Institute. Department of Agriculture, Thailand.

Anonymous, (1997c). *Office of Rice Quality Control*. Thai Chamber of Commerce, Thailand.

Arashi, K. (1975). Kinsei inasaku gijutsu shi: sono ritchi seitaiteki kaiseki. (In *Japanese: History of Rice Cultivation in Modern Times: Geological and Ecological View Point*). Nobunkyo, Tokyo, Japan, pp.468–490.

Baqui, M. A., Ham, M. E. Jones, D., & Straingfellow, R. (1997). *The Export Potential of Traditional Varieties of Rice from Bangladesh*. Bangladesh Rice Research Institute, Gazipur, Bangladesh.

BBS (Bangladesh Bureau of Statistics). (1994). Statistical year book of Bangladesh. Bangladesh Bureau of Statistics, Statistics Division, Ministry of Planning, Dhaka, Bangladesh.

Bhasin, V. K. (2000). India and the emerging global rice trade. In: *Aromatic Rices* (Singh, R. K., Singh, U. S., & Khush, G. S., eds.), Oxford & IBH Publ., New Delhi, India, pp. 257–276.

Bisne, R., & Sarawgi, A. K. (2008). Agro-morphological and quality characterization of badshah bhog group from aromatic rice germplasm of Chhattisgarh. *Bangladesh Journal of Agricultural Research, 33*(3), 479–492.

Boriboon, S. (1996). Khao Dawk Mali 105, Problems, Research efforts, and Future prospects. In: *Report of the INGER Monitoring Visit on Fine Grain Aromatic Rice in India,* Iran, Pakistan and Thailand, 21th September—10th October, 1996. pp.102–111.

Buu, B. C. (2000). Aromatic races of Vietnam, In: *Aromatic Rices* (Singh, R. K., Singh, U. S., & Khush, G. S., eds.), Oxford & IBH Publ., New Delhi, India. pp. 188–190.

CBS (Central Bureau of Statistic). (2004). Indonesian statistic. Indonesian Central Bureau of Statistic, Jakarta. 35.

Chang, T. T. (1976). The origin, evolution, cultivation, dissemination and diversification of Asian and African rices. *Euphytica, 25*, 425–441.

Chaudhary, R. C. (2003). Speciality rices of the world: effect of WTO and IPR and on its production trend and marketing. *Journal of Food Agriculture and Environment, 1*, 34–41.

Chuanyuang, Y., & Shuzhen, G. (1994). Scented rice in Jiangxi province, China. *International Rice Research Newsletter, 19*, 8–9.

Das, T., & Baqui, M. A. (2000). Aromatic rices of Bangladesh. In: *Aromatic Rices* (Singh, R. K., Singh, U. S., & Khush, G. S., eds.), Oxford & IBH Publ. New Delhi, India. pp. 186–187.

Efferson, J. N. (1985). Rice quality in world markets, in: *Proceedings of the Conference on Grain Quality and Marketing,* International Rice Research Conference, 1–5 June 1985, Los Banos, Laguna, Philippines, pp.1–29.

Fukuoka, S., Suu, T. D., Ebana, K., Trinh, L. N., Nagamine, T., & Okuno, K. (2006). Diversity in phenotypic profiles in landrace populations of Vietnamese rice: a case study of agronomic characters for conserving crop genetic diversity on farm. *Genetic Resources and Crop Evolution, 53*, 753–761.

Fushimi, T., & Ishitani, T. (1994). Scented rice and its aroma compounds. *Koryo* (Japan Perfumery and Flavorings Association), *183*, 73–80.

Gay, F., Mestres, C., Phung, C. V., Lang, N. T., Thinh, D. K., Laguerre, M., Boulanger, R., & Davrieux, F. (2004). Promising new technologies for classifying aromatic rices. *Omonrice, 12*, 157–161.

Haryanto, T. A. D., Suwarto, S., & Yoshida, T. (2008). Yield stability of aromatic upland rice with high yielding ability in Indonesia. *Plant Production Science, 11*(1), 96–103.

Hien, N. L., Sarhadi, W. A., Oikawa, Y., & Hirata, Y. (2007). Genetic diversity of morphological responses and the relationships among Asia aromatic rice (*Oryza sativa* L.) cultivars. *Tropics, 16*(4), 342–355.

Islam, M. R., Mustafi, B. A. A., & Hossain, M. (1996). Socic-economic aspects of fine quality rice cultivation in Bangladesh (Rice Res. Prioritization), *BRRI/IRRI.*

Itani, T. (1993). History, cultivation and breeding of aromatic rice varieties (in Japanese with English abstract). *Bull. Hiroshima Prefect University, 5*, 267–281.

Itani, T. (2002). Agronomic characteristics of rice cultivars collected from Japan and other countries (in Japanese with English abstract). *Japanese Journal of Crop Science, 71*, 68–75.

Itani, T., Tamaki, M., Hayata, Y., Fushimi, T., & Hashizume, K. (2004). Variation of 2-acetyl-1-pyrroline concentration in aromatic rice grains collected in the same region in Japan and factors affecting its concentration. *Plant Production Science, 7*(2), 178–183.

Kamath, S., Stephen, J. K. C., Suresh, S., Barai, B. K., Sahoo, A. K., Reddy, K. R., & Bhattacharya, K. R. (2008). Basmati rice: Its characteristics and identification. *Journal of the Science of Food and Agriculture, 88*(10), 1821–1831.

Kato, Y., Kamoshita, A., & Yamagishi, J. (2006b). Growth of three rice (*Oryza sativa* L.) cultivars under upland conditions with different levels of water supply. 2. Grain yield. *Plant Production Science, 9*, 435–445.

Kato, Y., Kamoshita, A., Yamagishi, J., & Abe, J. (2006a). Growth of three rice (*Oryza sativa* L.) cultivars under upland conditions with different levels of water supply. Nitrogen content and dry matter production. *Plant Production Science, 4*, 422–434.

Khush, G. S. (2000). Taxonomy and origin of rice. In: *Aromatic Rices* (Singh, R. K., Singh, U. S., & Khush, G. S., eds.), Oxford & IBH Publ., New Delhi, India. pp. 5–14.

Kiani, G., Nematzadeh, G. A., Ghareyazie, B., & Sattari, M. (2012). Pyramiding of cry1Ab and FGR genes in two Iranian rice cultivars Neda and Nemat. *Journal of Agriculture, Science and Technology, 14*, 1087–1092.

Kondo, H. (1987). *Visit the Forgotten Rice* (in Japanese). Japan book Publication Society, Tokyo, Japan, pp. 5–81.

Kunlu, Z., & Fuming, L. (1995). Xiangyou 63, a quasi-aromatic hybrid rice with good quality and high yield. *International Rice Research Notes, 20*, 9–10.

Kunlu, Z. (1995). Xiangyou 63, a quasi-aromatic hybrid rice with good quality and high yield. *Hunan Agricultural Research Newsletter, 2*, 4–6.

Lafitte, H. R., & Courtois, B. (2002). Interpreting cultivar × environment interactions for yield in upland rice: assigning value to drought-adaptive traits. *Crop Science, 42*, 1409–1420.

Miyagawa, S., & Nakamura, S. (1984). Regional differences in varietal characteristics of scented rice (in Japanese with English abstract). *Japanese Journal of Crop Science, 53*, 494–502.

Moumeni, A., Samadi, B. Y., Wu, J., & Leung, H. (2003). Genetic diversity and relatedness of selected Iranian rice cultivars and disease resistance donors assayed by simple sequence repeats and candidate defense gene markers. *Euphytica, 131*, 275–284.

Nakamura, Y., Kameshima, M., Mizobuchi, M., Uga, H., & Morita, Y. (1996). A new scented rice variety 'Kouikuka 37'. *Bulletin Kochi Agricultural Research Center, 5*, 38–49.

Nawar, W. W. (1996). Lipids. In: *Food Chemistry* (Fennema, O. R., ed.). Dekker, New York, pp. 225–319.

Nayak, A. R., Reddy, J. N., & Pattnaik, A. K. (2002). Quality evaluation of some Thailand and Vietnam scented rice. *Indian Journal of Plant Genetic Resources, 15*(2), 125–127.

Nematzadeh, G. A., & Kiani, G. (2007). Agronomic and quality characteristics of high-yielding rice lines. *Pakistan Journal of Biology and Science, 10*, 142–144.

Nematzadeh, G. A., Karbalaie, M. T., Farrokhzad, F., & Ghareyazie, B. (2000). Aromatic races of Iran, In: *Aromatic Rices* (Singh, R. K., Singh, U. S., & Khush, G. S., eds.), Oxford & IBH Publ., New Delhi, India. pp. 191–199.

Nghia, N. H., Buu, B. C., Trinh, L. N., & Thao, L. V. (2001a). Speciality rice in Viet Nam: breeding, production and marketing. In: *Speciality Rices of the World. Breeding, Production and Marketing* (Duffy, R., Chaudhary, R. C., & Tran, D. V., eds.). FAO of the UN Rome. Science Publishers, Inc. Plymouth, UK. pp.175–190.

Nghia, N. H., Buu, B. C., Trinh, L. N., & Thao, L. V. (2001b). Improvement of aromatic rice in Viet Nam. In: *Speciality Rices of the World. Breeding, Production and Marketing* (Duffy, R., Chaudhary, R. C., & Tran, D. V., eds.). FAO of the UN Rome. Science Publishers, Inc. Plymouth, UK. pp. 191–200.

Oikawa, T., Sasaki, T., Suzuki, K., Abe, S., Matsunaga, K., Wakin, S., & Tanno, K. (1991). A new variety of aromatic rice "Miyakaori". *Bulletin Miyagi Prefect Furakavu Agriculture Experiment Station, 1*, 50–62.

Oka, H. I. (1958). Intervarietal and classification of cultivated rice. *Indian Journal of Genetics and Plant Breeding, 18*, 79–89.

Okoshi, M., Matsuno, K., Okuno, K., Ogawa, M., Itani, T., & Fujimura, T. (2016). Genetic diversity in Japanese aromatic rice (*Oryza sativa* L.) as revealed by nuclear and organelle DNA markers. *Genetic Resources and Crop Evolution, 63*, 199–208.

Okuno, K., Katsuta, M., Nakayama, H., Ebana, K., & Fukuoka, S. (1998). International collaboration on plant diversity analysis. In: *Plant Genetic ReSources: Characterization and Evaluation.* NIAR-MAFF, Japan, pp. 158–169.

Ootsuka, K., Takahashi, I., Tanaka, K., Itani, T., Tabuchi, H., Yoshihashi, T., Tonouchi, A., & Ishikawa, R. (2014). Genetic polymorphisms in Japanese fragrant landraces and novel fragrant allele domesticated in northern Japan. *Breeding Science, 64*, 115–124.

Partoatmodjo, A., Alidawati, & Harahap, Z. (1994). Bengawan solo, a short duration aromatic rice in Indonesia. *International Rice Research Notes, 19*, 19–20.

Ramaiah, K., & Rao, M. V. B. N. (1953). Rice breeding and genetics. *ICAR Science Monograph 19.* Indian Council of Agricultural Research, New Delhi, India.

Rohilla, R., Singh, V. P., Singh, U. S., Singh, R. K., & Khush, G. S. (2000). Crop husbandry and environmental factors affecting aroma and other quality traits. In: *Aromatic Rices* (Singh, R. K., Singh, U. S., & Khush, G. S., eds.), Oxford & IBH Publ., New Delhi, India. pp. 201–316.

Sarhadi, W. A., Hien, N. L., Nhi, P. T. P., & Hirata, Y. (2006). Characterization of Afghanistan aromatic rice specific nature. *Breeding Research, 8*(Suppl. 2), 230.

Sarhadi, W. A., Hien, N. L., Zanjani, M., Yosofzai, W., Yoshihashi, T., & Hirata, Y. (2008). Comparative analysis for aroma and agronomic traits of native rice cultivars from central Asia. *Journal of Crop Science and Biotechnology, 11*, 17–22.

Sarhadi, W. A., Ookawa, T., Yoshihashi, T., Madadi, A. K., Yosofzai, W., Oikawa, Y., & Hirata, Y. (2009). Characterization of aroma and agronomic traits in Afghan. native rice cultivars. *Plant Production Science, 12*, 63–69.

Sarhadi, W. A., Shams, S., Bahram, G. M., & Sadeghi, M. B. (2015). Comparison of yield and yield components among different varieties of rice. *Globle Journal of Biology, Agriculture & Health Science, 4*(1), 158–163.

Sarkarung, S., Somrith, B., & Chitrakorn, S. (2000). Aromatic rices of Thailand. In: *Aromatic Rices* (Singh, R. K., Singh, U. S., & Khush, G. S., eds.), Oxford & IBH Publ., New Delhi, India. pp. 180–187.

Seno, D. S. H., Artika, M., Nurcholis, W., Santoso, T. J., Trijatmiko, K. R., Padmadi, B., Praptiwi, D., Ching, J. M., & Masud, Z. A. (2013). Identification of badh2 mutation type among Indonesian fragrant rice varieties. *Journal of Biology, Agriculture and Healthcare, 3*(17), 2224–3208.

Shahidullah, S. M., Hanafi, M. M., Ashrafuzzaman, M., Ismail, M. R., Salam, M. A., & Khair, A. (2010). Biomass accumulation and energy conversion efficiency in aromatic rice genotypes. *Comptes Rendus Biologies, 333*, 61–67.

Singh, R. K., & Singh, U. S. (1998). Indigenous scented rices of India: a survival issues. In: *Sustainable Agriculture for Food, Energy and Industry* (Bassam, N. E., Behl, R. K., & Prohnow, B., eds.). *Proceeding of International Conference* held in Braunschweig (Germany). pp. 676–681.

Singh, R. K., Khush, G. S., Singh, U. S., Singh, A. K., & Singh S. (2000c). Breeding aromatic rice for high yield, improved aroma and grain quality. In: *Aromatic Rices* (Singh, R. K., Singh, U. S., & Khush, G. S., eds.). Oxford & IBH Publ., New Delhi, India. pp. 71–105.

Singh, R. K., Singh, U. S., & Khush, G. S. (1997). Indigenous aromatic rices of India: Present scenario and needs. *Agriculture Situation in India LVI, 8*, 491–496.

Singh, R. K., Singh, U. S., & Khush, G. S. (2000b). Prologue. In: *Aromatic Rices* (Singh, R. K., Singh, U. S., & Khush, G. S., eds.). Oxford & IBH Publ., New Delhi, India. pp. 1–3.

Singh, R. K., Singh, U. S., & Khush, G. S. (2000e). Aromatic Rices of other countries. In: *Aromatic Rices* (Singh, R. K., Singh, U. S., & Khush, G. S., eds.). Oxford & IBH Publ., New Delhi, India. pp. 179–200.

Singh, R. K., Singh, U. S., Khush, G. S. Rohilla, R., Singh, J. P., Singh, G., & Shekhar, K. S. (2000d). Small and medium grained aromatic rices of India. In: *Aromatic Rices* (Singh, R. K., Singh, U. S., & Khush, G. S., eds.). Oxford & IBH Publ., New Delhi, India. pp. 155–178.

Singh, R. K., Singh, U. S., Khush, G. S., & Rohilla, R. (2000a). Genetics and biotechnology of quality traits in aromatic rices. In: *Aromatic Rices* (Singh, R. K., Singh, U. S., & Khush, G. S., eds.). Oxford & IBH Publ., New Delhi, India. pp. 47–69.

Singh, V. P. (2000). The basmati rice of India. In: *Aromatic Rices* (Singh, R. K., Singh, U. S., & Khush, G. S., eds.). Oxford & IBH Publ., New Delhi, India. pp. 135–154.

Tama, R. A. Z., Begum, I. S., Alam, M. J., & Islam, S. (2015). Financial profitability of aromatic rice production in some selected areas of Bangladesh. *International Journal of Innovation and Applied Studies, 12*(1), 235–242.

Tava, A., & Bocchi, S. (1999). Aroma of cooked rice (*Orvza sativa*): Comparison between commercial Basmati and Italian line 135–3. *Cereal Chemistry, 76*, 526–529.

Totok, A. D. H., & Utari, R. S. (2005). Yield trial of F6 pure lines from crossing of Mentikwangi x Poso in comparison with their parents. Research report, *Faculty of Agriculture*, Jenderal Soedirman University, pp. 22–30.

Totok, A. D. H. (2004). Growth, yield and rice quality of F5 genotypes from crossing of Mentikwangi X Poso for developing of aromatic upland rice. *Journal of Rural Development*, Jenderal Soedirman University, *2*, 122–128.

Totok, A. D. H., Daryanto, S., & Soesanto, L. (2005). Constructing of high yielding and aromatic upland rice variety for improving of its production and economic value. *Journal of Agroland, Faculty of Agriculture*, Tadulako University, *3*, 93–99.

Vazirzanjani, M., Sarhadi, W. A., New, J. J., Amirhosseini, M. K., Siranet, R., Trung, N. Q., Kawai, S., & Hirata, Y. (2011). Characterization of aromatic rice cultivars from Iran and surrounding regions for aroma and agronomic traits. *SABRAO Journal of Breeding and Genetics, 43*(1), 15–26.

Verma, D. K., & Srivastav, P. P. (2016). Extraction technology for rice volatile aroma compounds In: *Food Engineering Emerging Issues, Modeling, and Applications* (eds. Meghwal, M., & Goyal, M. R.). In book series on Innovations in Agricultural and Biological Engineering, Apple Academic Press, USA. pages 245–291.

Verma, D. K., & Srivastav, P. P. (2017). Proximate composition, mineral content and fatty acids analyses of aromatic and non-aromatic Indian rice. *Rice Science, 24*(1), 21–31.

Verma, D. K. (2011). Physico-chemical and cooking characteristics of azad basmati (CSAR 839–3) – *a Newly Evolved Variety of Basmati Rice* (*Oryza sativa* L.). *MSc. (Agri.) Thesis*, Chandra Shekhar Azad University of Agriculture and Technology, Kanpur (UP) – 208002, INDIA.

Verma, D. K., Mohan, M., & Asthir, B. (2013). Physicochemical and cooking characteristics of some promising basmati genotypes. *Asian Journal of Food Agro-Industry*, *6*(2), 94–99.

Verma, D. K., Mohan, M., Prabhakar, P. K., & Srivastav, P. P. (2015). Physico-chemical and cooking characteristics of Azad basmati. *International Food Research Journal*, *22*(4), 1380–1389.

Verma, D. K., Mohan, M., Yadav, V. K., Asthir, B., & Soni, S. K. (2012). Inquisition of some physico-chemical characteristics of newly evolved basmati rice. *Environment and Ecology*, *30*(1), 114–117.

VTPAEPC (Vietnam Trade Promotion Agency Export Promotion Center), (2008). Report on vietnamese rice sector (Bao cao nganh hang gao Viet Nam). Hanoi, 12/2008. Online Access on 23/01/2016.

Yang, S. C., Chang, W. L., Chao, C. N., & Chen, L. C. (1988). Development of aromatic rice variety Tainung Sen 20. *Journal of Agriculture Research in China, 37,* 349–359.

Yang, S. Y., Zou, Y. B., Liang, Y. Z., Xia, B., Liu, S. K., Ibrahim, M., Li, D. Q., Li, Y. Q., Chen, L., Zeng, Y., Liu, L., Chen, Y., Li, P., & Zhu, J. W. (2012). Role of soil total nitrogen in aroma synthesis of traditional regional aromatic rice in China. *Field Crops Research*, *125*, 151–160.

CHAPTER 3

EXTRACTION METHODS FOR CHEMICAL CONSTITUTES ASSOCIATED WITH RICE AROMA, FLAVOR, AND FRAGRANCE: A SHORT OVERVIEW

DEEPAK KUMAR VERMA,[1] DIPENDRA KUMAR MAHATO,[2] and PREM PRAKASH SRIVASTAV[3]

[1]Research Scholar, Agricultural and Food Engineering Department, Indian Institute of Technology, Kharagpur–721302, West Bengal, India, Mobile: +91-7407170260, +91-9335993005, Tel.: +91-3222-281673, Fax: +91-3222-282224, E-mail: deepak.verma@agfe.iitkgp.ernet.in, rajadkv@rediffmail.com

[2]Senior Research Fellow, Indian Agricultural Research Institute, Pusa Campus, New Delhi–110012, INDIA, Mobile: +91-9911891494, +91-9958921936, E-mail: kumar.dipendra2@gmail.com

[3]Associate Professor, Agricultural and Food Engineering Department, Indian Institute of Technology, Kharagpur–721302, West Bengal, India, Mobile: +91-9434043426, Tel.: +91-3222-283134, Fax: +91-3222-282224, E-mail: pps@agfe.iitkgp.ernet.in

CONTENTS

3.1 INTRODUCTION

Rice aroma, flavor, and fragrance are important quality characteristics for consumer perception (Nayak et al., 2002; Verma and Srivastav, 2016). In the consumer's perception, there are two types of components involved: (1) odor active component and (2) taste active component. The perception of rice aroma, flavor, and fragrance is considered as chemically derived phenomenon. The acceptability and desirability of rice aroma, flavor, and fragrance are highly affected by the cultural background and ethnic diversity of the consumers. For example, the majority of the Indian, indigenous aromatic cultivars of rice are small and medium grained (Singh et al., 2000a; Verma et al., 2012, 2013, 2015). In the USA, long-grain aromatic rice separates easily, and kernels are four times longer than its breadth that is most common and familiar during cooking (Rohilla et al., 2000). China residents prefer to eat semi-aromatic rice compared to pure aromatic rice (Singh et al., 2000c), whereas the Iranian consumers prefer generally two important quality factors: grain shape and aroma (Moumeni et al., 2003). Indonesian people prefer their rice to have a perfumed, nutty flavor aroma with light, fluffy texture after cooking (Partoatmodjo et al., 1994; Haryanto et al., 2008; Seno et al., 2013). Japanese people prefer to cultivate aromatic rice for daily use, special guests, festivals, or religious celebrations in different part of the country since about thousand years (Arashi, 1975; Kondo, 1987; Itani, 2002; Itani et al., 2004; Okoshi et al., 2016). On the basis of the discussion on cultural background and ethnic diversity of the consumer's acceptability and desirability of rice aroma, flavor, and fragrance, distinct differences were found, and specifically, the choice of rice preparation is different in different parts of the world. These rice aroma, flavor, and fragrance are also affected by the culture, specific environment, and ethnicity of the people. Changes in the aroma, flavor, and fragrance of rice, induced by processing, are also important factors that affect their diversity and characteristics.

Presently, it is important to assess the aroma compounds for the extraction of volatile aroma compounds in rice by extraction technology. There is no single technology that is optimal for aroma extraction in rice. Several traditional and modern methods for rice aroma chemical extraction (Table 3.1) coupled with analytical techniques were studied and have been used for extraction and quantification of rice aroma compounds at different levels of concentration ranges from 1–10 ppb level to 2 ppm. These extraction technologies made a cocktail of over 450 compounds—alcohols, aldehydes, esters, heterocyclic, hydrocarbons, ketones, and organic acids, as described previously by many researchers studying various aromatic and nonaromatic cultivars of rice (Widjaja et al., 1996a; Lin et al., 2010).

In addition of all these rice volatile aroma compounds, 2-AP is a particular chemical that has been shown to be the predominantly active volatile aroma compound in aromatic rice at several hundred parts per billion (ppb) level of concentrations and emits a popcorn-like aroma. The presence of this particular rice volatile aroma compound is reported by Yoshihashi (2002) in various parts of the rice plant except roots and is also reported in very trace amounts in nonaromatic rice (Buttery et al., 1983; Grimm et al., 2001; Park et al., 2010). Zhou et al. (1999) reviewed the methods, viz., direct extraction, distillation, and headspace, for volatile aroma compound analysis and their contribution to flavor in cereals, among which headspace complemented by solid phase microextraction has become one of the best isolation and extraction methods. The typical concentration levels (in ppb)

TABLE 3.1 Traditional and Modern Extraction Methods for Rice Volatile Aroma Compounds

S. N.	Traditional Method	S. N.	Modern Method
1.	Purge and trap	1.	Solid phase micro extraction
2.	Simultaneous steam distillation	2.	Supercritical fluid extraction with CO_2
3.	Micro steam distillation		
4.	Direct solvent extraction		

of different volatile aroma compounds from rice were alcohols 1869, aldehydes 5952, disulfides 79, heterocyclic compounds 1220, hydrocarbons 548, ketones 234, phenols 534, and terpenes 257. The discussed methods have been employed for extraction of aroma volatiles that affect considerably the quality and quantity of rice (Sriseadka et al., 2006; Stashenko and Martínez, 2007). Weber et al. (2000) reported that one of the major limitations is the lack of a quantitative assay in the improvement of aromatic rice for aroma through breeding because there is no single method to quantify relative concentration of different compounds or activity of some critical enzymes. Many researchers worked on the best analytical technique for extraction and isolation of rice volatile aroma compounds. To obtain higher extraction yield (referred as high efficiency) and potency of the extract (referred as efficacy), researchers have made considerable efforts to find efficient methods for extraction of volatile compounds for rice aroma.

3.2 TRADITIONAL AND MODERN METHODS FOR EXTRACTION

Rice has trace amounts of aroma constituents. The aroma components are found among carbohydrates, fat, proteins, salts, and water. For these reasons, there is a need of a specific extraction method for rice to be analyzed. There are many traditional extraction techniques such as liquid/liquid solvent extraction, static headspace sampling, dynamic headspace sampling, solid-phase microextraction, and Soxhlet extraction as well as modern extraction techniques such as supercritical fluid distillation, simultaneous steam distillation and molecular distillation, solid phase microextraction, and supercritical fluid extraction with CO_2 to extract aroma from rice. Purge and trap (P&T), simultaneous steam distillation, micro steam distillation, direct solvent extraction, solid phase microextraction, and supercritical fluid extraction with CO_2 are some of the methods that will be discussed.

3.2.1 PURGE AND TRAP

In the 1960s, P&T was used for the study of body fluids. Later, in mid-to-late 1970s, it became a widely applied technique for monitoring volatile organic compounds (VOCs) in drinking water. Today, it is routinely applied for the analysis of VOCs in foods, soil, and water (Seibel, 2015).

A measured amount of sample is placed in a sealed vessel. The sample is purged with inert gas, causing VOCs to be swept out of the sample. The VOCs are retained in an analytical trap, which allows the purge gas to pass through to vent. The VOCs are then desorbed by heating the trap, injected into the GC by backflushing the trap with a carrier gas, and separated and detected by normal gas chromatography–mass spectrometry (GC-MS) operation as shown in Figure 3.1 (Boswell, 1999). The P&T technique provides headspace samples by purging it with an inert gas. It can be applied to both solid and liquid matrices. Extracted compounds are trapped and concentrated in a tube filled with an appropriate sorbent material such as Tenax (Manzini et al., 2011). Connecting the P&T concentrator to the GC inlet is one of the major challenges in P&T GC-MS. The challenges are related to the flow rate requirements of the P&T concentrator, the capillary column, and the mass spectrometer.

On the other hand, the headspace P&T technique dynamically captures the volatile sample fraction in a solid adsorbent, which is

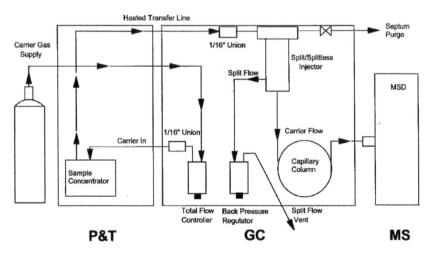

FIGURE 3.1 Purge and trap method coupled to GC-MS (Reprinted from Boswell, C. E. (1999). Fast and Efficient Volatiles Analysis by Purge and Trap GC/MS. WTQA – 15th Annual Waste Testing & Quality Assurance Symposium. pp. 190–194. https://clu-in.org/download/char/dataquality/EBoswell.pdf.).

recovered by subsequent thermal or solvent desorption. According to Contarini and Povolo (2002), the P&T technique is more suited to the determination of compounds with lower molecular weight than SPME. Despite the status and widespread application of P&T with GC/MS for volatile organics, various issues appear in consistent operation over the desired concentration ranges and at the required detection limits (Hollis and Prest, 2016).

3.2.2 SIMULTANEOUS STEAM DISTILLATION

Simultaneous steam distillation/extraction (SDE) is one of the traditional methods to extract aroma compounds from rice and was first described by Likens and Nickerson; so, SDE is also known as the Likens–Nickerson steam distillation (Likens and Nickerson, 1964; Nickerson and Likens, 1966; Siegmund et al., 1996). The SDE method produces solvent extract as the resulting final product (Blanch et al., 1993; Parliment, 1997), and this resulting extract do not contain any nonvolatiles (Siegmund et al., 1996).

To carry out this procedure, the sample and solvent are placed into a type of glassware used for refluxing. The solvent is an organic solvent such as methanol or di-ethyl ether. The sample mixture is heated up in a sand bath, and the vapors formed escape through a water-cooled tube. The extracted compounds are collected in a flask connected to the water-cooled glass tube. Steam distillation may be accomplished in several ways. The product may be simply put in a rotary evaporator (if liquid or initially slur-ried in water if solid), and a distillate is collected. This distillate is solvent extracted to yield an aroma isolate suitable for GC analysis. Chaintreau (2001) has also discussed the evolutionary forms of SDE like atmospheric pressure system and vacuum systems. An atmospheric pressure system (shown in Figure 3.2) works at atmospheric pressure. The steam distilla-tion at atmospheric pressure forms thermal artifacts (Shimoda et al., 1995). Therefore, composition of the extracts is close to that of an essential oil, but atmospheric pressure SDE is not applicable to heat-sensitive products. With this restriction, SDE gives the most representative GC-profiles over a wide range of volatiles as compared to other isolation means (Reineccius, 1993). To avoid thermal modification of flavors and fragrances, a device

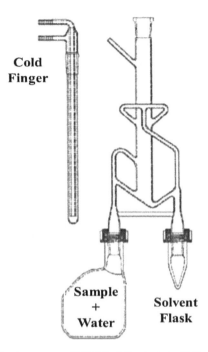

Cold
Finger

Sample
+
Water

Solvent
Flask

FIGURE 3.2 Atmospheric pressure SDE (Godefroot et al., 1981; Rijks et al., 1983).

working under vacuum (V-SDE) as shown in Figure 3.3 is used (Maignial et al., 1992). The vacuum systems must have joints that are air-tight, and all parts of the apparatus must be under rigid temperature control. This method provides high recoveries without thermal artifacts, and a preparative version allows large-scale operations (Pollien and Chaintreau, 1997; Pollien et al., 1998).

In a typical set-up (Figure 3.4), sample is added to distilled water in a larger round-bottom flask and stirred constantly throughout the distillation to avoid bumping. The smaller flask is filled with solvent (typically dichloromethane or freon), and both flasks are heated simultaneously. The sample is heated to 100°C and solvent to its boiling point, 45°C for dichloromethane. The vapors rise in each wing of the apparatus and combine centrally, and the volatiles are transferred between the condensing liquids. The water and the solvent then get collected separately.

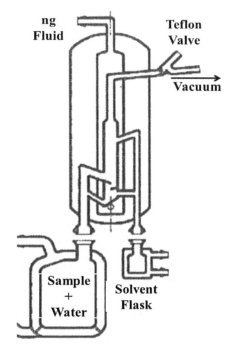

FIGURE 3.3   Vacuum SDE (Maignial et al., 1992).

The main advantage of steam distillation techniques is that the resulting extracts do not contain any nonvolatiles. Some disadvantages are that the glassware tends to be very expensive and delicate. The method itself, when carried out, can also be time consuming. A highly water-soluble sample such as strawberry will have volatile compounds that are very difficult to extract with this method. Target compounds that can easily be oxidized and degraded are also difficult to extract. Even though simultaneous SDE is a fairly simple procedure to carry out (Peppard, 1999), it still has limitations.

3.2.3   MICRO STEAM DISTILLATION

High level of water pollution due to industrial waste along with increasing pollution has created water scarcity for agricultural as well as drinking

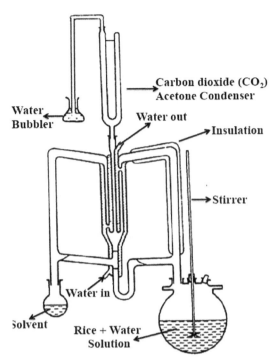

FIGURE 3.4 A typical set up of simultaneous steam distillation/extraction (SDE). (Reprinted from Verma, D. K. and Srivastav, P. P. (2016). Extraction Technology for Rice Volatile Aroma Compounds In: Food Engineering Emerging Issues, Modeling, and Applications (eds. Meghwal, M. and Goyal, M. R.). With permission from Apple Academic Press.).

purpose. Distillation is the process to remove all the contaminants from the water. The water is heated to vapor phase, which is then condensed to water leaving behind all the impurities.

The main components of the system are vertical tube evaporators (VTEs), condenser, mist eliminators, or feed entrainment separators, distillate withdrawal capillaries, forward feed and brine transferring capillaries, and pump (Figure 3.5). The unit is insulated to minimize heat loss to the surroundings (Vyas, 2003; Mudgal and Anurag, 2009). Primary steam from a baby boiler or a solar energy source is fed to the tube side of the first VTE which gets condensed by transferring its latent heat of condensation to the evaporating feed water on the shell side. The fresh steam formed on the shell side of this VTE carries brine mist along with it. To

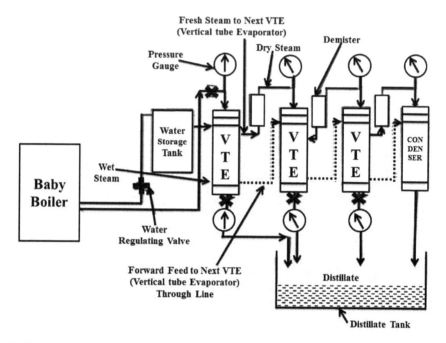

FIGURE 3.5   Micro-steam distillation unit. (Source: Modified from Sen, P. V., Bhuwanesh, K., Ashutosh, K., Engineer, Z., Hegde, S., Sen, P. K. and Lal, R. (2013). Micro-scale multi-effect distillation system for low steam inputs. *Procedia Engineering, 56,* 63–67.).

remove the mist, it is passed through an inter-effect mist separator before being fed to the next effect, and the process is repeated. Due to cyclonic effect, heavier water particles settle down in the mist separator, and steam moves upward due to lower density. The steam is further dewetted using de-mister screens. The steam generated in the last effect is condensed in a falling feed water film condenser. This liquid film is actually the feed water that is preheated in the process. The feed water progress is simplest if it passes from effect one to effect two, to effect three, and so on, as in these circumstances, the feed will flow without pumping. This "forward feed" arrangement also helps in conservation of energy. Fresh brine from water storage tank is used for cooling the condenser and returned to the tank. In this way, all vapors get condensed in the condenser (Sen et al., 2013).

3.2.4 DIRECT SOLVENT EXTRACTION

The direct solvent extraction method also known as liquid–liquid extraction method was initially developed by Landen (1982). Liquid–liquid extraction is an important separation technology (Figure 3.6), with a wide range of applications in the modern process industry. The extraction process is based on different solubilities of components in two immiscible, or partially miscible, liquids. The components that need to be recovered are extracted from the feed stream with the help of a solvent. Both liquids have to be thoroughly contacted and subsequently separated from each other again.

In one of the methods explained by Lee et al. (1998), samples of 0.4 g were accurately weighed into a 125-mL round-bottom glass bottle. Hot deionized water (80°C, 4 mL) was added to the sample and then mixed with a spatula. Ten milliliters of isopropanol was added to the mixture. Approximately, 5 g of anhydrous magnesium sulfate was added followed by 25 mL of extracting solvent (hexane:ethyl acetate, 90:10, v/v) containing

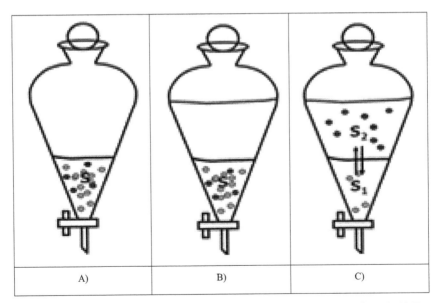

FIGURE 3.6 Liquid-liquid extraction method shown by the separator funnel. A) Two substances dissolved in water, B) add the second immiscible (insoluble) solvent and shake, and C) wait for partitioning then drain off bottom.

0.01% BHT. The mixture was homogenized with a homogenizer for 1 min at medium speed, and the homogenizer was rinsed with 5 mL extracting solvent. The mixture was filtered through a medium porosity glass filter using a vacuum bell jar filtration apparatus. The vacuum was released, and the filter cake was broken with a spatula and washed with 5 mL extracting solvent. The filter cake was transferred to the same 125-mL round-bottom glass bottle for the repeat extraction. Five milliliters of isopropanol and 30 mL of the extracting solvent were added to the mixture followed by homogenization and filtration. The combined filtrate was transferred to a 100-mL volumetric flask and diluted to volume with the extracting solvent followed by filtration using a 0.45-pm nylon membrane filter. A 1.0-mL aliquot of the combined filtrates was evaporated with nitrogen gas and then made to the appropriate concentration of analytes with mobile phase.

3.3   SOLID PHASE MICROEXTRACTION

Solid phase microextraction (SPME) was introduced in 1990s and developed by Arthur and Pawliszyn (1990) and Pawliszyn (1995, 1997) as a solvent-free extraction technique. It was initially named after the first experiment conducted using SPME device for extraction on solid fused-silica fibers which was later renamed in relation to a liquid or gaseous donor phase (Pawliszyn, 1997). Prior to the development of SPME, there were two types of headspace (HS) analysis techniques, viz., static and dynamic (Teranishi et al., 1972). SPME represents a new approach that combines the advantages of both static and dynamic modes of HS techniques.

SPME employs a short-length fused silica fiber coated with an adsorptive material and placed at the end of a syringe (Figure 3.7). The fiber adsorbs the volatile compounds in the headspace vapor. This technique uses a fused silica fiber (1 or 2 cm in length) coated with a polymeric film to collect the volatiles from the sample. A range of polar, non-polar, and mixed fibers are available in the market. The fiber is contained within a needle, which is placed into a SPME holder for sampling and desorbing purposes (Figure 3.7). The sample is placed in a SPME vial and sealed with a septum cap. The fiber is introduced (through the needle) into the headspace above the sample or can be immersed in liquid samples. After a fixed sampling time, in which volatiles are absorbed on the fiber coating,

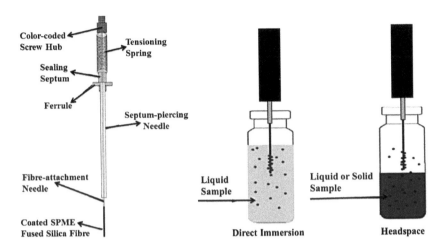

FIGURE 3.7   SPME needle with fiber for extraction.

the fiber is withdrawn and desorbed directly into a GC or GC-MS. The SPME fiber is reconditioned by heating in a GC injection port for 5–15 min. It takes time to determine the most suitable fiber type, as well as sampling time and temperature, to analyze a particular matrix. Once the optimal conditions are determined, as long as they are kept constant, the analyst should be able to obtain reproducible results.

SPME is one of the most widely used techniques among 204 extraction techniques used in various fields (Figure 3.7) and also gained popularity among scientists and researchers due to its superiority over other techniques (Arthur and Pawliszyn, 1990; Koziel et al., 1999; Peñalve et al., 1999; Mills and Walke, 2000; Ulrich, 2000; Augusto et al., 2003; Chang and Abdullah, 2010; Heydari and Haghayegh, 2014). It is based on analyte diffusion and is most widely used for the isolation of volatile and semi-volatile compounds from aqueous solutions (Harmon, 1997; Yu et al., 2012). SPME overcomes the difficulties of extraction and is used in the detection of flavoring volatile chemical compounds (Zhang et al., 1994) from various types of sample matrices by dividing them into a polymeric liquid coating (i.e., an immobilized stationary phase) from a gaseous or liquid sample matrix (Steffen and Pawliszyn, 1996). It has broad application in food, forensic, and toxicological studies; analysis of natural products; and in the study of volatile profile of many fruits and vegetables (Vas and Vekey, 2004; Castro et al., 2008; Flamini, 2010). Besides this, SPME has been employed to analyze the volatile aroma chemicals due to some benefits like comparatively simple, rapid, solvent-free, and less expensive, and consumes less time for extraction in comparison to other techniques like SDE (Zhang et al., 1994; Kataoka et al., 2000; Holt, 2001; Mehdinia et al., 2006; Djozan et al., 2009). Presently, many researchers consider it as one of the most brilliant inventions as it is simple, efficient, and eco-friendly method and widely used in sample preparation for analysis of rice aroma chemical compounds (Penalver et al., 1999; Koziel et al., 1999; Mills and Walke, 2000; Ulrich, 2000; Augusto et al., 2003).

3.3.1   SUPERCRITICAL FLUID EXTRACTION WITH CO_2

Supercritical fluid extraction (SCFE) is a rapid, non-selective method for sample preparation prior to the analysis of compounds in natural product matrices. Extraction yields are affected by the solubility of the solute in the fluid, diffusion through the matrix, and adsorption onto the matrix. Applications involve the extraction of carotenoids, lipids, flavor, fragrance compounds, etc. (Modey et al., 1996). The market potential and economics of supercritical extraction for the separation of essential oils and flavors from spices and herbs have been reviewed (Seidlitz and Lack, 1996). The use of CO_2 as a supercritical fluid, which is liquefied under pressure and heat, is to attain the properties of both liquid and gases that acts as solvent for the extraction of oil (Halim et al., 2012). This method requires high-pressure equipment. However, it is expensive and energy intensive. The advantage of this method is that it is nontoxic, easy to operate, and nonflammable (Xiaodan et al., 2012). From the above methods, the solvent extraction and SCFE methods are the most popular methods for oil extraction as compared to mechanical methods.

The SCFE method resembles Soxhlet extraction except that the solvent is in the supercritical state depicted in Figure 3.8, which is described by Sarmento et al. (2006). One of the solvents that can be used for this extraction method is supercritical CO_2. The major advantage of the method is that it does not need an organic solvent. The selectivity can be controlled by varying the density of CO_2. The compounds in this extraction have high diffusion coefficients that increase the rate of extraction and lessen

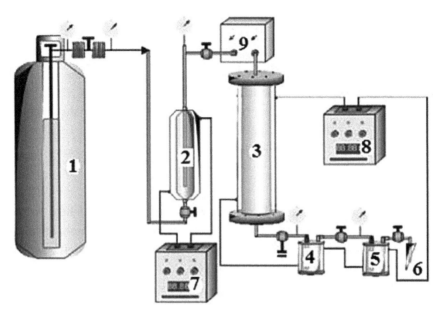

FIGURE 3.8 Schematic diagram of Supercritical Fluid Extraction (SCFE) set-up. 1) CO_2 cylinder, 2) surge tank, 3) extraction, 4) & 5) separators, 6) collector and gas measuring device, 7) & 8) thermostatic baths, and 9) isocratic pump (Source: Adapted from Sarmento, C. M. P., Ferreira, S. R. S. and Hense, H. (2006). Supercritical fluid extraction (SFE) of rice bran oil to obtain fractions enriched with tocopherols and tocotrienols. *Brazi. J. Chemi. Engg.*, *23*(2), 243–249.).

degradation of the solute. Another advantage is the low cost of CO_2. It is also nonflammable and devoid of oxygen, which prevents the sample from any degradation caused by oxidation. A disadvantage of SCFE is the problem with the fat content of the sample. A sample with high fat content cannot be used because the fat will be extracted along with the desired compounds by this method (Peppard, 1999).

3.4 SUMMARY AND CONCLUSION

Currently, simultaneous distillation extraction (SDE), solid phase micro-extraction (SFME), and supercritical fluid extraction (SFE) are the most important and common procedures coupled with GC-MS reported in the literature on aroma compounds. The impact of aroma chemical characteristics in rice can be identified by many classical techniques of the 21st century. The present chapter addressed and discussed a short overview of the traditional and modern extraction approaches for aroma chemical components, which are responsible for distinguished aroma, flavor, and fragrance in rice. There are number of references that exist on this chapter, which provides different perspectives.

ACKNOWLEDGMENT

Deepak Kumar Verma and Prem Prakash Srivastav are indebted to the Department of Science and Technology, Ministry of Science and Technology, Government of India, for an individual research fellowship (INSPIRE Fellowship Code No.: IF120725; Sanction Order No. DST/INSPIRE Fellowship/2012/686; Date: 25/02/2013).

KEYWORDS

- analytical techniques
- chemical constitutes
- extraction methods
- flavor
- fragrance
- GC analysis
- quantification
- rice aroma
- solvent extraction
- SPME steam distillation

REFERENCES

Adahchour, M., Beens, J., Vreuls, R. J. J., Batenburg, A. M., Rosing, E. A. E., & Brinkman, U. A. T. (2002). Application of solid-phase micro-extraction and comprehensive two-dimensional gas chromatography (GC × GC) for flavor analysis. *Chromatographia, 55*(5/6), 361–367.

Arashi, K. (1975). Kinsei inasaku gijutsu shi: sono ritchi seitaiteki kaiseki. (In Japanese: *History of Rice Cultivation in Modern Times: Geological and Ecological View Point*). Nobunkyo, Tokyo, Japan, pp. 468–490.

Berlardi, R., & Pawliszyn, J. (1989). The application of chemically modified fused silica fibers in the extraction of organics from water matrix samples and their rapid transfer to capillary columns. *Water Pollution Res. J. Canada, 24*, 179.

Blanch, G. P., Tabera, J., Herraiz, M., & Reglero, G. (1993). Preconcentration of volatile components of foods: optimization of the steam distillation–solvent extraction at normal pressure. *J. Chrom., 628*, 261–268.

Boswell, C. E. (1999). Fast and efficient volatiles analysis by purge and trap GC/MS. WTQA-15th *Annual Waste Testing & Quality Assurance Symposium*. pp. 190–194. https://clu-in.org/download/char/dataquality/EBoswell.pdf.

Chaintreau, A. (2001). Simultaneous distillation-extraction: from birth to maturity. Review, *Flavor Fragrance J., 16*(2), 136.

Contarini, G., & Povolo, M. (2002). Volatile fraction of milk: Comparison between purge and trap and solid phase microextraction techniques. *Journal of Agricultural and Food Chemistry, 50*(25), 7350–7355.

Godefroot, M., Sandra, P., & Verzele, M. (1981). New method for quantitative essential oil analysis. *Journal of Chromatography A., 203*, 325–335.

Halim, R., Danquah, M. K., & Webley, P. A. (2012). Extraction of oil from microalgae for biodiesel production: A review. *Biotechnology Advances, 30*, 709–732.

Haryanto, T. A. D., Suwarto, S., & Yoshida, T. (2008). Yield stability of aromatic upland rice with high yielding ability in Indonesia. *Plant Production Science, 11*(1), 96–103.

Hollis, J. S., & Prest, H. (2016). Volatile organic compound analysis using purge and trap success with VOC analysis using the agilent 5975C mass selective detector. URL: https://www.agilent.com/cs/library/applications/5991–0029EN.pdf. Accessed on: 14–05–2016.

Itani, T. (2002). Agronomic characteristics of rice cultivars collected from Japan and other countries (in Japanese with English abstract). *Japanese Journal of Crop Science, 71*, 68–75.

Itani, T., Tamaki, M., Hayata, Y., Fushimi, T., & Hashizume, K. (2004). Variation of 2-acetyl-1-pyrroline concentration in aromatic rice grains collected in the same region in japan and factors affecting its concentration. *Plant Production Science, 7*(2), 178–183.

Kapfer, G. F., Kruger, H., Nitz, S., & Drawert, F. (1990). Gewinnung von fermentativ hergestellten aromen durch adsorption und CO_2 Hochdruckextraktion, *ZFL Lebensmitteltechnik, 41*, 51–55.

Kondo, H. (1987). *Visit the Forgotten Rice* (in Japanese). Japan book Publication Society, Tokyo, Japan, pp. 5–81.

Landen, W. O. (1982). Application of gel permeation chromatography and non-aqueous reverse phase chromatography to high performance liquid chromatographic determination of retinyl palmitate and α-tocopheryl acetate in infant formulas. *Journal of American Oil Chemistry International, 65*, 810–813.

Lee, J., Landen, Jr. W. O., Phillips, R. D., & Eitenmiller, R. R. (1998). Application of direct solvent extraction to the LC quantification of vitamin E in peanuts, peanut butter, and selected nuts. *Peanut Science, 25*, 123–128.

Likens, S. T., & Nickerson, G. B. (1964). Detection of certain hop oil constituents in brewingproducta. *Am. Soc. Brew. Chem. Proc.*, 5–13.

Maignial, L., Pibarot, P., Bonetti, G., Chaintreau, A., & Marion, P. (1992). Simultaneous distillation extraction under static vacuum: isolation of volatiles at room temperature. *Journal of Chromatography, 606*, 87–94.

Manzini, S., Durante, C., Baschieri, C., Cocchi, M., Sighinolfi, S., & Totaro, S. (2011). Optimization of a dynamic headspace-thermal desorption-gas chromatography/mass spectrometry procedure for the determination of furfurals in vinegars. *Talanta, 85*(2), 863–869.

Modey, W. K., Mulholland, D. A., & Raynor, M. W. (1996). Analytical supercritical fluid extraction of natural products, *Phytochemical Analysis, 7*, 1–15.

Moumeni, A., Samadi, B. Y., Wu, J., & Leung, H. (2003). Genetic diversity and relatedness of selected Iranian rice cultivars and disease resistance donors assayed by simple sequence repeats and candidate defense gene markers. *Euphytica, 131*, 275–284.

Mudgal, A. (2009). Multi effect distillation (MED) of water as a rural micro enterprise. *PhD Thesis*, IIT Delhi, India.

Nickerson, G. B., & Likens, S. T. (1966). Gas chromatographic evidence for the occurrence of hop oil components in beer, *J. Chromatog., 21*, 1–3.

Okoshi, M., Matsuno, K., Okuno, K., Ogawa, M., Itani, T., & Fujimura, T. (2016). Genetic diversity in Japanese aromatic rice (*Oryza sativa* L.) as revealed by nuclear and organelle DNA markers. *Genetic Resources and Crop Evolution, 63*, 199–208.

Parliment, T. H. (1997). Solvent extraction and distillation techniques. In: Marsili, R. (ed.), *Techniques for Analyzing Food Aroma*. Marcel Dekker, Inc., New York, pp. 1–27.

Partoatmodjo, A., Alidawati, & Harahap, Z. (1994). Bengawan solo, a short duration aromatic rice in Indonesia. *International Rice Research Notes, 19*, 19–20.

Peppard, T. L. (1999). How chemical analysis supports flavor creation. *Food Technology, 53*, 46–51.

Pollien, P., & Chaintreau, A. (1997). Simultaneous distillation-extraction: theoretical model and development of a preparative unit. *Analytical Chemistry, 69*, 3285–3292.

Pollien, P., Ott, A., Fay, L. B., Maignial, L., & Chaintreau, A. (1998). Simultaneous distillation-extraction: Preparative recovery of volatiles under mild conditions in batch or continuous operations. *Flavor and Fragrance Journal, 13*(6), 412–423.

Reineccius, G. (1993) Biases in analytical flavor profiles introduced by isolation method. In: *Flavor Measurement,* Ho & Manley, New York.

Rijks, J. A., Curver, J., Noy, T., & Cramers, C. (1983). Possibilities and limitations of steam distillation-extraction as a pre-concentration technique for trace analysis of organics by capillary gas chromatography. *Journal of Chromatography, 279*, 395–407.

Rohilla, R., Singh, V. P., Singh, U. S., Singh, R. K., & Khush, G. S. (2000). Crop husbandry and environmental factors affecting aroma and other quality traits. In: *Aromatic Rices* (Singh, R. K., Singh, U. S., & Khush, G. S., eds.), Oxford & IBH Publ., New Delhi, India. pp. 201–316.

Seibel, B. (2015). *Fundamentals of Purge and Trap Posted.* URL: http://blog.teledynetekmar.com/fundamentals-of-purge-and-trap. Accessed on 14–05–2016.

Seidlitz, H., & Lack, E. (1996). Separation of essential oils & flavors from spices and herbs. Market potential and economics, *Chemical Industry Digest, 9*, 91–97.

Sen, P. V., Bhuwanesh, K., Ashutosh, K., Engineer, Z., Hegde, S., Sen, P. K., & Lal, R. (2013). Micro-scale multi-effect distillation system for low steam inputs. *Procedia Engineering, 56*, 63–67.

Seno, D. S. H., Artika, M., Nurcholis, W., Santoso, T. J., Trijatmiko, K. R., Padmadi, B., Praptiwi, D., Ching, J. M., & Masud, Z. A. (2013). Identification of badh2 mutation type among Indonesian fragrant rice varieties. *Journal of Biology, Agriculture and Healthcare, 3*(17), 2224–3208.

Shimoda, M., Shigematsu, H., Shiratsuchi, H., & Osajima, Y. (1995). Comparison of the odour concentrates by SDE and adsorptive column method from green tea infusion. *Journal of Agricultural and Food Chemistry, 43*, 1616–1620.

Siegmund, B., Leitner, E., Mayer, I., Farkas, P., Sadecka, J., Pfannhauser, W., & Kovac, M. (1996). Investigation of the extraction of aroma compounds using the simultaneous distillation extraction according to Likens-Nickerson method. *Dtsch. Lebensm. Rundsch. (German), 92*(9), 286–290.

Singh, R. K., Khush, G. S., Singh, U. S., Singh, A. K., & Singh S. (2000c). Breeding aromatic rice for high yield, improved aroma and grain quality. In: *Aromatic Rices* (Singh, R. K., Singh, U. S., & Khush, G. S., eds.). Oxford & IBH Publ., New Delhi, India. pp. 71–105.

Singh, R. K., Singh, U. S., Khush, G. S., & Rohilla, R. (2000a). Genetics and biotechnology of quality traits in aromatic rices. In: *Aromatic Rices* (Singh, R. K., Singh, U. S., & Khush, G. S., eds.). Oxford & IBH Publ., New Delhi, India. pp. 47–69.

Verma, D. K., & Srivastav, P. P. (2016). Extraction technology for rice volatile aroma compounds In: *Food Engineering Emerging Issues, Modeling, and Applications* (eds. Meghwal, M., & Goyal, M. R.). In book series on Innovations in Agricultural and Biological Engineering, Apple Academic Press, USA. pp. 245–291.

Verma, D. K., Mohan, M., & Asthir, B. (2013). Physicochemical and cooking characteristics of some promising basmati genotypes. *Asian Journal of Food Agro-Industry, 6*(2), 94–99.

Verma, D. K., Mohan, M., Prabhakar, P. K., & Srivastav, P. P. (2015). Physico-chemical and cooking characteristics of Azad basmati. *International Food Research Journal, 22*(4), 1380–1389.

Verma, D. K., Mohan, M., Yadav, V. K., Asthir, B., & Soni, S. K. (2012). Inquisition of some physico-chemical characteristics of newly evolved basmati rice. *Environment and Ecology, 30*(1), 114–117.

Vyas, S. K. (2003). Small scale multi effect distillation (MED) system for rural drinking water supply. *PhD Thesis*, IIT Delhi, India.

Williams, A., Ryan, D., Guasca, A. O., Marriott, P., & Pang, E. (2005). Analysis of strawberry volatiles using comprehensive two-dimensional gas chromatography with headspace solidphase microextraction. *Journal of Chromatography B., 817*, 97–107.

Xiaodan, W., Rongsheng, R., Zhenyi, D., & Yuhuan, L. (2012). Current status and prospects of biodiesel production from microalgae. *Energies, 5*, 2667–2682.

PART II

TRADITIONAL EXTRACTION TECHNOLOGY OF RICE AROMA CHEMICALS

SIMULTANEOUS DISTILLATION EXTRACTION (SDE): A TRADITIONAL METHOD FOR EXTRACTION OF AROMA CHEMICALS IN RICE

DEEPAK KUMAR VERMA,[1] DIPENDRA KUMAR MAHATO,[2] and PREM PRAKASH SRIVASTAV[3]

[1]Research Scholar, Agricultural and Food Engineering Department, Indian Institute of Technology, Kharagpur–721302, West Bengal, India, Mobile: +91-7407170260, +91-9335993005, Tel.: +91-3222-281673, Fax: +91-3222-282224, E-mail: deepak.verma@agfe.iitkgp.ernet.in, rajadkv@rediffmail.com

[2]Senior Research Fellow, Indian Agricultural Research Institute, Pusa Campus, New Delhi–110012, India, Mobile: +91-9911891494, +91-9958921936, E-mail: kumar.dipendra2@gmail.com

[3]Associate Professor, Agricultural and Food Engineering Department, Indian Institute of Technology, Kharagpur–721302, West Bengal, India, Mobile: +91-9434043426, Tel.: +91-3222-283134, Fax: +91-3222-282224, E-mail: pps@agfe.iitkgp.ernet.in

CONTENTS

4.1   INTRODUCTION

Simultaneous distillation extraction (SDE) is one of the most important extraction methods and more popular among all the known methods for rice aroma chemical analysis (Verma and Srivastav, 2016). This valuable extraction method is also known as Likens–Nickerson steam distillation (Siegmund et al., 1996), which was first reported in 1964 by Likens and Nickerson (1964) and described in 1966 (Nickerson and Likens, 1966). Later, the SDE method was modified by Maarse and Kepner (1970) and Schultz et al. (1977). The SDE method has been considered as one of the most cited methods until now, and it is used for the extraction and isolation of volatile aroma chemical components (Chaintreau, 2001; Pino et al., 2001, 2002). Previously, many researchers have successfully applied the SDE method for the extraction and analysis of various rice volatile aroma chemicals from different cultivars of rice as depicted in Table 4.1 (Buttery et al., 1983, 1986; Laksanalamai and Ilangantileke, 1993; Petrov et al., 1996; Widjaja et al., 1996a, 1996b; Tava and Bocchi, 1999; Bhattacharjee et al., 2003; Wongpornchai et al., 2003; Sunthonvit et al., 2005; Park et al., 2010).

4.2   PRINCIPLE AND INSTRUMENTATION

The SDE method simplifies the experimental procedures by merging vapor distillation and solvent extraction as shown in Figure 4.1 (Chaintreau, 2001). In the SDE method, the final product is a solvent extract (Blanch et al., 1993; Parliament, 1997). Various parts of the SDE apparatus, such as carbon dioxide (CO_2)–acetone condenser, water outlet, insulation, stirrer, solution of rice and water, and water bubbler, are shown in Figure 4.1.

TABLE 4.1   List of Extracted Major Rice Volatile Aroma Compounds Using SDE Techniques

Aroma Compounds	Aroma Compounds	Aroma Compounds
(E)-2, (E)-4-decadienal [d, e]	2-pentadecanone [e, f]	Indole [e, g]
(E)-2, (E)-4-heptadienal [e]	2-pentylfuran [d, e]	1-octen-3-ol [d, e]
(E)-2, (E)-C heptadienal [e]	2-phenylethanol [d, e]	Methyl heptanoate [e]
(E)-2, (E)-C nonadienal [e]	2-tridecanone [e]	n-decanal [e]
(E)-2-decenal [e, j]	2-undecanone [e]	n-heptanol [e, j]
(E)-2-heptenal [d, f]	4-vinylguaiacol [d, e, g]	n-hexanal [d, e, g]
(E)-2-hexenal [e]	4-vinylphenol [d, e]	n-nonanal [d, e, j]
(E)-2-nonenal [e, g]	6-methyl-5-hepten-2-one [e]	n-nonanol [e]
(E)-2-octenal [d, e].	Acetophenone [e]	n-octanal [e, f, j]
(E)-hept-2-enal [f]	Alk-2-enals [d]	n-octanol [e, j]
2,4,6-trimethylpyridine (TMP) or Collidine [e]	Alka-2,4-dienals [d]	n-pentanol [e, f]
2-acetyl-1-pyrroline (2-AP) [a–i]	Alkanals [d]	n-undecanal [e]
2-heptanone [e, j]	Benzaldehyde [e, f, g]	Oct-1-en-3-ol [g]
2-hexanone [e]	Butan-2-one-3-Me [j]	Pentanal [g, j]
2-methyl-3-furanthiol (2-MF) [i]	Decanol [j]	Phenylacetaldehyde [e]
2-nonanone [e]	Hexadecanol [f]	Pyridine [e]
2-octanone [j]	Hexanol [f, g]	Undecane [j]

Sources: [a]Buttery, et al. (1983), [b]Buttery, et al. (1986), [c]Laksanalamai and Ilangantileke, (1993), [d]Widjaja, et al. (1996a), [e]Widjaja, et al. (1996b), [f]Petrov, et al. (1996), [g]Tava and Bocchi, (1999), [h]Sunthonvit, et al. (2005), [i]Park, et al. (2010), [j]Bhattacharjee, et al. (2003).

4.3 MERITS AND DEMERITS

The merits of the SDE technique involve simple apparatus in one-step extraction operation that can be handled for a range of samples for extraction of volatile compounds and concentrates the compounds. It further enables the detection of trace component from obtained extracts which are free from non-volatile compounds because the volatiles are separated from non-volatiles by the process of distillation (Chaintreau, 2001). This technique is quite safe because gas chromatography liners and columns are decontaminated as there is absence of non-volatile or high-boiling materials during extraction. It is performed under reduced pressure at room temperature

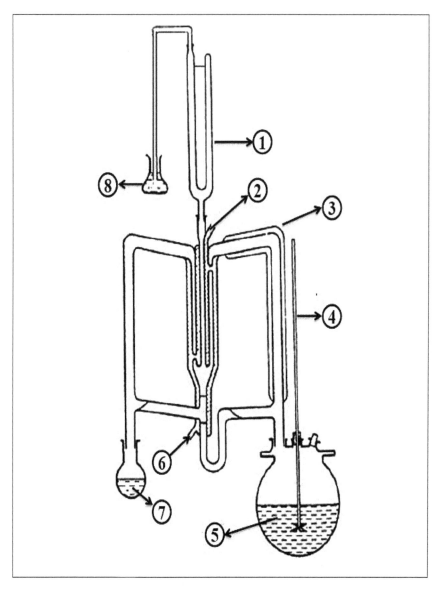

FIGURE 4.1 Simultaneous distillation extraction apparatus. Carbon dioxide (CO_2) – Acetone Condenser, 2) Water out, 3) Insulation, 4) Stirrer, 5) Solution of rice & water, 6) Water in, 7) Solvent, and 8) Water bubbler. (Source: Reprinted from Sunthonvit, N., Srzednicki, G., & Craske, J. (2005). Effects of high temperature drying on the flavor components in Thai fragrant rice varieties. *Dry. Techno.*, *23*, 1407–1418 © 2005. With permission from Taylor & Francis).

(25°C) in laboratory (Maignial et al., 1992). This technique consumes less time and a little volume of solvent (due to continuous recycling) to allow the extraction of aroma compounds (Chaintreau, 2001). Thus, the extraction of volatile compounds is performed without further concentrating the solvents. The extraction technique is performed to achieve higher concentration recoveries of aroma compounds under certain conditions. The SDE technique reduces the problems of artifact, and breakdown products are reduced or generate fewer products which build-up the solvents and further concentrates. The SDE technique has a high-extraction efficiency that is always associated with high reproducibility. Due to its efficiency, this technique has been employed for quantification of aroma compounds in various matrices (Chaintreau, 2001).

The demerits of the SDE technique include the following: it is not best extraction technique as it yields comparatively lower quantitative extracts than solvent extraction methods. The artifact formation in the SDE technique is maximum possible because of the thermal degradation of the rice sample. The extraction of acids and alcohols is very poor as they do not appear in the final extract due to their hydrophilic/polar nature. This technique may introduce silicone contaminants due to use of anti-foaming agents for avoiding foaming. The SDE technique is not suitable for fresh rice samples, as the extraction of aroma compounds is more akin to cooked rice (previously thermally processed). The major difficulties in the SDE technique are to balance two boiling flasks, constant pressure, and minimize evaporation of the boiling solvent.

4.4   CASE STUDY ON THE SDE METHOD FOR THE EXTRACTION OF RICE AROMA CHEMICALS

There are considerably large numbers of different rice cultivars grown around the world that possess aroma. The knowledge of the volatile aroma components of rice is obtained by a number of research works using SDE. In 1983, the steam distillation continuous extraction method was used by Buttery et al. (1983) to determine 2-AP from 10 different rice varieties. The determined levels of concentration in the steam volatile oils of rice varieties were based on the dry weight and varied from less than 0.006 to 0.09 ppm for Calrose and Malagkit Sungsong rice variety, respectively. The same method was again used with an acid-phase solvent extraction in

1986 by Buttery et al. (1986) for extraction and quantification of a potent rice volatile compound, 2-AP, from 200 g of bland rice sample. The aim of the study was to develop a simple and practical method for extraction and isolation of rice volatile aroma chemicals, and the findings showed that bland rice sample with this developed method contains known added concentration level of 2-AP and found the best-suited method for the purpose of quantitative analysis of 2-AP volatile aroma chemical from rice samples. A comparative study was made by Laksanalamai and Ilangantileke (1993) using the SDE method to identify the aroma chemical compound 2-AP in fresh and aged (shelf-stored) Khao Dawk Mali (KDML) 105 (a well-accepted Thai aromatic rice variety) in non-aromatic rice and pandan leaves.

In 1996, Widjaja et al. (1996a) conducted a comparative study on volatile compounds of fragrant and non-fragrant rice and reported that (*E*)-2, (*E*)-4-decadienal, (*E*)-2-octenal, (*E*)-Z heptenal, 2-AP, 2-pentylfuran, 2-pentylfuran, 2-phenylethanol, 2-phenylethanol, 4-vinylguaiacol, 4-vinylphenol, alk-2-enals, alka-2,4-dienals, alkanals, 1-octen-3-ol, *n*-hexanal, and *n*-nonanal were the most important volatile compounds out of 70 identified compounds, but other identified volatile chemical components were also responsible for contributing to total rice aroma profile. In the same year, Widjaja et al. (1996b) conducted another study in storage condition to know the variation in the volatile chemical compounds and reported an increase in the level of total volatile aroma chemical components in all three forms of rice, namely paddy, brown, and white fragrant rice. Petrov et al. (1996) isolated 100 compounds by simultaneous steam distillation-solvent extraction method from scented rice and non-scented rice in cooking water; among them, 78 were identified including (*E*)-hept-2-enal, 2-AP, 6,10,14-trimethyl-pentadecan-2-one, benzaldehyde, hexadecanol. hexanol, octanal, pentadecan-2-one, and pentanol.

In commercial basmati and Italian rice, Tava and Bocchi (1999) identified and quantified volatile compounds that were collected by steam distillation as alcohols, aldehydes, disulfides, heterocyclic, hydrocarbons, ketones, phenols, and terpenes in rice samples. Hexanal was the most abundant compound present in both commercial samples, followed by 2-AP, 4-vinylguaiacol, benzaldehyde, hexanol, indole, oct-1-en-3-ol, pentanal, and *trans*-2-nonenal. The presence of 2-AP was reported at 570 and 2,350 ppb level of concentration in Basmati and B5-3 rice sample, respectively. Bhattacharjee et al. (2003) conducted a comparative study of aroma profile from commercial brand of Basmati rice using two extraction techniques (viz. Likens–Nickerson and supercritical fluid extraction (SFE)) to find out the best method. The Likens–Nickerson method was reported to be the best for smaller sample size, shorter time of extraction, and negligible possibility of artifacts as compared to the SFE technique, because of its superiority for recovery of volatiles chemical constitutes from rice.

Sunthonvit et al. (2005) used SDE combined with the distillation-extraction apparatus (modified Likens and Nickerson apparatus) for the extraction of volatiles from Thai fragrant rice varieties and reported a total of 94 volatile compounds, including 17 acids, 21 alcohols, 19 aldehydes, 9 heterocyclic compounds, 8 hydrocarbons, 14 ketones, and 6 miscellaneous compounds. 2-AP was found as the major volatile aroma chemical component. Park et al. (2010) studied potent aroma chemical component of cooked Korean non-aromatic rice using SDE techniques and characterized 16 aroma volatiles chemicals with log3 flavor dilution >1. The detected volatile aroma chemical components of rice were 2-MF and 2-AP, which were considered as potent active aroma chemical components. The volatile aroma chemical 2-MF was first time reported for its aroma potential in cooked Korean nonaromatic rice.

4.5 CONCLUSION, FUTURE SCOPE, AND RESEARCH OPPORTUNITIES

Presently, SDE involves steam distillation followed by solvent extraction. It is one of the most popularly used extraction methods. It has always played a key role in solvent extraction. This method has two important features. First, it can exploit the differences between volatile compounds and non-volatile compounds present in the rice matrix. The SDE technique has been selected to improve extraction of rice volatile aroma compounds as well as to limit or exclude the nonvolatile components that have no importance and only cause hinderance during analysis. Second, the time duration for extraction is a strong parameter for analysis. Generally, rice extraction consumes 1 to 4 hours (Chaintreau, 2001; Bhattacharjee et al., 2003). Due to solvent nature and extraction time, the extraction yield is always best in the SDE technique. This is the main reason that researchers are trying to find out the best suitable experimental conditions to recover the selected volatile aroma components of rice.

4.6 SUMMARY

SDE is the most important and common traditional procedures coupled with gas chromatography–mass spectrometry (GC-MS) reported in the literature for extraction of aroma compounds. The impact of aroma chemical characteristics of rice can identified by many classical techniques of the 21st century. The present chapter addressed and discussed in detail the SDE extraction approach for rice aroma compounds, principle and instrumentation, merits and demerits of the SDE method, factors influencing method performance, case study on research progress, and strength and weakness of rice aroma research. There are a number of references that exist on the topic entitled "Simultaneous Distillation Extraction (SDE): A Traditional Method for Extraction of Aroma Chemicals in Rice," which provides a different perspective.

ACKNOWLEDGMENT

Deepak Kumar Verma and Prem Prakash Srivastav are indebted to the Department of Science and Technology, Ministry of Science and Technology, Government of India, for an individual research fellowship (INSPIRE Fellowship Code No.: IF120725; Sanction Order No. DST/INSPIRE Fellowship/2012/686; Date: 25/02/2013).

KEYWORDS

- 2-AP
- aroma chemicals
- basmati
- Likens–Nickerson

- **pandan leaves**
- **rice**
- **SDE method**
- **solvent extraction**
- **traditional method**
- **vapor distillation**
- **volatile aroma**

REFERENCES

Bhattacharjee, P., Ranganathan, T. V., Singhal, R. S., & Kulkarni, P. R. (2003). Comparative aroma profiles using supercritical carbon dioxide and Likens-Nickerson extraction from a commercial brand of Basmati rice. *J. Sci. Food Agric., 83*, 880–883.

Blanch, G. P., Tabera, J., Herraiz, M., & Reglero, G. (1993). Preconcentration of volatile components of foods: optimization of the steam distillation–solvent extraction at normal pressure. *J. Chromato., 628*, 261–268.

Buttery, R. G., Ling, L. C., & Mon, T. R. (1986). Quantitative analysis of 2-acetyl-1-pyrroline in rice. *J. Agric. Food Chemi., 34*(1), 112.114.

Buttery, R. G., Ling, L. C., Juliano, B. O., & Turnbagh, J. G. (1983). Cooked Rice aroma and 2-acetyl-l-pyrroline. *J. Agric. Food Chem., 31*, 823–826.

Chaintreau, A. (2001). Simultaneous distillation–extraction: from birth to maturity – review. *Flav. Fragra. J., 16*, 136–148.

Laksanalamai, V., & Ilangantileke, S. (1993). Comparison of aroma compound (2-acetyl-1-pyrroline) on leaves from pandan (*Pandanum amaryllifolius*) and Thai fragrant rice (Khao Dawk Mali-105). *Cereal Chemi., 70*, 381–384.

Likens, S. T., & Nickerson, G. B. (1964). Detection of certain hop oil constituents in brewingproducta. *Am. Soc. Brew. Chem. Proc.,* 5–13.

Maarse, H., & Kepner, R. (1970). Changes on composition of volatile trpenes in Douglas fir needls during maturation. *J. Agric. Food Chem., 18*, 1095–1101.

Maignial, L., Pibarot, P., Bonetti, G., Caintreau, A., & Marion, J. P. (1992). Simultaneous distillation-extraction under static vacuum: isolation of volatile compounds at room temperature. *J. Chromato., 606*, 87–94.

Nickerson, G. B., & Likens, S. T. (1966). Gas chromatographic evidence for the occurrence of hop oil components in beer. *J. Chromatog., 21*, 1–3.

Park, J. S., Kim, K., & Baek, H. H. (2010). Potent aroma-active compounds of cooked Korean non aromatic rice. *Food Sci. Biotechnol., 19*(5), 1403–1407.

Parliament, T. H. (1997). Solvent extraction and distillation techniques. *In: Techniques for Analyzing Food Aroma* (Marsili, R., ed.). Marcel Dekker, Inc., New York, pp. 1–27.

Petrov, M., Danzart, M., Giampaoli, P., Faure, J., & Richard, H. (1996). Rice aroma analysis: discrimination between a scented and a non-scented rice. *Sci. Des Alim., 4*(16), 347–360.

Pino, J. A., Marbot, R., & Bello, A. (2002). Volatile compounds of psidium salutare (HBK.) Berg. Fruit. *J. Agric. Food Chem., 50*, 5146–5148.

Pino, J. A., Marbot, R., & Vazquez, C. (2001). Volatile compounds of psidium salutare (HBK.) Berg. Fruit. Characterizarion of volatiles in strawberry guava (*Psidium cattleianum* Sabine) fruits. *J. Agric. Food Chem., 49*, 5883–5887.

Schultz, T. H., Flath, R. A., Mon, T. R., Eggling, S. B., & Teranishi, R. (1977). Isolation of volatile compounds from a model system. *J. Agric. Food Chem., 25*, 446–449.

Sunthonvit, N., Srzednicki, G., & Craske, J. (2005). Effects of high temperature drying on the flavor components in Thai fragrant rice varieties. *Dry. Techno., 23*, 1407–1418.

Tava, A., & Bocchi, S. (1999). Aroma of cooked rice (*Oryza sativa*): comparison between commercial basmati and Italian line B5–3. *Cer. Chemi., 76*(4), 526–529.

Verma, D. K., & Srivastav, P. P. (2016). Extraction technology for rice volatile aroma compounds In: *Food Engineering Emerging Issues, Modeling, and Applications* (eds. Meghwal, M., & Goyal, M. R.). In book series on Innovations in Agricultural and Biological Engineering, Apple Academic Press, USA. pp. 245–291.

Widjaja, R., Craske, J. D., & Wootton, M. (1996a). Comparative studies on volatile components of non-fragrant and fragrant rices. *J. Sci. Food Agric., 70*(2), 151–161.

Widjaja, R., Craske, J. D., & Wootton, M. (1996b). Changes in volatile components of paddy, brown and white fragrant rice during storage. *J. Sci. Food Agric., 71*, 218–224.

Wongpornchai, S., Sriseadka, T., & Choonvisase, S. (2003). Identification and quantitation of the rice aroma compound, 2-acetyl-1-pyrroline, in bread flowers (*Vallaris glabra* Ktze). *J. Agric. Food Chem., 51*(2), 457–462.

PART III

NEW AND MODERN APPROACHES IN EXTRACTION OF RICE AROMA CHEMICALS

SOLID PHASE MICRO EXTRACTION (SPME): A MODERN EXTRACTION METHOD FOR RICE AROMA CHEMICALS

DEEPAK KUMAR VERMA,[1] DIPENDRA KUMAR MAHATO,[2] SUDHANSHI BILLORIA,[3] and PREM PRAKASH SRIVASTAV[4]

[1]*Research Scholar, Agricultural and Food Engineering Department, Indian Institute of Technology, Kharagpur–721302, West Bengal, India, Mobile: +91-7407170260, +91-9335993005, Tel.: +91-3222-281673, Fax: +91-3222-282224, E-mail: deepak.verma@agfe.iitkgp.ernet.in, rajadkv@rediffmail.com*

[2]*Senior Research Fellow, Indian Agricultural Research Institute, Pusa Campus, New Delhi–110012, India, Mobile: +91-9911891494, +91-9958921936, E-mail: kumar.dipendra2@gmail.com*

[3]*Research Scholar, Agricultural and Food Engineering Department, Indian Institute of Technology, Kharagpur–721302, West Bengal, India, Mobile: +91-8768126479, E-mail: sudharihant@gmail.com*

[4]*Associate Professor, Agricultural and Food Engineering Department, Indian Institute of Technology, Kharagpur–721302, West Bengal, India, Mobile: +91-9434043426, Tel.: +91-3222-283134, Fax: +91-3222-282224, E-mail: pps@agfe.iitkgp.ernet.in*

CONTENTS

5.1   INTRODUCTION

Solid phase micro extraction (SPME) was initially named after the first experiment conducted using SPME device for extraction on solid fused-silica fibers, which was later renamed in relation to a liquid or gaseous donor phase (Pawliszyn, 1997). Prior to the development of SPME, there were two types of headspace (HS) analysis techniques, viz., static and dynamic (Teranishi et al., 1972) used for analysis of food flavor chemical compounds. In addition to these well-established HS techniques in analysis of food flavor (Kolb and Ettre, 1997), SPME was the third emerging technique that represented a new approach that combines the advantages of both static and dynamic HS techniques. SPME was introduced in 1990s (Yang and Peppard, 1994) and developed by Arthur and Pawliszyn, and Pawliszyn (Arthur and Pawliszyn, 1990: Pawliszyn, 1995, 1997) as a solvent-free extraction technique that can be used in both laboratory and on-site.

SPME is a reliable extraction method for quantitative examination that is characterized by high sensitivity when compared with the dynamic HS method. It has been reported that both SPME and dynamic HS method were used for the analysis of different food products like cola (Elmore et al., 1997), milk (Marsili, 1999a; Contarini and Povolo, 2002), butter (Povolo and Contarini, 2003), and chees (Mallia et al., 2005). With regard to flavor compound isolation, SPME is preferred over simultaneous distillation extraction (SDE) methods as in that isolation is performed at low temperatures and usually for a short duration of time, which prevents breakdown of thermally labile compounds (Jelen, 2003). SPME is based on analyte diffusion and is most widely used for the isolation of volatile and semi-volatile compounds from aqueous solutions (Harmon, 1997; Yu et al., 2012). It requires only a small sample volume coupled with gas chromatography and mass spectrometry (GC-MS) to provide high sensitivity. It is relatively and rapidly emerging as a new robust solvent-free, simple, rapid, low cost, easy to operate, sensitive, and reliable technique among all the other techniques (Zhou et al., 1999; Adahchour et al., 2002; Djozan and Ebrahimi, 2008; Djozan et al., 2009; Kaseleht et al., 2011). SPME overcomes the difficulties of extraction and is used in the detection of flavoring volatile chemical compounds (Zhang et al., 1994) from various types of sample matrices by dividing them into a polymeric liquid coating (i.e., an immobilized stationary phase) from a gaseous or liquid sample matrix (Steffen and Pawliszyn, 1996). It has broad application in food, forensic, and toxicological studies; analysis of natural products, and in the study of volatile profile of many fruits and vegetables (Vas and Vekey, 2004; Castro et al., 2008; Flamini, 2010). Besides this, SPME has been employed to analyze the volatile aroma chemicals due to some benefits like comparatively simple, rapid, solvent-free, less expensive, and consumes less time for extraction in comparison to other techniques like SDE (Zhang et al., 1994; Kataoka et al., 2000; Holt, 2001; Mehdinia et al., 2006; Djozan et al., 2009). Presently, many researchers have considered it as one of the most brilliant inventions as it is simple, efficient, and eco-friendly method and is widely used in sample preparation for analysis of rice aroma chemical compounds (Koziel et al., 1999; Penalver et al., 1999; Mills and Walke, 2000; Ulrich, 2000; Augusto et al., 2003). The volatile aroma compounds from rice sample identified to date are given in Table 5.1.

5.2 PRINCIPLE AND INSTRUMENTATION

5.2.1 PRINCIPLE

The SPME technique includes very simple setup, and there is no requirement of any additional instrumentation other than a GC-MS with injection

TABLE 5.1   List of Extracted Major Rice Volatile Aroma Compounds Using SPME Techniques

Aroma Compounds	Aroma Compounds	Aroma Compounds
(1S,Z)-calamenene[1]	2-methyl-5-decanone[1]	Farnesol[f,1]
(2-aziridinylethyl) amine[c]	2-methyl-5-isopropyl-furan[g]	Formic acid hexate[g]
(2E)-dodecenal[f]	2-methyl-butanal[e]	Geranyl acetone[f,i,j,k,1]
(E)-3-Octene-2-one[e]	2-methyl-dodecane[g]	Glycerin[1]
(E) 6, 10-dimethyl-5,9 undecadien-2-one[e]	2-methyl-furan[g]	Heptacosane[g]
(E)-14-hexadecenal[c]	2-methyl-hexadecanal[a]	Heptadecane[a,c,e,f,g,k,1]
(E)-2-decenal[e,f,j,k,1]	2-Methyl-I-propenyl benzene[e]	Heptadecyl-cyclohexane[g]
(E)-2-heptanal[b,e]	2-methyl-naphthalene[f]	Heptanal[a,c,d,g,h,i,j,k,1]
(E)-2-heptenal[f,j,k,1]	2-methyl-tridecane[f]	Heptane[e]
(E)-2-hexenal[d,e,k,1]	2-methyl-undecanal[f]	Heptanol[h]
(E)-2-nonen-1-ol[a]	2-methyl-undecanol[g]	Hepten-2-ol<6-methyl-5->[f]
(E)-2-nonenal[a,e,f,i,j,k,1]	2-n-butyl furan[c]	Heptene-dione[g]
(E)-2-octen-1-ol[k,1]	2-nonanone[c,d,g,k,1]	Heptylcyclohexane[a]
(E)-2-octenal[f,h,j,k,1]	2-nonenal[h]	Hexacosane[c,f,g]
(E)-2-undecenal[f,j,k,1]	2-ocetenal (E)[a]	Hexadecanal[g,k,1]
(E)-3-nonen-2-one[k,1]	2-octanone[g,1]	Hexadecane 2,6-bis-(t-butyl)-2,5[e]
(E)-4-nonenale[k,1]	2-octenal[e,g]	Hexadecanoic acid[d,e,j,k,1]
(E,E)-2,4-heptadienal[f,k,1]	2-pentadecanone[d,f,g,j,k,1]	Hexadecyl ester, 2,6-difluoro-3-methyl benzoic acid[a]
(E,E)-2,4-nonadienal[e,f,i,k,1]	2-pentylfuran[a,b,c,e,f,g,h,i,j,k,1]	Hexane[e,h]
(E,E)-2,4-octadienal[f]	2-phenyl-2-methyl-aziridine[f]	Hexanoic acid[b,e,k,1]
(E,E)-farnesyl acetone[k,1]	2-propanol[d]	Hexanoic acid dodecate[g]
(E,Z)-2,4-decadienal[f,j,k,1]	2-tetradecanone[d]	Hexanol[a,h]
(S)-2-methyl-1-dodecanol[d]	2-tridecanone[g,j,k,1]	Hexatriacontane[a,g]
(Z)-2-octen-1-ol[k,1]	2-undecanone[e,g,j,k,1]	Hexyl furan[f]
(Z)-3-Hexenal[e]	3-(t-Butyl)-phenol[e]	Hexyl hexanoate[k]
[(1-methylethyl)thio] cyclohexane[a]	3,5,23-trimethyl-tetracosane[g]	Hexylpentadecyl ester-sulphurous acid[a]

TABLE 5.1 (Continued)

Aroma Compounds	Aroma Compounds	Aroma Compounds
<6-methyl->Heptan-2-ol [f]	3,5,24-trimethyl-tetracon-tane [a]	Indane [e]
<cis-2-tert-butyl->Cyclohexanol acetate [f]	3,5-dimethyl-1-hexene [a]	Indene, 2,3-dihydro-1,1,3 trimethyl-3-phenyl [e]
1-(1H-pyrrol-2-yl)-etha-none [a]	3,5-dimethyl-octane [g]	Indole [a, e, f, g, j, k, l]
1, 2 –Bis, cyclobutane [h]	3,5-di-tert-butyl-4-hydroxy-benzaldehyde [a]	I-S, cis-calamenelonol 2 [e]
1,1-dimethyl-2-octyl-cyclobutane [g]	3,5-heptadiene-2-one [g]	Isobutyl hexadecyl ester oxalic acid [a]
1,2,3,4-Tetramethyl benzene [e]	3,5-octadien-2-ol [c]	Isobutyl nonyl ester oxalic acid [a]
1,2,3-trimethyl-benzene [g]	3,7,11,15-tetramethyl-2-hexadecimalen-1-ol [g]	Isobutyl salicylate [f]
1,2,4-trimethyl-benzene [c, e, f]	3,7,11-trimethyl-1-dodeca-nol [g]	Isolongifolene [e]
1,2-dichloro-benzene [h]	3,7-dimethyl-hexanoic acid [g]	Isopropylmyristate [d]
1,2-dimethyl-benzene [e]	3-dimethyl-2-(4-chlorphenyl)-thioacrylamide [a]	Limonene [f]
1,3,5-trimethyl-benzene [h, f]	3-ethyl-2-methyl-heptane [g]	Ethylbenzene [c]
1,3-diethyl-benzene [c]	3-Hexanone [e]	l-Pentanol [e]
1,3-dimethoxy-benzene [f]	3-hydroxy-2,4,4-trimethyl-pentyl iso-butanoate [l]	Methoxy-phenyl-oxime [a, g]
1,3-dimethyl-benzene [c, f]	3-methyl butanal [e, h]	Methyl (E)-9-octadeceno-ate [j, k, l]
1,3-dimethyl-naphthalene [f]	3-methyl pentane [h]	Methyl (Z,Z)-11,14-eicosa-dienoate [j, k, l]
1,4-dimethyl-benzene [e]	3-methyl-1-butanol [d, e]	Methyl (Z,Z)-9,12-octa-decadienoate [j, k, l]
1,4-dimethyl-naphthalene [f]	3-methyl-2-butyraldehyde [g]	Methyl butanoate [e]
10-heptadecene-1-ol [g]	3-Methyl-2-heptyl acetate [e]	Methyl decanoate [e]
10-pentadecene-1-ol [g]	3-methyl-hexadecane [g]	Methyl dodecanoate [e]
17-hexadecyl-tetratriac-ontane [g]	3-methyl-pentadecane [f]	Methyl ester tridecanoic acid [c]
17-pentatricontene [a]	3-methyl-tetradecane [f]	Methyl hexadecanoate [e, j, k, l]

TABLE 5.1 (Continued)

Aroma Compounds	Aroma Compounds	Aroma Compounds
1-chloro-3,5-bis(1,1-dimethylethyl)2-(2-propenyloxy) benzene [a]	3-nonene-2-one [g]	Methyl isocyanide [c]
1-chloro-3-methyl butane [a]	3-octdiene-2-one [g]	Methyl isoeugenol [f]
1-chloro-nonadecane [a]	3-octene-2-one [g]	Methyl linolate [e]
1-decanol [l]	3-tridecene [a]	Methyl octanoate [e]
1-docosene [a]	4,6-dimethyl-dodecane [c]	Methyl oleate [e]
1-dodecanol [j, k, l]	4-[2-(methylamino) ethyl]-phenol [c]	Methyl pentanoate [e]
1-ethyl-2-methyl-benzene [g]	4-acetoxypentadecane [a]	Methyl tetradecanoate [e]
1-ethyl-4-methyl-benzene [e]	4-cyclohexyl-dodecane [a]	Methyl-naphthalene [g]
1-ethyl-naphthalene [f]	4-ethyl-3,4-dimethyl-cyclohexanone [f]	Myrcene [c]
1-heptanol [d, e, f, k, l]	4-ethylbenzaldehyde [d]	N,N-dimethyl chloestan-7-amine [a]
1-hexacosene [a]	4-hydroxy-4-methyl-2-pentanone [g]	N,N-dinonyl-2-phenylthio ethylamine [a]
1-hexadecanol [a, j, k, l]	4-methyl benzene formaldehyde [g]	Naphthalene [a, b, e, f]
1-hexadecene [e]	4-methyl-2-pentyne [a]	n-decane [f]
1-hexanol [d, e, j, k, l]	4-vinyl-guaiacol [i]	n-dodecanol [g]
1-methoxy-naphthalene [f]	4-vinylphenol [j, k, l]	n-eicosanol [f, g]
1-methyl-2-(1-methylethyl)-benzene [c, e]	5-amino-3,6-dihydro-3-imino-1(2H) pyrazine acetonitrile [a]	n-heneicosane [f]
1-methyl-4-(1-methylethylidene)-cyclohexene [c]	5-butyldihydro-2(3H)-furanone [e]	n-heptadecanol [g]
1-methylene-1H-indene [c]	5-ethyl-4-methyl-2-phenyl-1,3-dioxane [a]	n-heptadecylcyclohexane [a]
1-nitro-hexane [c, f]	5-ethyl-6-methyl-(E)3-hepten-2-one [e]	n-heptanal [f]
1-nonanol [d, f, k, l]	5-ethyl-6-methyl-2-heptanone [g]	n-heptanol [g]
1-octadecene [f]	5-ethyldihydro-2(3H)-furanone [e]	n-hexadecane [g]

TABLE 5.1 (Continued)

Aroma Compounds	Aroma Compounds	Aroma Compounds
1-octanol [d, e, f, i, j, k, l]	5-*iso*-Propyl-5*H*-furan-2-one [k, l]	*n*-hexadecanoic acid [c]
1-octene-3-ol or 1-octen-3-ol [a, c, f, g, i, j, k, l]	5-methy-3-hepten-2-one [a]	*n*-hexanal [a, b, c, d, e, f, g, h, i, j, k, l]
1-pentanol [d, h, k, l]	5-methyl-2-hexanone [a]	*n*-hexanol [g, f]
1-tetradecyne [a]	5-methyl-pentadecane [g]	Nitro-ethane [c]
1-tridecene [e]	5-methyl-tridecane [g]	*n*-nonadecane [f]
1-undecanol [l]	5-pentyldihydro-2(3*H*)-furanone [e]	*n*-nonadecanol [a]
2,2,4-trimethyl-3-carboxyisopropyl, isobutyl ester pentanoic acid [a]	5-Propyldihydro-2(3H)-furanone [e]	*n*-nonanal [b, c, d, e, f, g, h, i, k, l,]
2,2,4-trimethyl-heptane [e]	6,10,13-trimethyl-tetradecanol [g]	*n*-nonanol [c, e, g]
2,2,4-trimethyl-pentane [c]	6,10,14-trimethyl-2-pentadecanone [e, g, j, k, l]	*n*-octanal [a, c, d, f, g, h, i, j, k, l]
2,2-dihydroxy-1-phenyl-ethanone [a]	6,10,14-trimethyl-pentadecanone [d]	*n*-octane [e, f]
2,3,5-trimethyl-naphthalene [f]	6,10-dimethyl-2-undecanone [g, k, l]	*n*-octanol [f, g]
2,3,6-trimethyl-naphthalene [f]	6,10-dimethyl-5,9 undecadien-2-one [a]	*n*-octyl-cyclohexane [f]
2,3,6-trimethyl-pyridine [e]	6,10-dimethyl-5,9-undecandione [g]	Nonadecane [a, e, f, g]
2,3,7-trimethyl-octanal [g]	6-dodecanone [k, l]	Nonane [g]
2,3-butandiol [e]	6-methyl-2-heptanone [g]	Nonanoic acid [j, k, l]
2,3-dihydrobenzofuran [g]	6-methyl-3,5-heptadiene-2-one [g, j, k, l]	Nonene [a]
2,3-dihydroxy-succinic acid [g]	6-methyl-5-ene-2-heptanone [g]	Nonnenal [g]
2,3-octanedione [g, l]	6-methyl-5-heptanone [g]	Nonyl-cyclohexane [g]
2,4,4-trimethylpentan-1,3-diol di-*iso*-butanoate [l]	6-methyl-5-hepten-2-one [a, e, j, k, l]	*n*-tetradecanol [f]
2,4,6-trimethyl-decane [g]	6-methyl-octadecane [a]	*n*-tricosane [f]
2,4,6-trimethylpyridine (TMP) or Collidine [e, m]	7,9-di-tert-butyl-1-oxaspiro-(4,5)deca-6,9-diene-2,8-dione [a]	*n*-undecanol [f]

TABLE 5.1   (Continued)

Aroma Compounds	Aroma Compounds	Aroma Compounds
2,4,7,9-tetramethyl-5-decyn-4,7-diol [a]	7-chloro-4-hydroxyquinoline [a]	Octacosane [g]
2,4-bis(1,1-dimethylethyl)-phenol [a]	7-methyl-2-decene [l]	Octadecane [c, e, f, g]
2,4-di(*tert*butyl) phenol [j, k, l]	7-tetradecene [a]	Octadecyne [a]
2,4-diene dodecanal [g]	9-methyl-nonadecane [f]	Octanal [e]
2,4-hexadienal [h]	Acetic acid [g]	Octanoic acid [k, l]
2,4-hexadiene aldehyde [g]	Acetic acid tetradecate [g]	Octylformate [f]
2,4-nonadienal [g]	Acetone [e, g, b]	*O*-decylhydroxamine [a]
2,4-pentadiene aldehyde [g]	Acetonitrile [c]	Oxalic acid-cyclohexyl methyl nonate [g]
2,4-pentandione [e]	Acetophenone [c]	Palmitate [g]
2,5,10,14-tetramethyl-pentadecane [e]	Alkyl-cyclopentane [g]	Pentacontonal [a]
2,5,10-trimethyl-pentadecane [e]	Anizole [f]	Pentacosane [f, g]
2,5-dimethyl-undecane [g]	Azulene [c]	Pentadecanal [a]
2,6,10,14-tetramethyl-heptadecane [g]	Benzaldehyde [b, c, d, e, f, j, k, l]	Pentadecane [c, e, f, g, k, l]
2,6,10,14-tetramethyl-hexadecane [e, f, g]	Benzene acetaldehyde [e, g]	Pentadecanoic [c]
2,6,10,14-tetramethyl-pentadecane [f]	Benzene [c]	Pentadecyl-cyclohexane [g]
2,6,10-trimethyl-dodecane [c, g]	Benzene formaldehyde [g]	Pentanal [a, e, g, h, j, k, l]
2,6,10-trimethyl-pentadecane [c, f, g]	Benzothiazole [a, d, e]	Pentanoic acid [e]
2,6,10-trimethyl-tetradecane [g]	Benzyl alcohol [e]	Pentylhexanoate [k]
2,6-bis(t-*butyl*)-2,5 cyclohexadien-I-one [e]	Bezaldehyde [h]	Phenol [l]
2,6-bis(t-*butyl*)-2,5-cyclohexadien-1, 4-dione [e]	Bicyclo[4.2.0]octa-1,3,5-triene [c]	Phenyl acetic acid-4-tridecate [g]
2,6-di(*tert*-butyl)-4-methylphenole [j, k, l]	Bis-(I-methylethyl) hexadecanoate [e]	Phthalic acid [f]

TABLE 5.1 (Continued)

Aroma Compounds	Aroma Compounds	Aroma Compounds
2,6-Diisopropyl-naphthalene [e]	Butanal [e]	Phytol [g]
2,6-dimethoxy-phenol [f]	Butandiol[e]	Propane [g]
2,6-dimethyl-aniline [f]	Butanoic acid [e]	Propiolonitrile [a]
2,6-dimethyl-decane [g]	Butylatedhydroxy toluene [a, e, f, h]	Propyl acid [g]
2,6-dimethyl-heptadecane[g]	Citral [g]	p-xylene [c]
2,6-dimethyl-naphthalene [c, f]	Cubenol [l]	Pyridine [e]
2,7-dimethyl-octanol [h]	Cyclodecanol [a]	Pyrolo[3,2-d]pyrimidin-2,4(1H,3H)-dione [a]
2-acetyl-naphthalene [f]	Cyclosativene[f]	Styrene [c, e]
2-acetyl-1-pyrroline (2-AP) [a, c, d, e, f, i, m]	d-dimonene [d]	Tetracosane [f, g]
2-butyl-1,2-azaborolidine [a]	Decanal [c, d, e, f, h, i, j, k, l]	Tetradecanal [g]
2-butyl-1-octanol [a, d, g]	Decane [e, g, j, l]	Tetradecane [b, c, e, f, g, h]
2-butyl-2-octenal [b, e, g]	Decanoic acid [l]	Tetradecanoic acid [d, j, k, l]
2-butylfuran [e, f, k, l]	Decyl aldehyde [g]	Tetradecanol [g]
2-butyl-octanol [h]	Decyl benzene [a]	Tetradec-I-ene [e]
2-chloro-3-methyl-1-phenyl-1-butanone [a]	Diacetyl[c]	Tetrahydro-2,2,4,4-tetramethyl furan [a]
2-chloroethyl hexyl ester isophthalic [a]	Dichlorobenzene [b]	Toluene [c, d, e, g]
2-decanone [k, l]	Diethyl phthalate [a]	trans-2-octenal [c]
2-decen-1-ol [a]	Diethyl phthalate [c, f, j, k, l]	trans-caryophylene[e]
2-decenal [g]	Di-iso-butyl adipate [l]	Triacontane [g]
2-dodecanone [d, g]	Dimethyl disulphide [c, e]	Tricosane [g]
2-ethyl-1-decanol [g]	Dimethyl sulphide [e]	Tridecanal[g, l]
2-ethyl-1-dodecanol [d]	Dimethyl trisulphide [e]	Tridecane [c, e, f, g, k, l]
2-ethyl-1-hexanol [a, e]	D-limonene [c]	Trimethylheptane [e]
2-ethyl-2-hexenal [g]	Docosane [f, g]	Tritetracontane [a]
2-ethyl-decanol [h]	Dodecanal [b, g, k, l]	Turmerone [f]
2-heptadecanone [j, k, l]	Dodecane [c, e, f, g, h]	Undecanal [e, j, k, l]
2-heptanone [c, e, g, h, l]	Dotriacontane [a, g]	Undecane [b, c, e, f, g]

TABLE 5.1   (Continued)

Aroma Compounds	Aroma Compounds	Aroma Compounds
2-heptenal [g]	Eicosane [c, e, g]	Undecanol [f]
2-heptene aldehyde [g]	Eicosanol [a]	Undecyl-cyclohexane [g]
2-heptylfurane [k]	Ethanol [e, f, j, k, l]	Vanillin [k, l]
2-hexadecanol [a]	Ethenyl cyclohexane [e]	Z-10-pentadecen-1-ol [a]
2-hexyl-1-decanol [a]	Ethyl (E)-9-octadecenoate [j, k, l]	α-cadinene [l]
2-hexyl-1-octanol [a, c, e]	Ethyl benzene [e]	α-cadinol [l]
2-hexyl-decanol [a]	Ethyl decanoate [e]	α-terpineol [e]
2-hexylfurane [k, l]	Ethyl dodecanoate [e]	β-bisabolene [f]
2-methoxy phenol [a]	Ethyl hexadecanoate [e, j, k, l]	β-terpineol [e]
2-methoxy-4-vinylphenol [f, j, k, l]	Ethyl hexanoate [e]	γ-nonalacton [k]
2-methyl-butanol [e]	Ethyl linoleate [e]	γ-nonalactone [j, l]
2-methyl decane [a]	Ethyl nonanoate [e]	γ-terpineol [e]
2-methyl-1,3-pentanediol [a]	Ethyl octanoate [e]	δ-cadinene [l]
2-methyl-1-hexadecanol [a]	Ethyl oleate [e]	δ-cadinol [l]
2-methyl-2,4-diphenyl-pentane [e]	Ethyl tetradecanoate [e]	2-methyl-3-octanone [k, l]

Sources: [a]Bryant and McClung, (2011); [b]Bryant, et al. (2006); [c]Ghiasvand, et al. (2007); [d]Goufo, et al. (2010); [e]Grimm, et al. (2011); [f]Khorheh, et al. (2011); [g]Lin, et al. (2010); [h]Monsoor, et al. (2004); [i]Tananuwonga and Lertsirib, (2010); [j]Zeng, et al. (2008a); [k]Zeng, et al. (2008b); [l]Zeng, et al. (2009).

port as shown in Figure 5.1 (Hawthorne et al., 1992; Steffen and Pawliszyn, 1996). As shown in Figure 5.2, a small quantity of the extracting phase connected with a solid support is placed in contact with the matrix of sample for a fixed period of time till the concentration equilibrium is recognized between matrix of the sample and the extraction phase. After the equilibrium condition is attained, exposing the fiber further does not collect any more analytes. Figure 5.2 shows the cross-section of SPME device.

When the fiber assembly is introduced into the headspace, the transport of analytes from the sample matrix into the coating begins, and the extraction ability differs for the liquid and solid coatings. Both adsorptive and absorptive equilibrium extraction takes place. The extraction process

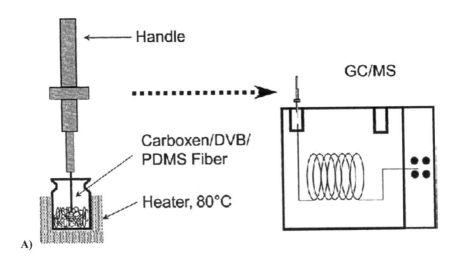

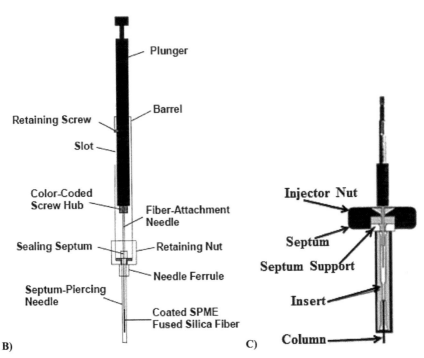

FIGURE 5.1   Solid phase micro extraction. A) Block diagram of SPME analysis, B) Commercial SPME device, and C) SPME-GC interface (Adapted from Pawliszyn J. (2001) Solid Phase Microextraction. In: Rouseff R.L., Cadwallader K.R. (eds) Headspace Analysis of Foods and Flavors. Advances in Experimental Medicine and Biology, vol 488. Springer.)

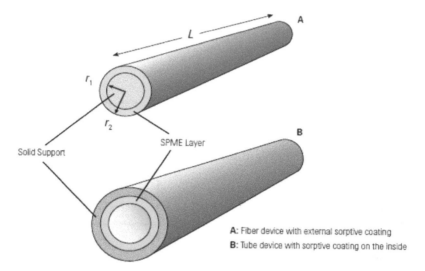

FIGURE 5.2 Cross-section of an SPME device (Adapted from Hinshaw, J. (2012). Solid-Phase Microextraction. LCGC North America, *30*(10), 904–910).

initiates by the adsorption of analytes at the interface of extraction phase–matrix followed by diffusion of analytes into the bulk of the extraction phase. Extraction in the SPME is completed only when the distribution equilibrium of the analyte concentration is attained between matrix of the sample and the fiber coating. At equilibrium condition, the extracted volume is constant (for limits of experimental error) and is independent of extraction time as depicted by the following equation (Louch et al., 1992):

$$n = \frac{\left(K_{fs}V_{f}V_{s}C_{o}\right)}{K_{fs}V_{f} + Vs} \tag{1}$$

where, n = mass of analyte extracted by the coating; K_{fs} = fiber coating–sample matrix distribution constant; V_{f} = fiber coating volume; V_{s} = sample volume; and C_{0} = initial concentration of a given analyte in the sample.

SPME has also been used for rapid extraction and transfer of a sample to the instrument for the analysis. Presently, SPME has a broad range of application in diverse fields like analysis of the biodegradation pathways of industrial contaminants (Pawliszyn, 1999); analyte distribution in a complex multiphase system (Poerschmann et al., 1997), and speciate different forms of analytes in a matrix (Mester and Pawliszyn, 1999). In case, if volume of the sample is very large, then equation (1) is simplified as:

$$n = (K_{fs}V_{f}C_{o}) \tag{2}$$

Equation (2) clearly shows that the quantity of extracted analyte (n) is independent of the volume of the sample (V_{s}), which means that sample collection prior to analysis is not required. In this case, the sample is collected directly from ambient air, water, or the production stream. Elimination of the sampling step hastens the process by avoiding the errors related to the analyte losses through breakdown or adsorption on the sampling container walls. This could be used for developing portable field devices for commercial purpose (Pawliszyn, 2000).

5.2.2 INSTRUMENTATION

SPME device is a combination of two steps that occurs separately. The first step is related to partitioning the analytes of desired target between the matrix of sample and the fiber surface, while the other step is related to the direct desorption of absorbed analytes into injection port of GC as depicted in Figure 5.1A (Djozan and Ebrahimi, 2008). SPME being a simple, solvent-free extraction technique uses a phase-coated fused silica fiber, which is exposed to the headspace above the liquid or solid sample. After adsorbing of analytes to the fiber, they are desorbed in the GC and transferred to a capillary column (Arthur et al., 1992; Zhang et al., 1994). SPME concretes analytes to its fiber and leaves behind unwanted solvent and nonvolatiles. Though it can be applied to both GC as well as liquid chromatography (LC), GC is preferred over LC. The SPME process involves major two steps (Adahchour et al., 2002): (1) partitioning of analytes between the extraction phase and the sample matrix; and (2) desorption of concentrated extracts into an analytical instrument.

5.2.2.1   Additional Hardware Associated with SPME

5.2.2.1.1   *Fiber bakeout station (FBS)*

Fiber bakeout station (FBS) is employed for pre-conditioning of fiber outside the GC to prevent the contamination of the column. This is done after the fiber has been desorbed in the inlet port. This is important in those cases where the inlet desorption temperature is low in the method for the analysis. Hence, a proper conditioning is required to prevent the fiber bleed from the previous sample run.

5.2.2.1.2. *Single magnet mixer*

Single magnet mixer (SMM) is a heated stirring device. It consists of a magnet placed inside the sample vial. It is applicable for more liquid SPME than headspace SPME (HS-SPME). The SMM is gentle to the fiber in the immersed condition. The stirring provides better responses for semi-volatiles as compared to shaking.

5.3   DIFFERENT TYPES OF SPME DEVICE

5.3.1   *FIBER SPME*

Fiber SPME consists of two components, viz., a fiber holder and a fiber assembly (Vas and Vekey, 2004). The fiber assembly is a 1- to 2-cm long retractable SPME fiber (Vas and Vekey, 2004) with a built-in coated fiber (Figure 5.2) (Zhang and Pawliszyn, 1993; Lord and Pawliszyn, 2000). The SPME needle with fiber inside is inserted into the headspace of a septum-sealed vial, and the fiber is exposed out into the vial for adsorption to take place until equilibrium of concentrations is reached between matrix of the sample and the extraction phase. Extreme sensitivity is achieved, and a proportional relationship is calculated between the quantity of the extracted analyte and the concentration in the sample as the adsorption process is proportional to the concentration up to certain limits after which it saturates (Ai, 1997). Hence, the SPME method is a sort of qualitative analysis rather than quantitative one as the calibration curve cannot be obtained using it.

5.3.2 IN-TUBE SPME (IT-SPME)

In-tube SPME (IT-SPME) was introduced by Eisert and Pawliszyn (1997a) for application with high performance liquid chromatography (HPLC) or liquid chromatography–mass spectrometry (LC-MS, or alternatively HPLC-MS) as SPME fiber could barely withstand aggressive HPLC solvent conditions (Eisert and Pawliszyn, 1997a; Lord and Pawliszyn, 2000). It has an open tubular fused-silica capillary with the extraction phase either as an inner surface coating or a sorbent bed, which overcomes the drawbacks of SPME fiber like fragility, low sorption capacity and bleeding from thick-film coatings. IT-SPME has higher mechanical stability as compared to the SPME fiber and can be automated with the HPLC (Nerin et al., 2009). One of the drawbacks of in-tube SPME is clogging of the capillary, which could be eliminated by working with samples without interfering phases like particles or macromolecules (Nerin et al., 2009). The IT-SPME methods are based on either extraction coatings where the coating is applied as an internal extraction phase, immobilized in the capillary wall or the extraction fillings where the extraction phase is a sorbent packing (Figure 5.3) (Silva et al., 2014).

5.3.3 COOLED COATED FIBER (CCF) DEVICE

An internally cooled coated fiber device (CCF) has been established to facilitate the adsorption process by the fiber coatings from the complex matrices (Zhang and Pawliszyn, 1995; Chen and Pawliszyn, 2006). For desorption of

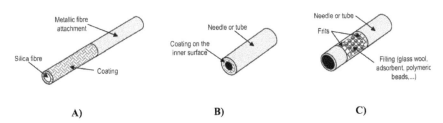

FIGURE 5.3 Classical solid phase micro extraction (SPME) fiber and in-tube fibers A) "Classical" SPME fiber, B) ITE with extraction coating, and C) ITE with extraction filling. (Reprinted from Nerin, C., Salafranca, J., Aznar, M. and Batlle, R. (2009). Critical review on recent developments in solventless techniques for extraction of analytes. *Analytical and Bioanalytical Chemistry, 393*, 809–833. © 2008. With permission from Springer.).

analytes from solid particles, heating is often done to get the volatile compounds collected in the headspace, which are trapped by the fiber. On the other hand, absorption of the analytes through fiber coating is known as an exothermic process, due to which decrease in partition coefficients is observed (Nerin et al., 2009). Hence, to accelerate the mass transfer process and enhance the distribution constants of analytes (Nerin et al., 2009), the fiber coatings are equipped with the cooling facility as shown in Figure 5.4. CCF was first miniaturized and automated by Chen and Pawliszyn (2006). It provides more sensitivity and higher sample throughput than conventional HS-SPME (Carasek et al., 2007). However, one of the major drawbacks includes lack of selectivity, i.e., along with the desired analytes, the interferences are also extracted onto the fiber. Neverthless, it has been successfully used to extract analytes from various environmental matrices (Ghiasvand et al., 2006) and for food analysis (Carasek and Pawliszyn, 2006).

5.3.4   NONFIBER SPME TECHNIQUES

Nonfiber SPME techniques are of two types: (1) static method and (2) dynamic method. The static method includes stir-bar sorptive extraction

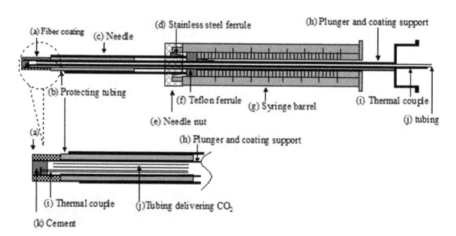

FIGURE 5.4   Internally cooled SPME device. (Reprinted with permission from Chen, Y., & Pawliszyn, J. (2006). Miniaturization and automation of an internally cooled coated fiber device. *Analytical Chemistry, 78*, 5222–5226. © 2006. American Chemical Society.).

(SBSE) and thin-film micro extraction (TFME), while the dynamic method includes in-needle and in-tip SPME (Padron et al., 2014; Silva et al., 2014). The overview of the technique is shown in Figure 5.5.

5.3.4.1   Stir-Bar Sorptive Extraction (SBSE)

Stir-bar sorptive extraction (SBSE) is assisted by magnetic stir bar which is coated with polydimethylsiloxane (PDMS). It is either stirred in or positioned above an aqueous sample (Kataoka, 2010; Padron et al., 2014) and removed after the extraction process. In SBSE-GC, the desorption of the analytes is done by introducing the bar into the heated GC injection port or by employing it in a small vial and backextracting with a few microliters

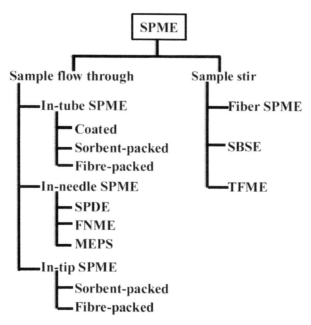

FIGURE 5.5   Types of non-fiber SPME techniques (Merkle et al., 2015). SPDE: solid phase dynamic extraction; FNME: fiber-packed needle micro extraction; MEPS: micro extraction by packed syringe; SBSE: stir-bar sorptive extraction; TFME: thin-film micro extraction. (Modified from Merkle, S., Kleeberg, K. K., & Fritsche, J. (2015). Recent developments and applications of solid phase microextraction (SPME) in food and environmental analysis—A Review. *Chromatography, 2*(3), 293–381.).

of an appropriate liquid solvent, while in LC, the mobile phase is added directly to the stir bar. The PDMS of SBSE is 50- to 250-fold thicker than the SPME fiber and hence has a higher concentration capacity. Despite this, SPME fiber is widely accepted as achieving full automation with SBSE is difficult (Kataoka, 2010; Padron et al., 2014).

5.3.4.2   Thin-Film Micro Extraction (TFME)

In thin-film micro extraction (TFME), a flat film is used for the extraction of analytes. It has high surface-to-volume ratio so that the extraction efficiency is enhanced and the extraction time is reduced (Jiang and Pawliszyn, 2012; Mirnaghi et al., 2013). The extraction of analytes by TFME includes conditioning of blades followed by their exposure to the sample matrix (Jiang and Pawliszyn, 2012; Mirnaghi et al., 2013). The 96-well plates are agitated by placing them on an orbital shaker, washed, and then moved to desorption solvent, which is finally injected to GC. The different steps in TFME like conditioning, extraction, washing, and desorption are controlled by an automatic robotic workstation. TFME is applicable to sample matrices of solid, liquid, or gas. Presently, the extraction plates are even coated with PDMS (Jiang and Pawliszyn, 2012; Mirnaghi et al., 2013).

5.3.4.2.1   In-needle SPME (In-SPME)

In-needle SPME (In-SPME) methods make use of needle for extraction in place of tube. In 2001, a needle trap (NT) device with quartz wool was developed to trap particulate matter and aerosols in air (Koziel et al., 2011), and later, a sorbent-packed NT device for the analysis of volatile organic compounds (VOCs) in gaseous samples was developed by Wang et al. (2005). Similarly, Saito et al. (2006) developed in-needle extraction device for VOC analysis using a copolymer of methacrylic acid and ethylene glycol dimethacrylate, which had higher extraction performance and thermal stability. The major advantage is that the extracted analytes are very stable and can be analyzed even after several days of storage at room temperature (Lipinski, 2001; Ampuero et al., 2004; SPDE™, 2005; Kataoka, 2010).

5.3.4.2.2 Micro Extraction by Packed Syringe (MEPS)

Micro extraction by packed syringe (MEPS) is a miniature form of solid-phase extraction (SPE) where a packed bed is utilized for the extraction. It can be directly connected on-line to GC equipment (Lipinski, 2001; Ampuero et al., 2004; SPDE™, 2005; Kataoka, 2010). The MEPS sorbent consist of a reverse phase (C-2, C-8, and C-18), a normal phase (silica), a restricted access material (RAM), and molecular imprinted polymers (MIPs) and can be used repeatedly till 100 times. After the conditioning of the packed bed, sample solution is passed through the syringe followed by washing of sorbent. The analytes are eluted with organic solvent, which is injected into the GC, or with the LC mobile phase for analysis (Lipinski, 2001; Ampuero et al., 2004; SPDE™, 2005; Kataoka, 2010).

5.4 FIBER IN SPME: TYPE AND SELECTION

Selection of fiber is one of the most important factors that affect the result of SPME extraction technology as the extracted amounts of volatile chemical compounds depends upon selected fiber. The chemistry of rice volatile aroma compounds may determine their behaviors of adsorption and desorption on a particular type of fiber. Analytes varied in polarities may require different fiber chemistry. Presently, various types of SPME fibers are available in the market with different variety of thicknesses coated with polymers ranging from the nonpolar (e.g., PDMS) to the more polar (e.g. CW) (Table 5.2). These SPME fibers depend on the desired target volatile aroma compounds that are extracted from the sample matrix (Arthur and Pawliszyn, 1990; Pawliszyn, 1997; Furton et al., 2000; Namiesnik et al., 2000). This also provides additional advantages as it allows the choice of absorption or adsorption characteristics for the extraction of volatile aroma compounds.

According to available literatures initially, only three types of commercial fibers of a different composition and polarities are present, viz., 1) PDMS; 2) CAR/PDMS; and 3) DVB/CAR/PDMS in SPME technology, which are commonly used to extract aroma volatiles compounds from the rice samples. A fiber of 100 μm PDMS coating is opted for rice volatile

TABLE 5.2 Available Commercial SPME Fibers in Market

S. N.	Commercial Available Fibres	Used in Volatile Aroma Extraction	
		In General	In Rice
1.	Carbowax/Divinylbenzene (CW/DVB)	Used	Not Used
2.	Carbowax/Templated Resin (CW/TPR)	Used	Not Used
3.	Polydimethylsiloxane (PDMS)	Used	Used*
4.	Carboxen/Polydimethylsiloxane (CAR/PDMS)	Used	Used*
5.	Divinylbenzene/Carboxen/Polydimethylsiloxane (DVB/CAR/PDMS)	Used	Used*
6.	Polyacrylate (PA)	Used	Not Used
7.	Polydimethylsiloxane/Divinylbenzene (PDMS/DVB)	Used	Not Used

Note: *Grimm et al. (2001), Monsoor, et al. (2004), Bryant, et al. (2006), Ghiasvand, et al. (2007), Zeng, et al. (2008a, b), Zeng, et al. (2009), Lin, et al. (2010), Goufo, et al. (2010), Tananuwonga and Lertsirib, (2010), Bryant and McClung, (2011), Bryant, et al. (2011), Khorheh, et al. (2011).

aroma compounds. The 30-μm PDMS-coated fiber is much suitable for rapid equilibration, and 7-μm PDMS-coated fiber is best for high boiling component sample (e.g., polyaromatic hydrocarbons) as it is chemically bonded to the fused silica support; this fiber is also best for those operations that require higher temperature to thermally desorb in the injection port of the gas chromatograph.

5.4.1 POLYDIMETHYLSILOXANE (PDMS)

Polydimethylsiloxane (PDMS) is a commercially available and one of the most commonly used coating materials with physically and chemically stable silicone rubber. It has lowest glass transition temperatures ($Tg \approx 125°C$) due to which it provides unique flexibility with a shear elastic modulus ($G \approx 250$ KPa). It also has a little change in the shear elastic modulus versus temperature (1.1 KPa/°C) and has high compressibility with almost no change in G versus frequency. PDMS fiber is most preferable for use as micro-machined mechanical and chemical sensors (like accelerometers and

ISFET [ion-selective field effect transistor] devices as ion selective membrane) due to its low curing temperature, very low drift with time and temperature, high flexibility, and flexibility for changes in functional groups. The type and the thickness of fiber are selected based on the target of analytes to be extracted, equilibrium time, and sensitivity of the method. Other applications of PDMS includes as an adhesive in wafer bonding, as a cover material in tactile sensors, and as the mechanical decoupling zone in sensor packaging (Lotters et al., 1996).

5.4.2   CARBOXEN/POLYDIMETHYLSILOXANE (CAR/PDMS)

Carboxen/polydimethylsiloxane (CAR/PDMS) SPME fibers have strong adsorptive capability. It has been used for the extraction of aroma chemical components from raw homogenates of turkey breast muscle. The aroma components were found to be significantly affected by various parameters like sample temperatures, sample volume, and the headspace. This fiber has versatile nature for the selection of operating parameters for enhanced sensitivity for both high and low volatile compounds. Even at low pressure, the trapping ability of the CAR/PDMS fiber is better than that of porous polymer materials such as PDMS/DVB. Among the different coatings used with SPME, CAR/PDMS was found most efficient for detecting flavoring compounds in traditional smoke-cured bacon with maximum number of compounds being detected (Dietz et al., 2006). The CAR/PDMS fiber has excellent potential and is simple and effective means of extracting volatile aroma components from raw foods (like aromatic rice) as well as other sample matrices where the application of heat is not required (Brunton et al., 2001). Marsili (1999b, 2000) evaluated off-flavors in milk and reported that CAR/PDMS combination of SPME fiber detected more different types of volatiles than other SPME fibers. The CAR/PDMS coating comprise a special case as it contains a mixed carbon of carboxen 1006 adsorbent with ~1000 m^2g^{-1} surface area phased with small micropores. Hence, the extraction processes wish to modify into two different physicochemical mechanisms (Gorecki et al., 1999). PDMS fiber is used for the extraction of nonpolar analytes. CAR/PDMS fiber is preferred over PDMS for better extraction efficiency due to mutual potential effect of adsorption and distribution to the stationary phase (Pawliszyn, 1999; Kataoka et al., 2000). The remaining types (CW/DVB, CW/TPR, CAR/PDMS and PDMS/DVB) are mixed coatings that stay as a mono-layer on the surface of the fiber (Chen and Pawliszyn, 1995).

5.4.3 DIVINYLBENZENE/POLYDIMETHYLSILOXANE (DVB/PDMS)

The analysis of volatiles by SPME fibers coupled with GC-MS is most effective since its development in the 1990s (Arthur and Pawliszyn, 1990). Several different types of combinations of stationary phases and film thickness of SPME fibers are being manufactured by Supelco, Bellefonte, PA. It has been observed that the CAR or DVB are used as a stationary phase for low-molecular-weight compounds, while for the high-molec-ular-weight compounds, PDMS is preferred as a stationary phase. The film thickness of PDMS/DVB is thinner than that of CAR/PDMS; so, the adsorption and desorption equilibrium is attained faster for organic vola-tile compounds. The effect of PDMS/DVB SPME fiber is moderate. It is used to extract alcohol, carboxylic acids, ester, heterocycle, phenol, and sulfide compounds. However, PDMS/DVB is not feasible for extracting alkene compounds.

5.4.4 DIVINYLBENZENE/CARBOXEN/ POLYDIMETHYLSILOXANE (DVB/CAR/PDMS)

The SPME fiber divinylbenzene/carboxen/polydimethylsiloxane (DVB/ CAR/PDMS) coating consists of three different kinds of materials with a film thickness similar to CAR/PDMS. The efficiency of DVB/CAR/PDMS for aldehyde, ketone, and organic acids is less than that of PDMS/DVB (Yu et al., 2012). The DVB/CAR/PDMS fiber has intermediary polarity and shows better performance with different polarities and is reported to be most efficient in trapping the headspace (Ceva-Antunes et al., 2006). The DVB/CAR/PDMS fiber combination has been applied to analyze vol-atile aroma compounds in rice (Grimm et al., 2001; Laguerre et al., 2007; Zeng et al., 2008a, 2008b). This combination of DVB/CAR/PDMS fiber has been reported to trap a wide range of volatile aroma compounds with different polarities, e.g., alcohols, aldehydes, esters, ketones, and terpenic hydrocarbons, than other SPME fibers; hence, it is important for rice volatile aroma compounds (Mondello et al., 2005; Ceva-Antunes et al., 2006).

Ghiasvand et al. (2007) used three commercial fibers, viz. PDMS, CAR/PDMS, and DVB/CAR/PDMS for investigating flavor profiling in the headspace of the Iranian fragrant rice samples at the beginning of the experiments by using cold-fiber SPME–GC–TOF–MS. Each fiber offered particular advantages. For instance, PDMS is well efficient to perform for a wide range of nonpolar analytes as it has a nonpolar coating. Hence, the chromatogram in GC that utilizes PDMS fiber coating for the analysis of rice volatile aroma compounds is relatively poor for the first 15 min in beginning and turns richer at peak at more than 15 min retention time period. This ability of PDMS fibers proves that they have more efficiency for extraction of nonpolar volatile aroma compounds. On the other hand, CAR-PDMS fiber reveals contrary effect on compression with PDMS coating as chromatogram is richer in peaks at the beginning only. Relative to the PDMS and CAR/PDMS fibers, the DVB/CAR/PDMS is most suitable for the extraction of target analytes shown in Figure 5.6 (Ghiasvand et al., 2007).

5.5 MODES OF SPME TECHNIQUES

The volatiles analytes of rice sample can be extracted by three modes from different kinds of media either in liquid or gas phase. They are:

5.5.1 DIRECT EXTRACTION (LIQUID PHASE)

This method (Figure 5.7A) involves direct immersion of coated fiber into a gaseous or liquid sample without headspace and direct transport of analytes from matrix of the sample to the extraction phase. If the sample is liquid, certain agitation is mandatory to reduce the extraction time. Agitation is required to some extent for transportation of analytes from bulk to the adjacent fiber for the rapid extraction. This direct extraction mode of SPME aids to calculate required time period for the extraction of analytes

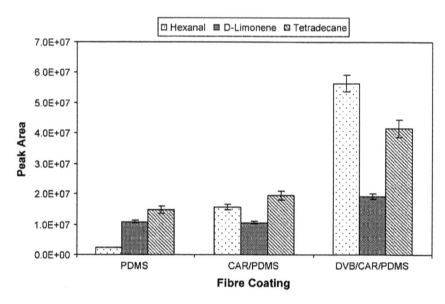

FIGURE 5.6 Comparative presentation of extraction efficiencies of fibers for volatile compounds in the headspace of rice samples. (Reprinted with permission from Ghiasvand, A. R., Setkova, L., & Pawliszyn, J. (2007). Determination of flavor profile in Iranian fragrant rice samples using cold-fibre SPME–GC–TOF–MS. Flavor and Fragrance Journal, 22(5), 377–391. © 2007 John Wiley and Sons.).

from the sample. For volatile chemical compounds in gaseous samples, the naturally occurring air flow is often sufficient to reach the equilibrium (Balasubramanian and Panigrahi, 2011).

5.5.2 HEADSPACE (HS) EXTRACTION (GAS PHASE)

This mode of HS extraction of SPME involves direct contact of fiber into the gaseous and relatively clean water sample as analytes shows high affinity for the fiber and interfering contaminants do not exist. The SPME fiber is exposed to the vapor phase above the liquid or solid sample in the headspace. It protects the fiber from damaging substances present in matrix of the sample and a possible adjustment of pH conditions. Significant differences have been observed in the analysis by direct immersion SPME (DI-SPME) and HS-SPME techniques (Nilsson et al., 1997). The extraction

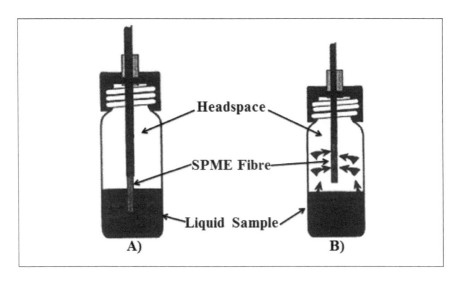

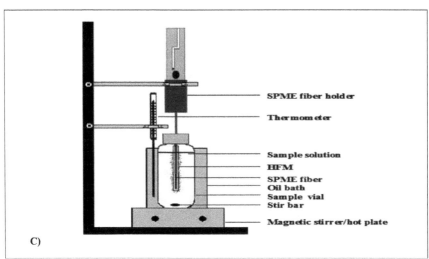

FIGURE 5.7 Modes of SPME sampling. They are: A) direct extraction (solid or liquid phase), B) headspace extraction (gas phase), and C) SPME technique for membrane protection sampling. (Figures A and B adapted from Wercinski, S. A. S., & Pawliszyn, J. (1999). Solid Phase Microextraction Theory. In: Solid Phase Microextraction: A Practical Guide (Wercinski, S. A. S., ed.). CRC Press. Figure C is reprinted with permission from Basheer, C., & Lee, H. K. (2004). Hollow fiber membrane-protected solid phase microextraction of triazine herbicides in bovine milk and sewage sludge samples. *Journal of Chromatography A, 1047*, 189–194]. © 2014 Elsevier.).

kinetics of HS-SPME are governed by Henry's law (Bene et al., 2002) according to which Henry's constant of a given matrix is directly proportional to the concentration of the chemical components in the headspace. Therefore, higher the Henry's constant, higher is the concentration and rapid is the extraction from the headspace. There occurs diffusion between the fiber and the sample without any interference in static HS sampling, whereas in dynamic HS sampling, there is an air movement device like air sampling pumps to move the headspace air (Razote et al., 2002). The static HS sampling does not require careful calibration processes and expensive air sampling pumps, which makes it a preferable method over dynamic HS sampling (Parreira et al., 2002). However, the demerits are seen as the effect of temperature and pressure on the efficiency of the static sampling process and also the effects from sampling containers (Razote et al., 2002). Despite these consequences, static HS sampling has many advantages, viz., selectivity, sensitivity, simplicity, and ease of automation over the dynamic HS sampling techniques (Jung and Ebeler, 2003). Further, to collect and pre-concentrate the headspace gas, the headspace air is moved to SPME filament or another extraction trap chamber (Razote et al., 2004). In the headspace mode (Figure 5.7B), analytes are extracted from the gas phase equilibrated with sample. Presently, HS-SPME is widely used for the analysis of rice aroma compounds.

5.5.3 MEMBRANE PROTECTED EXTRACTION (MPE)

Membrane protection extraction (MPE) is the third type of extraction technique. The first two methods cannot control the interfering compounds; hence, to overcome this shortcoming, this method was developed. It is applied for matrices with interfering compounds like proteins, humic acids, and fatty material (Basheer and Lee, 2004). These high-molecular-weight compounds usually interfere with the proper sampling process; hence, extraction is done using selectively permeable membranes for analytes of interest depicted in Figure 5.7C (Basheer and Lee, 2004). MPE is slow as compared to other extraction techniques, but efficiently extracts compounds with low volatility (Balasubramanian and Panigrahi, 2011).

In either of the cases, the SPME fiber is positioned to appropriate levels in the headspace. The fiber coatings act like a sponge, concentrating the analytes by absorption/adsorption processes. The extraction is based on the concept of either gas–liquid or liquid–liquid partitioning (Ulrich, 2000). The kinetics of SPME extraction process always depends on various factors such as agitation of the sample, film thickness, sampling time, etc. Finally, after the sampling, the fiber is inserted into the GC injection port for the analysis of the analytes.

5.6 FACTORS AFFECTING RICE AROMA IN THE SPME METHOD

There are several factors that affect the sensitivity of the SPME technique. However, these factors highly affect the quantitative accuracy with SPME techniques. These factors, viz., analyte characteristics, coating material of the fiber, extraction time, ionic strength of the matrix, sample matrix, temperature of the sample, type of fiber, etc., are really very important to investigate the determination of optimum sample materials and operating condition for the isolation and extraction of rice volatile aroma compounds.

5.6.1 ANALYTE CHARACTERISTICS

The extraction of analytes from the matrices depends upon the type of matrices. For complex matrices like soil and sediments, the extraction depends on the nature of the medium. For example, soil matrices with weak analyte/matrix interactions release analytes easily into the headspace (Zhang et al., 1994). While the case for the matrices with high adsorptive nature is different (Eisert and Pawliszyn, 1997b) and tend to show strong analyte/matrix interactions, thereby making HS sampling difficult. Hence, in the case of rice and other matrices, a small amount of water is added to release the analytes from the matrix into the headspace (Zhang et al., 1994).

5.6.2 COATING MATERIAL OF THE FIBER

The principle and chemistry of fiber-coated SPME and that of column for GC are the same (Hiroyuki et al., 2000). The types of fiber coating determine the efficiency of SPME analyses. The different sorts of SPME as polar, non-polar, and the combination of (CW/DVB, CAR/PDMS, or PDMS/DVB) (shown in Table 5.3) have their respective applications for the majority of polar, organic volatiles (Spietelun et al., 2013). When the extraction efficiencies of the three fiber coatings, viz., PA, PDMS and CAR/PDMS, were equalated at 40°C on reaching equilibrium time, it was found that the PA-coated fiber extracted the least esters; the PDMS-coated fiber extracted the most esters (except diethyl succinate, ethyl lactate); and the CAR/PDMS-coated fiber extracted the most alcohols, diethyl succinate, ethyl lactate, and the second most esters. As alcohols, diethyl succinate, and ethyl lactate are more polar than the other target analytes, the CAR/PDMS coating offers the most polar phase among the three types of fiber coatings and is also most sensitive for both diethyl succinate and lactate (Popp and Paschke, 1997; Henryk et al., 1998). In 2014, five different types of SPME fiber viz. PDMS, CAR/DVB, PDMS/DVB, CAR/PDMS,

TABLE 5.3 Types of Fiber Coatings and Their Selection

Absorption SPME Fibres	Polarity	Material	Coating method	Application
Carbowax/Divinylbenzene (CW/DVB)	Polar	Stable flex	Cross-linked	Alcohols/ polars
Carbowax/Templated Resin (CW/TPR) (HPLC only)	Polar	Templated resin	Cross-linked	Surfactants
Polydimethylsiloxane (PDMS)	Non-polar	Fused silica	Non-bonded	General volatiles
Carboxen/Polydimethylsiloxane (CAR/PDMS)	Bipolar	Fused silica, stable flex	Cross-linked	Low MW volatiles
Divinylbenzene/Carboxen/Polydimethylsiloxane (DVB/CAR/PDMS)	Bipolar	Stable flex	Cross-linked	Odors and flavors
Polyacrylate (PA)	Polar	Fused silica	Cross-linked	Polar semi-volatiles
Polydimethylsiloxane/ divinylbenzene (PDMS/ DVB)	Bipolar	Fused silica, stable flex	Cross-linked	Polar volatiles

Modified from Vas, G., & Vekey, K. (2004). Solid-phase microextraction: A powerful sample preparation tool prior to mass spectrometric analysis. *Journal of Mass Spectrometry, 39*, 233–254. With permission. © 2004. John Wiley and Sons.

and DVB/CAR/PDMS were used by Mahattanatawee and Rouseff (2014) to evaluate their efficacy, and among these fibers, DVB/CAR/PDMS was reported to be better for extraction and concentration of the headspace volatiles of cooked rice.

5.6.3   EXTRACTION TIME

Extraction time determines the extraction efficiency of analytes by SPME fiber. It also depends on other factors like temperature and partition coefficient of the analytes. In fact, the time depends on the equilibration of analytes among sample matrix, the headspace, and the fiber coating. The concentration of the analytes is constant in all the three phases only after the equilibrium is attained (Zhang and Pawliszyn, 1993). Hence, the extraction time is the time duration taken for the analytes to reach the state of equilibrium that can be carefully controlled and even used for the quantization of analytes (Whiton and Zoecklein, 2000). Generally, equilibration time is longer than the GC analysis time. By taking equilibrium time and GC method into consideration, the analysis time can be reduced significantly (Song et al., 1997). It has also been found that more extraction time is required for less volatile and/or more polar compounds of the matrix, but less extraction time for more volatile compounds in order for better extraction efficiency. By manipulating the temperature, sample size and agitating the sample, the time parameter can be optimized. Optimized conditions for equilibrium temperature, extraction temperature, and sample weight were used, and 60 min for equilibration and 50 min for extraction were found to be optimum for aromatic rice (Lin et al., 2010).

5.6.4   IONIC STRENGTH OF THE MATRIX

The extraction efficiency also depends on the ionic strength of the matrix. It affects the release of specific analytes into the headspace. So, the addition of salt or adjustment of pH is done to alter the SPME yields (Eisert and Pawliszyn, 1997b). In some cases, addition of salt increases the extraction efficiency, for example, with polar compounds and volatiles but it is reverse for analytes with high distribution constants. The compounds of acidic and basic nature are more effectively extracted under acidic and alkaline conditions, respectively. Hence a combination of salt and pH is taken into consideration for better extraction of the analytes from the headspace (Supelco, 1998).

5.6.5   SAMPLE MATRIX

The analytes are present in the sample matrix, and hence, their release into the headspace also depends upon the type of matrix. If the matrix binds the analytes strongly than the partition coefficient is low and adsorption to the fiber ultimately decreases. The extraction of analytes depends upon the partitioning between the matrix and the fiber coating. The chemical potential difference of the analytes acts as a driving force for moving analytes from the matrix to the fiber coating in the HS-SPME. Besides this, the concentration of analytes is directly proportional to its original concentration in the matrix and inversely proportional to partition coefficient (Spietelun et al., 2013). The headspace/water partition coefficients are directly associated with the Henry's law constants of the analytes determined by their volatility and hydrophobicity (Zhang et al., 1994). The work of Mathure et al. (2011) and Lin et al. (2010) in different cultivars of aromatic rice stated that sample quantity and volume adjustment should have received attention for better SPME efficiency.

5.6.6   TEMPERATURE OF THE SAMPLE

Temperature plays one of the major roles in the extraction of analytes by the SPME method. With the increase in temperature, the vapor-pressure of the sample increases, which is inversely proportional to the partition coefficient. Therefore, maximum quantities of analytes are collected in the headspace associated with the decrease in partition coefficient (Kolb et al., 1992). Temperature maintains the equilibrium of analytes between the matrix and the fiber coating. It has also been observed that with the increase in temperature for the higher boiling polar compounds like alcohols and ethyl lactate, the response was increased, but the concentration of volatiles in the headspace reduced. As the adsorption by fiber is an exothermic process, increase in temperature even reduces the distribution constant and the extraction efficiency. At the same time, the passage of analytes from the sample to the headspace is facilitated by the temperature. Hence, the optimum extraction and pre-incubation temperature based on the analytes of interest and sensitivity to enhance the volatility of target volatile compounds in headspace should be identified (Penalver et al., 1999; Wang et al., 2004; Panavaite et al., 2006). The extraction temperature for analysis of rice volatile compounds was optimized by Lin et al. (2010) and Mathure et al. (2011) and reported 80°C was the optimum temperature in the cultivars of aromatic rice.

5.6.7 TYPE OF FIBERS

The SPME fiber coatings are selected based on the type of analytes to be analyzed. Before analysis, the fiber is conditioned properly to avoid background noise by the fiber bleed in the chromatogram (Balasubramanian and Panigrahi, 2011). Some of the commercially available fiber coatings for food and environmental analysis are CAR, CW, DVB, PA, and PDMS along with various coating combinations, blends or copolymers (Table 5.3) as well as different film thicknesses, and fiber assemblies. According to the need for novel application (Table 5.4), the new coating materials and their combinations are being developed (Dietz et al., 2006). The most widely used fiber coatings are PA and PDMS. Mixed coatings with primary extraction phase as a porous solid are available as well (Gorecki et al., 1999). For a wide range of analytes with different polarities and molecular weights, CAR/PDMS and DVB/CAR/PDMS have been reported to provide the best extraction efficiencies. The SPME fiber PDMS can tolerate high temperature up to 300°C and has great stability (Balasubramanian and Panigrahi, 2011). These fibers are well suited for the analysis of non-polar analytes, while PA is used to extract polar analytes. The DVB fiber on the other hand has a polar porous solid coating used for extracting polar compounds such as disulfides and trisulfides (Cai et al., 2001). The fiber blends of non-polar (PDMS) and a polar material (DVB) are often employed for extraction of bipolar compounds such as alcohols, aldehydes, carboxylic acids, ethers, and ketones (Balasubramanian and Panigrahi, 2011).

TABLE 5.4   SPME Fiber Need for Novel Application

SPME fibre	Novel application
Carbowax/divinylbenzene	For polar analytes, especially for alcohols
Carbowax/templated resin	Developed for high-performance liquid chromatograpy applications, e.g., surfactants
Carboxen/poly(dimethylsiloxane)	Ideal for gaseous/volatile analytes, high retention for trace analysis
Divinylbenzene/Carboxen/poly(dimethylsiloxane)	Ideal for a broad range of analyte polarities, good for C3–C20 range
Poly(dimethylsiloxane)	Considered non-polar for non-polar analytes
Poly(dimethylsiloxane)/divinylbenzene	Ideal for many polar analytes, especially amines
Polyacrylate	Highly polar coating for general use, ideal for phenols

Significantly modified from Reineccius, G. A. (2007). Flavor-Isolation Techniques. In: *Flavors and Fragrances: Chemistry, Bioprocessing and Sustainability* (Berger, R. G., ed.). Springer-Verlag Berlin Heidelberg, Germany. pp. 409-426.

5.7   EXTRACTION PROCESS

Different extraction methods have been applied by researchers for the analysis of flavor compounds in rice cultivars. Some of them are as follows:

In a method developed by Grimm et al. (2001), the whole rice grains were analyzed where flour samples were taken immediately before analyses. Sample preparation consisted of placing 0.75 g of rice directly into a 2 mL vial. Water was added to the sample by spraying Milli-Q water onto the top of the rice kernels. 2,4,6-Trimethylpyridine (TMP; Sigma-Aldrich, St. Louis, MO) was employed as the internal standard. A 2-µl aliquot of a 1 ppm solution was added to each sample, thus effectively placing 2 ng of TMP in each vial. The standard was placed on the inside of the glass vial just below the neck. Following preparation, samples were placed in an autosampler tray and maintained at room temperature until analyzed. Samples were preheated for 25 min at 80°C prior to sampling. Collection of volatile compounds was accomplished using a 15-min adsorption period at 80°C during which the sample was shaken. The SPME fiber employed was a 1-cm 50/30 carboxen/DVB/PDMS fiber (Supelco, Bellefonte, PA).

A CTC SPME autosampler equipped with a heated sample shaker and a needle heater for thermal cleaning of the SPME fiber was employed (Leap Technologies, Carrboro, NC).

According to Wongpornchai et al. (2003), 1.2 g of sample was taken in a sealed 27-mL bottle fitted with a PTFE/ silicone septum (Restek Corp., Bellefonte, PA) and an aluminum cap. The sample bottle was left at room temperature (27°C) for 30 min. A SPME fiber (Supelco, Bellefonte, PA) of 1 cm in length, coated with poly (dimethylsiloxane) (PDMS) at 100 μm thickness, was mounted in the manual SPME holder (Supelco) and was preconditioned in a GC injection port set at 250°C for 1 h. By insertion through the septum of the sample bottle, the fiber was then exposed to the sample headspace for 10 min prior to desorption of the volatiles into the splitless injection port of the GC-MS instrument.

5.8 ADVANTAGES AND DISADVANTAGES OF SPME

5.8.1 ADVANTAGES

1. SPME technique is suitable for the extraction of volatile aroma compounds from liquids (water), gases, or solids rice matrix.
2. This one step extraction microtechnology is free from calibration as all analytes from rice sample are transferred to the extraction phase.
3. The most obvious advantages of SPME technique are solvent-free sampling (100%), comparatively low cost, reproducible (minimizes the use of solvents and their disposal), fast (reduces sample preparation time by 70%), and easy to use. Volatile aroma compounds from rice are isolated and concentrated without any hinderance.
4. Equilibrium methods in SPME technique are more selective because analysis can be take full advantage of the differences in extracting phase/matrix distribution constants to separate target volatile compounds from interferences and enhanced by the selection of the stationary phase that best suits the rice aroma compounds.
5. SPME is best suited sample preparation technique to quantify the rice aroma compound coupled with gas chromatography.

5.8.2 DISADVANTAGES

1. In SPME technique, the comparison of obtained data is only possible when the same fiber is used on all the samples.
2. The aroma profiling of collected volatiles from rice matrix in SPME technique always depends upon fiber type, thickness and length, sampling time, and temperature.
3. Mixed fibers have largely solved the difficulty of differentiation against polar components in SPME technique, whereas early fibers support differentiation against such components.

5.9 CASE STUDY ON THE SPME METHOD FOR RICE AROMA CHEMICALS

Several studies have been done for the analysis of rice volatile aroma compounds using SPME, and many components have been identified. Grimm et al. (2001) screened for 2-AP in rice headspace using SPME on 21 experimental rice varieties. The SPME method involves 100 μL water to 0.75 g of rice sample at 80°C temperature. Further, 25-min preheated rice headspace was exposed to DVB/CAR/PDMS fiber for 15 min followed by subsequent GC-MS analysis for 35 min. Several nanogram 2-AP was recovered from 0.75 g aromatic rice sample, whereas nonaromatic rice produced only trace amount of 2-AP. Single SPME headspace analysis recovered 0.3% of the total 2-AP in the sample.

In 2003, SPME was used to identify 2-AP for the first time by Wongpornchai et al. (2003) in headspace of fresh bread flowers (*Vallaris glabra* Ktze). Monsoor et al. (2004) analyzed volatile aroma component of commercially milled head and broken rice samples and observed that broken rice contains high concentration of volatile compounds than head rice. He also reported that the concentration of volatile components was increased after storage of both rice types, and pentanal, pentanol, hexanal, pentyfuran, octanal, and nonanal are major volatile components in head and broken rice samples.

In 2006, Bryant et al. (2006) experimented upon "Koshihikari" and "Basmati" rice for texture profile and volatile compound analysis and cooked both in different rice cookers. He reported the higher amount of dodecanal and hexanal in the sample prepared in Hitachi cooker and higher amount of acetone and nephthalene in the sample prepared in National cooker using the SPME/GC-MS method. A similar method was used by Bryant and McClung (2011) to identify volatile profile of aromatic and nonaromatic rice cultivars. They reported 93 volatile compounds, out of which 64 compounds were not reported previously. They differentiated aromatic rice from non-aromatic rice by the presence of 2-AP in former. Bryant et al. (2006) also studied the diversity of volatiles in aromatic and non-aromatic rice cultivars. They stated that further research may conclude in better understanding of flavoring compounds. Again in 2011, Bryant et al. (2011) used SPME with GC-MS and differentiated between nonaromatic rice (*Oryza sativa* L.) kernels and aromatic rice kernels. They detected the presence of 2-AP in aromatic rice germplasm by development of single kernel analysis method as aromatic rice cultivars were higher in demand. Determination of 2-AP is very important as it can be utilized in breeding methods to detect aromatic and nonaromatic rice in the market place.

Zeng et al. (2007) designed a new apparatus using SPME for direct extraction of flavor volatiles from rice during cooking and tested in the Japanese rice cultivar Akitakomachi. Zeng et al. (2008a) used modified HS-SPME/GC-MS for the analysis of flavor volatiles of 3 Japanese rice cultivars, viz. Nihonbare, Koshihikari and Akitaomachi and detected 46 components like aldehydes, ketones, alcohols, heterocyclic compounds, fatty acids, esters, phenolic compounds, hydrocarbons, etc. On cooking, the amount of low boiling volatiles was decreased, whereas the amount of key odorant compounds was increased. Three rice cultivars were compared on the basis of similarities and differences of their volatile compounds. Zeng et al. (2008b) again used modified HS-SPME with GC-MS for analyzing flavor volatiles in 3 rice cultivars during cooking with less amount of digestive protein. They detected 77 volatile compounds, out of which 13 components were new and not reported previously. These volatile compounds were indole, vanillin, (*E,E*)-2, 4-decadienal, (*E*)-2-nonenal, 2-pentyfuran, 2-methoxy-4-vinylphenol, etc. Zeng et al. (2009) identified a broad range of the total 96 flavor volatile compounds from cooked glutinous rice that could be extracted and detected by single HS-SPME/GC-MS by which 27 compounds were detected, viz. 2-methyl-3-octanone, 2-hexylfuran, 7-methyl-2-decane, 2-heptyfuran, (*E*)-4-nonenal, (*E*)-3-nonen-2-one, 2-methyl-5-decanone, (*Z*)-2-octen-1-ol, 6-dodecanone, 2,3-octanedione, 6,10-dimethyl-2-undecanone, dodecanal, alpha-cadinene, 1-decanol, delta-cadinene, 1-undecanol, tridecanal, (1*S,Z*)-calamenene, 3-hydroxy-2,4,4-trimethylpentyl isobutanoate, 2,4,4-trimethylpentan-1,3-diol di-iso-butanoate, hexadecanal, cubenol, delta cadinol, 5-iso-propyl-5H-furan-2-one, di-iso-butyl adipate, alpha-cadinol and glycerine. These detected volatile components in 3 glutinous rice cultivars belongs to various groups like aldehyde, ketone, alcohols and heterocyclic compounds, fatty acids, esters, phenolic compounds, hydrocarbons, etc. In 2010, Lin et al. (2010) presented a paper on volatile compounds in the 10th International Working Conference on Stored Product Protection. They experimented with 20 g sample of heated rice for 30 min at 80°C prior to headspace absorption, and then extracted for 30 min using HS-SPME with GCMS. They reported alcohols, aldehydes, ketones, esters, hydrocarbons, organic acids, as well as heterocyclic compounds as volatile compounds in indica and japonica rice and also reported that aldehydes were present in abundance, around 30 % of volatiles by weight.

Another new powerful device, cold-fiber solid phase micro extraction (CF–SPME) was developed in 2007 by Ghiasvand et al. (2007) for rapid detection of flavoring agents in sample of fragrant rice headspace (Figure 5.8). In CF–SPME volatile compounds concentration showed that two uncooked varieties of Indian fragrant rice and nine Iranian rices were successfully analyzed as dry kernels by utilizing fully automated CF-SPME-GC-TOF-MS without addition of water. Different experimental parameters were optimized including 2-AP as a key odorant based on four target analysis.

Tananuwonga and Lertsirib (2010) studied the changes in volatile aroma compounds of red organic fragrant rice cv. HomDaeng under different conditions of storage and evaluated effects of temperature at storage, time, and effects of packing by using SPME/GC-MS. The study detected total 13 key volatile compounds which includes 10 lipid oxidation products. In 2010, Goufo et al. (2010) conducted a study to investigate two fragrant rice cultivars, viz., Guixiangzhan and Peizaruanxiang, grown in South China. They used HS-SPME and static HS with GC-MS to find out

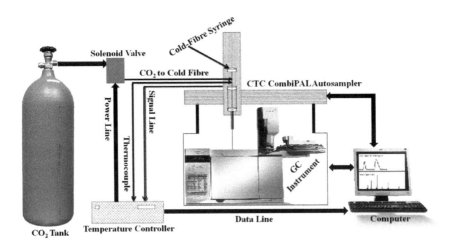

FIGURE 5.8 Schematic diagram of the automated cold-fiber headspace solid phase micro extraction system (Ghiasvand et al., 2007).

the factors that affects concentration of aroma compound, i.e., 2-AP and noted 5 times difference of 2-AP levels which is highest content (3.86 $\mu g.g^{-1}$) in Guixiangzhan compared to Thai KDML 105 rice. Despite the presence of 2-AP, other compounds were supposed to contribute the characteristic aroma of Peizaruanxiang. The findings showed that manipulation of pre and post-harvest treatment can largely improve the specific attributes of domestically produced cultivars. They reported that China could increase the share in domestic market of fragrant rice as well as in the international market (Goufo et al., 2010).

Maraval et al. (2010) analyzed 2-AP, a potent flavor compound in rice, and its ring-deuterated analog, 2-acetyl-1-d(2)-pyrroline (2-AP-d(2)). A stable isotope dilution assay (SIDA), involving HS-SPME combined with gas chromatography-positive chemical ionization-ion trap-tandem mass spectrometry (GC-PCI-IT-MS), was developed for 2-AP quantification. DVB/CAR/PDMS fiber was used for HS-SPME procedure and parameters affecting analytes recovery, such as extraction time and temperature, pH and salt, were studied. Limits of detection (LOD) and quantification (LOQ) for 2-AP were 0.1 and 0.4 $ng.g^{-1}$ of rice, respectively. The recovery of spiked 2-AP from rice matrix was almost complete. The developed method was applied to the quantification of 2-AP in aerial parts and grains of scented and non-scented rice cultivars

The HS-SPME method combined with GC-MS was employed by Khorheh et al. (2011) for assessment of flavor volatiles of Iranian rice, two modified Iranian rice and Hashemi rice cultivar using gelatinization process. The proposed combination provided a powerful system for easy and rapid screening of a wide range of flavors in fragrant rice samples. For optimization of different experimental parameters fiber composition effects, water content of rice sample and equilibrium time was evaluated. The result showed that the amounts of volatile compounds were increased during gelatinization. All free flavored volatiles, bound flavor components were liberated during gelatinization process were formed by thermal decomposition of non-volatile constituents present in rice. The identified volatile components during gelatinization process were 75, 55, and 66 from Hashemi, HD5 and HD6 Iranian rice cultivars, respectively, out of which 58 unique compounds were not detected previously. The chemical groups of identified volatile components in three Iranian rice cultivars belong to aldehydes, ketones, alcohols, heterocyclic compounds, fatty acids, esters, phenolic compounds, and hydrocarbon.

In a study conducted by Bryant and McClung (2011), seven aromatic and two non-aromatic cultivars were examined for their volatile profiles. Ninety-three volatile compounds were identified, 64 of which had not been previously reported in rice. Differences were found in the volatile compounds of aromatic and non-aromatic rice, besides 2-AP. Most of the volatile compounds were present in freshly harvested rice and rice following storage, with very few new compounds being identified only after storage. Dellrose, an aromatic cultivar, and Cocodrie, a non-aromatic cultivar, had the most complex volatile profiles (over 64 volatiles). Sixteen compounds were found only in the aromatic cultivars, and some volatiles were found to be unique to specific aromatic cultivars.

Mathure et al. (2011) developed another method for quantitative determination of 2-AP and other aroma volatiles using HS-SPME/GC-FID in scented rice. The sample extraction was done at 80°C for 30 min pre-incubation followed by 20-min adsorption from 1g rice containing 300μl of odor-free water were the optimum conditions for quantification. The optimized conditions were employed for quantitative analysis in 33 scented and two non-scented rice samples. Highest amount of 2-AP was recorded in Indrayani Brand 2 (0.552 ppm), followed by Kamod (0.418 ppm) and Basmati Brand 5 (0.411 ppm). Rice types (Basmati, Ambemohar, Kolam, Indrayani and local) significantly contributed to the variation in 2-AP, hexanal, nonanal, decanal, benzyl alcohol, vanillin, guaiacol, and indole.

Givianrad (2012) combined gas chromatography mass spectrometry with the HS-SPME method for flavor volatiles of three different rice cultivars including two modified Iranian rice cultivars and Hashemi rice cultivar during gelatinization. Altogether, 74, 55, and 66 components were identified for Hashemi, HD5 and HD6 rice samples, respectively, of which 56 unique compounds were not identified, previously. Subsequently, seven fragrance chemicals were detected, which were most frequently reported as contact allergens in the European Union.

Mahattanatawee and Rouseff (2014) characterized and identified aroma volatiles from three cooked fragrant rice types (Jasmine, Basmati, and Jasmati) using GC-MS. A total of 26, 23, and 22 aroma active volatiles were observed in Jasmine, Basmati, and Jasmati cooked rice samples, respectievly. 2-AP was aroma active in all three rice types, but the sulfur-based, cooked rice characteristic impact volatile, 2-acetyl-2-thiazoline was aroma active only in Jasmine rice. Five additional sulfur volatiles were found to have aroma activity: dimethyl sulfide, 3-methyl-2-butene-1-thiol, 2-methyl-3-furanthiol, dimethyl trisulfide, and methional. Other newly reported aroma active rice volatiles were geranyl acetate, β-damascone, β-damascenone, and α-ionone, contributing nutty, sweet floral attributes to the aroma of cooked aromatic rice.

5.10 CONCLUSION

SPME provide significant benefits to rice volatile aroma compounds research as an extraction technique prior to GC analysis. The quantitative and qualitative results obtained from SPME techniques are always better than traditional/conventional extraction techniques. Since past decades, the SPME technique has matured as a tool for the analysis of rice volatile aroma compounds to date. The strategies of SPME method is straightforward applications ranging from the quantitation of specific chemicals at low ppb levels to the complete identification of the volatile aroma compounds in rice sample. A number of publications listed due to sustainability of SPME techniques in the field of rice volatile aroma compounds since more than 10 years back. The researchers of this area are continuing to realize the benefits from SPME. As numerous findings are examples on SPME techniques for rice aroma study illustrated in this chapter, it is a potent extraction tool for the analysis of impact rice volatile aroma compounds and also superior than other traditional/conventional extraction techniques in terms of accuracy.

ACKNOWLEDGMENT

Deepak Kumar Verma and Prem Prakash Srivastav are indebted to the Department of Science and Technology, Ministry of Science and Technology, Government of India for an individual research fellowship (INSPIRE Fellowship Code No.: IF120725; Sanction Order No. DST/INSPIRE Fellowship/2012/686; Date: 25/02/2013).

KEYWORDS

- aroma chemicals
- dynamic HS
- extraction method
- fiber SPME
- flavoring volatile
- GC-MS
- head space
- polydimethylsiloxane
- rice aroma
- rice sample
- SPME

REFERENCES

Adahchour, M., Beens, J., Vreuls, R. J. J., Batenburg, A. M., Rosing, E. A. E., & Brinkman, U. A. T. (2002). Application of solid-phase micro-extraction and comprehensive two-dimensional gas chromatography (GC × GC) for flavor analysis. *Chromatographia, 55*(5/6), 361–367.

Ai, J. (1997). Solid phase microextraction for quantitative analysis in nonequilibrium situations. *Analytical Chemistry, 69,* 1230–1236.

Ampuero, S., Bogdanov, S., & Bosset, J. O. (2004). Classification of unifloral honeys with an MS-based electronic nose using different sampling modes: SHS, SPME and IN-DEX. *European Food Research and Technology, 218,* 198–207.

Arthur, C. L., & Pawliszyn, J. (1990). Solid phase microextraction with thermal desorption using fused silica optical fibers. *Analytical Chemistry, 62,* 2145–2148.

Arthur, C. L., Potter, D. W., Buchholz, K. D., Motlagh, S., & Pawliszyn, J. (1992). Solid-phase microextraction for the direct analysis of water: *Theory and Practice. LC-GC, 10,* 656–661.

Augusto, F., Lopes, A. L., & Zini, C. A. (2003). Sampling and sample preparation for analysis of aromas and fragrances. *Trends in Analytical Chemistry, 22,* 160–169.

Balasubramanian, S., & Panigrahi, S. (2011). Solid-phase microextraction (SPME) techniques for quality characterization of food products: A Review. *Food and Bioprocess Technology, 4,* 1–26.

Basheer, C., & Lee, H. K. (2004). Hollow fiber membrane-protected solid phase microextraction of triazine herbicides in bovine milk and sewage sludge samples. *Journal of Chromatography A., 1047,* 189–194.

Bene, A., Luisier, J. L., & Fornage, A. (2002). Applicability of a SPME method for the rapid determination of VOCs. *CHIMIA International Journal for Chemistry, 56*(6), 289–291.

Brunton, N. P., Cronin, D. A., & Monahan, F. J. (2001). The effects of temperature and pressure on the performance of Carboxen/PDMS fibres during solid phase microextraction (SPME) of headspace volatiles from cooked and raw turkey breast. *Flavor and Fragrance Journal, 16,* 294–302.

Bryant, R. J., & McClung, A. M. (2011). Volatile profiles of aromatic and non-aromatic rice cultivars using SPME/GC–MS. *Food Chemistry, 124*(2), 501–513.

Bryant, R. J., Jones, G., & Grimm, C. (2006). Texture profile and volatile compound analyses of 'Koshihikari' and 'Basmati' rice prepared in different rice cookers. B. R. Wells Rice Research Studies, *AAES Research Series, 550,* 370–376.

Bryant, R. J., McClung, A. M., & Grimm, C. (2011). Development of a single kernel analysis method for detection of 2-acetyl-1-pyrroline in aromatic rice germplasm. *Sensing and Instrumentation for Food Quality and Safety, 5*(5), 147–154.

Cai, J., Liu, B., & Su, Q. (2001). Comparison of simultaneous distillation extraction and solid-phase microextraction for the determination of volatile flavor components. *Journal of Chromatography A., 930,* 1–7.

Carasek, E., & Pawliszyn, J. (2006). Screening of tropical fruit volatile compounds using solid-phase microextraction (SPME) fibers and internally cooled SPME fiber. *Journal of Agricultural and Food Chemistry, 54,* 8688–8696.

Carasek, E., Cudjoe, E., & Pawliszyn, J. (2007). Fast and sensitive method to determine chloroanisoles in cork using an internally cooled solid-phase microextraction fiber. *Journal of Chromatography A., 1138*, 10–17.

Castro, R., Natera, R., Duran, E., & Garcia-Barroso, C. (2008). Application of solid phase extraction techniques to analyse volatile compounds in wines and other enological products. *European Food Research and Technology, 228*, 1–18.

Ceva-Antunes, P. M. N., Bizzo, H. R., Silva, A. S., Carvalho, C. P. S., & Antunes, O. A. C. (2006). Analysis of volatile composition of siriguela (*Spondias purpurea* L.) by solid phase microextraction (SPME). *LWT – Food Science and Technology, 39*, 437–443.

Chen, J., & Pawliszyn, J. (1995). Solid phase microextraction coupled to high-performance liquid chromatography. *Analytical Chemistry, 67*, 2530–2533.

Chen, Y., & Pawliszyn, J. (2006). Miniaturization and automation of an internally cooled coated fiber device. *Analytical Chemistry, 78*, 5222–5226.

Contarini, G., & Povolo, M. (2002). Volatile fraction of milk: comparison between purge and trap and solid phase microextraction techniques. *Journal of Agricultural and Food Chemistry, 50*, 7350–7355.

Dietz, C., Sanz, J., & Camara, C. (2006). Recent developments in solid-phase microextraction coatings and related techniques. *Journal of Chromatography A., 1103*, 183–192.

Djozan, D., & Ebrahimi, B. (2008). Preparation of new solid phase micro extraction fiber on the basis of atrazine –molecular imprinted polymer: Application for GC/MS screening of triazine herbicides in water, rice and onion. *Analytica Chimica Acta., 616*, 152–159.

Djozan, D., Makham, M., & Ebrahimi, B. (2009). Preparation and binding study of solid-phase microextractionfiber on the basis of ametryn-imprinted polymer: Application to the selective extraction of persistent triazine herbicides in tap water, rice, maize and onion. *Journal of Chromatography A., 1216*, 2211–2219.

Eisert, P., & Pawliszyn, J. (1997a). Automated in-tube solid-phase microextraction coupled to high-performance liquid chromatography. *Analytical Chemistry, 69*, 3140–3147.

Eisert, R., & Pawliszyn, J. (1997b). New trends in solid-phase microextraction. *Critical Review in Analytical Chemistry, 27*, 103–135.

Elmore, J. S., Erbahadir, M. A., & Mottram, D. S. (1997). Comparison of dynamic headspace concentration on Tenax with solid phase microextraction for the analysis of aroma volatiles. *Journal of Agricultural and Food Chemistry, 45*, 2638–2641.

Flamini, R. (2010). Volatile and aroma compounds in wines. In: *Mass Spectrometry in Grape and Wine Chemistry,* Flamini, R., & Traldi, P. (eds.). John Wiley & Sons, Inc. Hoboken, N. J., USA, pp. 117–162.

Furton, K. G., Almirall, J. R., Bi, M., Wang, J., & Wu, L. (2000). Application of solid-phasemicroextraction to the recovery of explosives and ignitable liquid residues from forensicspecimens. *Journal of Chromatography A., 885*, 419–432.

Ghiasvand, A. R. Hosseinzadeh, S., & Pawliszyn, J. (2006). New cold-fiber headspace solid-phase microextraction device for quantitative extraction of polycyclic aromatic hydrocarbons in sediment. *Journal of Chromatography A., 1124*, 35–42.

Ghiasvand, A. R., Setkova, L., & Pawliszyn, J. (2007). Determination of flavor profile in Iranian fragrant rice samples using cold-fibre SPME–GC–TOF–MS. *Flavor and Fragrance Journal, 22*(5), 377–391.

Givianrad, M. H. (2012). Characterization and assessment of flavor compounds and some allergens in three Iranian rice cultivars during gelatinization process by HS-SPME/GC-MS. *E-Journal of Chemistry, 9*(2), 716–728.

Gorecki, T., Yu, X., & Pawliszyn, J. (1999). Theory of analyte extraction by selected porous polymer SPME fibres. *Analyst, 124,* 643–649.

Goufo, P., Duan, M., Wongpornchai, S., & Tang, X. (2010). Some factors affecting the concentration of the aroma compound 2-acetyl-1-pyrroline in two fragrant rice cultivars grown in South China. *Frontiers of Agriculture in China, 4*(1), 1–9.

Grimm, C. C., Bergman, C., Delgado, J. T., & Bryant, R. (2001). Screening for 2-Acetyl-1-pyrroline in the Headspace of Rice Using SPME/GC-MS. *Journal of Agricultural and Food Chemistry, 49*(1), 245–249.

Grimm, C. C., Champagne, E. T., Lloyd, S. W., Easson, M., Condon, B., & McClung, A. (2011). Analysis of 2-Acetyl-l-Pyrroline in Rice by *HSSE/GCIMS. Cereal Chemistry, 88*(3), 271–277.

Harmon, A. D. (1997). Solid-phase microextraction for the analysis of flavors. In: *Techniques for Analyzing Food Aroma*, Marsili, R., (ed.). Marcel Decker, Inc. New York, USA, pp. 81–112.

Hawthorne, S. B., Miller, D. J., Pawliszyn, J., & Arthur, C. L. (1992). Solventless determination of caffeine in beverages using solid-phase microextraction with fused-silica fibers. *Journal of Chromatography, 603*(1–2), 185–191.

Henryk, H. J., Krystian, W., Erwin, W., & Edward, K. (1998). Solidphase microextraction for the analysis of some alcohols and esters in beer: comparison with static headspace method. *Journal of Agricultural and Food Chemistry, 46,* 1469–1473.

Hinshaw, J. (2012). Solid-Phase Microextraction. *LCGC North America, 30*(10), 904–910.

Hiroyuki, K., Heather, L., & Janusz, P. (2000). Applications of solidphase microextraction in food analysis. *Journal of Chromatography A., 380,* 35–62.

Holt, R. U. (2001). Mechanisms effecting analysis of volatile flavor components by solid-phase microextraction and gas chromatography. *Journal of Chromatography A., 937,* 107–114.

Jelen, H. H. (2003). Use of solid phase microextraction for profiling fungal volatile metabolites. *Letters in Applied Microbiology, 36,* 263–67.

Jiang, R., & Pawliszyn, J. (2012). Thin-film microextraction offers another geometry for solid-phase microextraction. *TrAC Trends in Analytical Chemistry, 39,* 245–253.

Jung, D. M., & Ebeler, S. E. (2003). Headspace solid-phase microextraction method for the study of the volatility of selected flavor compounds. *Journal of Agricultural and Food Chemistry, 51,* 200–205.

Kaseleht, K., Leitner, E., & Paalme, T. (2011). Determining aroma-active compounds in Kama flour using SPME–GC/MS and GC–olfactometry. *Flavor and Fragrance Journal, 26,* 122–128.

Kataoka, H. (2010). Recent developments and applications of microextraction techniques in drug analysis. *Analytical and Bioanalytical Chemistry, 396,* 339–364.

Kataoka, H., Lord, H. L., & Pawliszyn, J. (2000). Applications of solid-phase micro-extraction and gas chromatography. *Journal of Chromatography A., 800,* 35–62.

Khorheh, N. A., Givianrad, M. H., Ardebili, M. S., & Larijani, K. (2011). Assessment of Flavor Volatiles of Iranian Rice Cultivars during Gelatinization Process. *Journal of Food Biosciences and Technology, 1*, 41–54.

Kolb, B., & Ettre, L. S. (1997). Static headspace–gas chromatography: *Theory and Practice.* Wiley-VCH Inc., New York. pp. 335.

Kolb, B., Welter, C., & Bichler, C. (1992). Determination of partition coefficients by automatic equilibrium headspace gas chromatography by vapor phase calibration. *Chromatographia, 34*(5), 235–240.

Koziel, J. A., Odziemkowski, M., & Pawliszyn, J. (2001). Sampling and analysis of airborne particulate matter and aerosols using in-needle trap and SPME fiber devices. *Analytical Chemistry, 73*, 47–54.

Koziel, J., Jia, M. Y., Khaled, A., Noah, J., & Pawliszyn, J. (1999). Field air analysis with SPME device. *Analytical Chimica Acta., 400*, 153–162.

Laguerre, M., Mestres, C., Davrieux, F., Ringuet, J., & Boulanger, R. (2007). Rapid discrimination of scented rice by solid-phase microextraction, mass spectrometry, and multivariate analysis used as a mass sensor. *Journal of Agricultural and Food Chemistry, 55*, 1077–1083.

Lin, J. Y., Fan, W., Gao, Y. N., Wu, S. F., & Wang, S. X. (2010). Study on volatile compounds in rice by HS-SPME. 10th International working conference on stored product protection. *Julius-Kühn-Archiv, 425*, 125–134.

Lipinski, J. (2001). Automated solid phase dynamic extraction – Extraction of organics using a wall coated syringe needle. *Fresenius Analytical and Bioanalytical Chemistry, 369*, 57–62.

Lord, H., & Pawliszyn, J. (2000). Evolution of solid-phase microextraction technology. *Journal of Chromatography A., 885*, 153–193.

Lotters, J. C., Olthuis, W., Veltink, P. H., & Bergveld, P. (1996). The mechanical properties of the rubber elastic polymer polydimethylsiloxane for sensor applications. *Journal of Micromechanics and Microengineering, 7*(3), 145–147.

Louch, D., Motlagh, S., & Pawliszyn, J. (1992). Dynamics of organic compound extraction from water using solid phase microextraction with fused silica fibers. *Analytical Chemistry, 64*, 1187.

Mahattanatawee, K., & Rouseff, R. L. (2014). Comparison of aroma active and sulfur volatiles in three fragrant rice cultivars using GC–Olfactometry and GC–PFPD. *Food Chemistry, 154*, 1–6.

Mallia, S., Fernandez-Garcia, E., & Bosset, J. O. (2005). Comparison of purge and trap and solid phase microextraction techniques for studying the volatile compounds of three European PDO hard cheeses. *International Dairy Journal, 15*, 741–758.

Maraval, I., Sen, K., Agrebi, A., Menut, C., Morere, A., Boulanger, R., Gay, F., Mestres, C., & Gunata, Z. (2010). Quantification of 2-acetyl-1-pyrroline in rice by stable isotope dilution assay through headspace solid-phase microextraction coupled to gas chromatography-tandem mass spectrometry. *Analytica Chimica Acta., 675*(2), 148–155.

Marsili, R. T. (1999a). Comparison of solid phase microextraction and dynamic headspace methods for the gas chromatographic-mass spectrometric analysis of light-induced lipid oxidation products in milk. *Journal of Chromatographic Science, 37*, 17–23.

Marsili, R. T. (1999b). SPME–MS–MVA as an electronic nose for the study of offflavors in milk. *Journal of Agricultural and Food Chemistry, 47*, 648–654.

Marsili, R. T. (2000). Shelf-life prediction of processed milk by solid-phase microextraction, mass spectrometry, and multivariate analysis. *Journal of Agricultural and Food Chemistry, 48*, 3470–3475.

Mathure, S. V., Wakte, K. V., Jawali, N., & Nadaf, A. B. (2011). Quantification of 2-Acetyl-1-pyrroline and other rice aroma volatiles among Indian scented rice cultivars by HS-SPME/GC-FID. *Food Analytical Methods, 4*, 326–333.

Mehdinia, A., Mousavi, M. F., & Shamsipur, M. (2006). Nano-structured lead dioxide as a novel stationary phase for solid-phase microextraction. *Journal of Chromatography A., 1134*, 24–31.

Merkle, S., Kleeberg, K. K., & Fritsche, J. (2015). Recent developments and applications of solid phase microextraction (SPME) in food and environmental analysis – a review. *Chromatography, 2*(3), 293–381.

Mester, Z., & Pawliszyn, J. (1999). Electrospray mass spectrometry of trimethyllead and triethyllead with in-tube solid phase microextraction sample introduction. *Rapid Communications in Mass Spectrometry, 13*, 1999–2003.

Mills, G. A., & Walke, V. (2000). Headspace solid-phase microextraction procedures for gaschromatographic analysis of biological fluids and materials. *Journal of Chromatography A., 902*, 267–287.

Mirnaghi, F. S., Hein, D., & Pawliszyn, J. (2013). Thin-film microextraction coupled with mass spectrometry and liquid chromatography–Mass spectrometry. *Chromatographia, 76*, 1215–1223.

Mondello, L., Costa, R., Tranchida, P. Q., Chiofalo, B., Zumbo, A., & Dugo, P. (2005). Determination of flavor components in sicilian goat cheese by automated HS–SPME-GC. *Flavor and Fragrance Journal, 20*, 659–665.

Monsoor, M. A., & Proctor, A. (2004). Volatile Component Analysis of Commercially Milled Head and Broken Rice. *Journal of Food Science, 69*(8), 632–636.

Namiesnik, J., Zygmunt, B., & Jastrzebska, A. (2000). Application of solid-phase micro extraction for determination of organic vapours in gaseous matrices. *Journal of Chromatography A., 885*, 405–418.

Nerin, C., Salafranca, J., Aznar, M., & Batlle, R. (2009). Critical review on recent developments in solventless techniques for extraction of analytes. *Analytical and Bioanalytical Chemistry, 393*, 809–833.

Nilsson, T., Ferrari, R., & Facchetti, S. (1997). Inter-laboratory studies for the validation of solid-phase microextraction for the quantitative analysis of volatile organic compounds in aqueous samples. *Analytica Chimica Acta., 356*, 113–123.

Padron, M. E., Afonso-Olivares, C., Sosa-Ferrera, Z., & Santana-Rodriguez, J. J. (2014). Microextraction techniques coupled to liquid chromatography with mass spectrometry for the determination of organic micropollutants in environmental water samples. *Molecules, 19*, 10320–10349.

Panavaite, D., Padarauskas, A., & Vickackaite, V. (2006). Silicone glue coated stainless steel wire for solid phase microextraction. *Analytica Chimica Acta, 571*(1), 45–50.

Parreira, F. V., De Carvalho, C. R., & Cardeal, Z. L. (2002). Evaluation of indoor exposition to benzene, toluene, ethylbenzene, xylene and styrene by passive sampling with a solid-phase microextraction device. *Journal of Chromatographic Science, 40*, 122–126.

Pawliszyn, J. (1995). New directions in sample preparation for analysis of organic compounds. *Trends in Analytical Chemistry, 14*, 113–122.

Pawliszyn, J. (1997). Solid phase microextraction: *Theory and Practice*. Wiley-VCH. New York. pp. 247.

Pawliszyn, J. (1999). *Applications of Solid Phase Microextraction*. RSC, Cambridge, UK, pp. 609–621.

Pawliszyn, J. (2000). Solid-phase microextraction. In: *Extraction*. Academic Press, pp. 1416–1424.

Penalver, A., Pocurull, E., Borrull, F., & Marce, R. M. (1999). Trends in solid-phase microextraction for determining organic pollutants in environmental samples. *Trends in Analytical Chemistry, 18*(8), 557–568.

Poerschmann, J., Kopinke, F. D., & Pawliszyn, J. (1997). Solid phase microextraction to study the sorption of organation compounds onto particulate and dissolved humic organic matter. *Environmental Science & Technology, 31*, 3629–3636.

Popp, P., & Paschke, A. (1997). Solid phase microextraction of volatile organic compounds using carboxen-polydimethylsiloxane fibers. *Chromatographia, 46*, 419–424.

Povolo, M., & Contarini, G. (2003). Comparison of solid phase microextraction and purge-and-trap methods for the analysis of the volatile fraction of butter. *Journal of Chromatography A., 985*, 117–125.

Razote, E. B., Maghirang, R. G., Seitz, L. M., & Jeon, I. J. (2004). Characterization of volatile organic compounds on airborne dust in a swine finishing barn. *Transactions of the ASAE, 47*, 1231–1238.

Razote, E., Jeon, I., Maghirang, R., & Chobpattana, W. (2002). Dynamic air sampling of volatile organic compounds using solid phase microextraction. *Journal of Environmental Science and Health, Part B., 37*, 365–378.

Reineccius, G. A. (2007). Flavor-Isolation techniques. In: *Flavors and Fragrances: Chemistry, Bioprocessing and Sustainability* (Berger, R. G., ed.). Springer-Verlag Berlin Heidelberg, Germany. pp. 409–426.

Saito, Y., Ueta, I., Kotera, K., Ogawa, M., Wada, H., & Jinno, K. (2006). In-needle extraction device designed for gas chromatographic analysis of volatile organic compounds. *Journal of Chromatography A., 1106*, 190–195.

Silva, C., Cavaco, C., Perestrelo, R., Pereira, J., & Camara, J. S. (2014). Microextraction by packed sorbent (MEPS) and solid-phase microextraction (SPME) as sample preparation procedures for the metabolomic profiling of urine. *Metabolites, 4*(1), 71−97.

Song, J., Gardner, B. D., Holland, J. F., & Beaudry, R. M. (1997). Rapid analysis of volatile flavor compounds in apple fruit using SPME and GC/time-of-flight mass spectrometry. *Journal of Agricultural and Food Chemistry, 45*, 1801–1807.

SPDE™ (2015). The magic needle. flyer; chromtech GmbH: Idstein, Germany, 2006. URL: http://www.chromtech.de/download/20130130160105/CT-ProduktblattSP-DE-231205.pdf. Accessed on 31–03–2015.

Spietelun, A., Kloskowski, A., Chrzanowski, W., & Namiesnik, J. (2013). Understanding solid-phase microextraction: Key factors influencing the extraction process and trends in improving the technique. *Chemical Reviews, 113*, 1667−1685.

Steffen, A., & Pawliszyn, J. (1996). Analysis of flavor volatiles using headspace solid-phase microextraction. *Journal of Agricultural and Food Chemistry, 44*, 2187–2193.

Supelco, (1998). Solid phase microextraction: theory and optimisation of conditions. Bulletin 923, Sigma-Aldrich Co., Bellefonte, USA. pp. 1–8.

Tananuwonga, K., & Lertsirib, S. (2010). Changes in volatile aroma compounds of organic fragrant rice during storage underdifferent conditions. *Journal of the Science of Food and Agriculture, 90*, 1590–1596.

Teranishi, R., Mon, T. R., Robinson, A. B., Cary, P., & Pauling, L. (1972). Gas chromatography of volatiles from breath and urine. *Analytical Chemistry, 44*, 18–20.

Ulrich, S. (2000). Solid-phase microextraction in biomedical analysis. *Journal of Chromatography A., 902*(1), 167–194.

Vas, G., & Vekey, K. (2004). Solid-phase microextraction: A powerful sample preparation tool prior to mass spectrometric analysis. *Journal of Mass Spectrometry, 39*, 233–254.

Wang, A., Fang, F., & Pawliszyn, J. (2005). Sampling and determination of volatile organic compounds with needle trap devices. *Journal of Chromatography A., 1072*, 127–135.

Wang, L., Xu, Y., Zhao, G., & Li, J. (2004). Rapid analysis of flavor volatiles in apple wine using headspace solid-phase microextraction. *Journal of the Institute of Brewing, 110*(1), 57–65.

Wercinski, S. A. S., & Pawliszyn, J. (1999). Solid phase microextraction theory. In: *Solid Phase Microextraction: A Practical Guide* (Wercinski, S. A. S., ed.). CRC Press. Marcel Dekker, Inc. New York. pp. 1–26.

Whiton, R. S., & Zoecklein, B. W. (2000). Optimization of headspace solid-phase microextraction for analysis of wine aroma compounds. *American Journal of Enology and Viticulture, 51*, 379–382.

Wongpornchai, S., Sriseadka, T., & Choonvisase, S. (2003). Identification and quantitation of the rice aroma compound, 2-acetyl-1-pyrroline, in bread flowers (*Vallaris glabra* Ktze). *Journal of Agricultural and Food Chemistry, 51*(2), 457–462.

Yang, X., & Peppard, T. (1994). Solid-phase microextraction for flavor analysis. *Journal of Agricultural and Food Chemistry, 42*, 1925–1930.

Yu, Y. J., Lu, Z. M., Yu, N. H., Xu, W., Li, G. Q., Shi, J. S., & Xu, Z. H. (2012). HS-SPME/GC-MS and chemometrics for volatile composition of Chinese traditional aromatic vinegar in the Zhenjiang region. *Journal of the Institute of Brewing, 118*, 133–141.

Zeng, Z., Zhang, H., Chen, J. Y., Zhang, T., & Matsunaga, R. (2008a). Flavor volatiles of rice during cooking analyzed by modified headspace SPME/GC-MS. *Cereal Chemistry, 85*(2), 140–145.

Zeng, Z., Zhang, H., Chen, J. Y., Zhang, T., & Matsunaga, R. (2007). Direct Extraction of volatiles of rice during cooking using solid-phase microextraction. *Cereal Chemistry, 84*(5), 423–427.

Zeng, Z., Zhang, H., Zhang, T., & Chen, J. Y. (2008b). Flavor volatiles in three rice cultivars with low levels of digestible protein during cooking. *Cereal Chemistry, 85*(5), 689–695.

Zeng, Z., Zhang, H., Zhang, T., Tamogami, S., & Chen, J. Y. (2009). Analysis of flavor volatiles of glutinous rice during cooking by combined gas chromatography–mass spectrometry with modified headspace solid-phase microextraction method. *Journal of Food Composition and Analysis, 22*(4), 347–353.

Zhang, Z., & Pawliszyn, J. (1993). Headspace solid-phase microextraction. *Analytical Chemistry Analytical Chemistry, 65*, 1843–1852.

Zhang, Z., Yang, M. J., & Pawliszyn, J. (1994). Solid-phase micro-extraction. *Analytical Chemistry, 66*(17), 844–853.

Zhou, M., Robards, K., Glennie-Holmes, M., & Helliwell, S. (1999). Analysis of volatile compounds and their contribution to flavor in cereals. *Journal of Agricultural and Food Chemistry, 47*(16), 3941–3953.

CHAPTER 6

HEADSPACE-SOLID PHASE MICRO EXTRACTION COUPLED WITH GAS CHROMATOGRAPHY-MASS SPECTROSCOPY (HS-SPME-GCMS): AN EFFICIENT TOOL FOR QUALITATIVE AND QUANTITATIVE RICE AROMA ANALYSIS

RAHUL ZANAN,[1] VIDYA HINGE,[2] KIRAN KHANDAGALE,[3] and ALTAFHUSAIN NADAF[4]

[1]Young Scientist (SERB), Department of Botany, Savitribai Phule Pune University, Pune–411007, India, Mobile: +91-9689934273, Tel.: +91-20-25601439, Fax: +91-20-25601439
E-mail: rahulzanan@gmail.com

[2]Women Scientist (WOS-A), Department of Botany, Savitribai Phule Pune University, Pune–411007, India, Mobile: +91-7588611028
Tel.: +91-20-25601439, Fax: +91-20-25601439
E-mail: vidyahinge17@gmail.com

[3]Research Scholar, Department of Botany, Savitribai Phule Pune University, Pune–411 007, India, Mobile: +91-9860898859
Tel.: +91-20-25601439, Fax: +91-20-25601439
E-mail: kirankhandagale253@gmail.com

[4]Associate Professor, Department of Botany, Savitribai Phule Pune University, Pune–411007, India, Mobile: +91-7588269987,
Tel.: +91-20-25601439, Fax: +91-20-25601439
E-mail: abnadaf@unipune.ac.in

CONTENTS

6.1   INTRODUCTION

Aroma is one of the most important features of rice grain quality that has a significant role in the marketing of aromatic rice and its acceptance by consumers around the world (Sakthivel et al., 2009; Verma et al., 2012, 2013, 2015). Volatile chemistry reveals that the total mixture of several aroma volatiles in a specific proportion is associated with unique flavor and strength of aroma of diverse fragrant rice (Weber et al., 2000; Champagne, 2008; Mathure et al., 2012; Verma and Srivastav, 2016). Volatile analysis of aromatic rice cultivars revealed the presence of large number of volatile compounds in varying concentrations. More than 400 volatiles belonging to alcohols, ketones, esters, acids, pyridines, phenols, aldehydes, pyrazines, hydrocarbons, and other substances have been reported in aromatic rice by a number of researchers (Buttery et al., 1986; Jezussek et al., 2002; Yang et al., 2008; Mathure et al., 2011). Among these, the volatile 2-acetyl-1-pyrroline (2-AP) has been found to be a major aroma component that imparts characteristic aroma to basmati and other aromatic rice

(Buttery et al., 1983; Fitzgerald et al., 2008). It is having very low odor threshold value, and its amount is 10–15 times higher in aromatic rice than nonaromatic rice (Champagne, 2008; Yi et al., 2009).

Researchers have taken efforts for the detection of volatiles in several rice varieties. Initially, the detection of rice aroma was a tough job, and various crude methods were employed by researchers, such as chewing a few seeds (Dhulappanavar, 1976) and chewing a half of a single seed (Bemer and Hoff, 1986). Further, Reinke et al. (1991) reported tasting of individual grain as an ideal method for selection of aromatic rice varieties. Sood and Siddiq (1978) developed a sensory evaluation test for aromatic rice in which grains/ leaf sample are placed in 1.7% KOH and incubated at room temperature for 10 min in a Petri plate. The alkaline condition releases aroma volatiles that could be detected by inhaling them against standard compounds. Further, charm and aroma extract dilution analysis were developed for recording the presence or absence of individual odorants (Blank, 1997). This was followed by the osme analysis method (Grosch, 2001) and surface of nasal impact frequency method (Reineccius, 2006) for aroma analysis. In most of the analytical methods, samples need to distilled and/or evaporate for extraction of aroma volatiles and also require lot of hazardous organic solvents. Flavor is a complex mixture of compounds; hence, all these methods have limitation for identification and quantification of aroma volatiles. Researchers then developed new methodologies for aroma analysis. Several researchers used gas chromatography with olfactometry for separation of aroma compounds and identification of odor of individual compound (Buttery et al., 1988; Wadija et al., 1996). Recently, researchers have developed rapid, sensitive, and accurate method (headspace – solid phase micro-extraction (HS-SPME)) for the identification and quantification of aroma volatile. HS-SPME coupled with gas chromatography (GC) and/or gas chromatography–mass spectroscopy (GC-MS) are relatively easier method for the identification of aroma contributing volatiles.

6.2 METHODS USED FOR VOLATILE ANALYSIS

Generally, volatile compounds are present in trace amount and are complex mixture of various organic compounds with different molecular structures and polarities. Trapping or extraction of such volatiles is the first and crucial step in studying theses volatiles; hence, there is need to choose suitable sampling or volatile extraction technique. Steam distillation (SD), purge and trap (P&T), supercritical fluid extraction (SFE), and solid phase micro-extraction (SPME) are the four major sampling techniques routinely used by researchers for studying volatiles. Details of each method are given below.

6.2.1 STEAM DISTILLATION METHOD

Steam distillation is probably the very first technique used for isolation of plant volatiles due to its simplicity. Plant volatile organic compounds (VOCs) are first distilled in vapor phase and then are extracted in a suitable solvent. Schematic representation of steam distillation method is given in Figure 6.1. There are chances of changing the properties of heat-sensitive volatiles due to high temperature used during steam distillation. It was one of the popular methods of volatile sampling due to the large extraction capacity. Steam distillation and simultaneous steam and micro steam distillation with solvent extraction have been employed by many researchers for

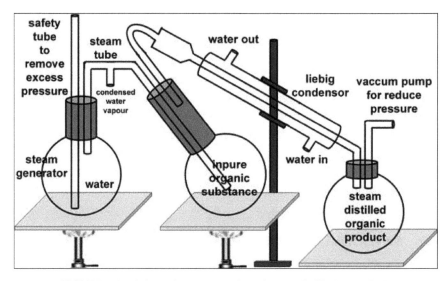

FIGURE 6.1   Schematic representation of steam distillation method.

isolation of rice aroma volatiles (Buttery et al., 1986; Tanchotikul and Hsieh, 1991; Lin et al., 1990; Widjaja et al., 1996; Mahatheeranont et al., 2001). Major drawbacks of this method involves the use of organic solvents, time consuming, laborious, needs large quantity of sample, and high temperature that may interfere with quality of volatiles (Theodoridis et al., 2000; Stashenko and Martínez, 2007). Volatile profile obtained from steam distillation depends upon the volatility of the compound and solubility in solvent (Reineccius, 2006).

6.2.2   PURGE AND TRAP SYSTEM

Purge and trap is one of the efficient and popular methods for extraction and concentration of volatile compounds for volatile analysis by GC or GC/MS. This technique is also called as dynamic headspace in which volatiles were transported from the sample by using inert gas as a carrier gas, and they are trapped in the solid sorbent such as Tenax. The trapped volatiles are then desorbed in GC/MS for analysis (Figure 6.2). In this method, volatiles with highest vapor pressure are preferentially removed. Trapping

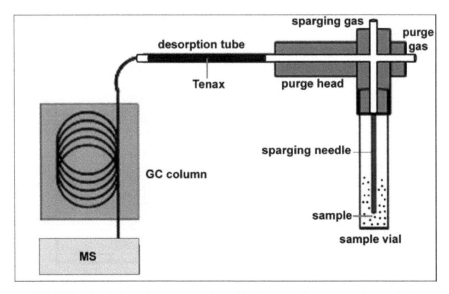

FIGURE 6.2   Schematic representation of the Purge and trap extraction method.

of volatiles depend on their polarity as Tenax has higher affinity for non-polar compounds (Reineccius, 2006). Buttery et al. (1988) and Wang and Ha (2012) used the purge and trap method to study aroma volatiles of rice.

6.2.3 SUPERCRITICAL FLUID EXTRACTION METHOD

Supercritical fluid extraction is the process of separation of single volatile from the sample matrix using supercritical fluids as the extracting solvent. These supercritical fluids are cleaner and safe compared to organic solvents used in steam distillation. Carbon dioxide (CO_2) is one of the commonly used supercritical fluid. Supercritical fluid extraction is an alternative to liquid extraction techniques and has proved to be an efficient for the study of plant aroma volatile studies (Figure 6.3). Bhattacharjee et al. (2003) generated aroma profile of commercial basmati rice by using supercritical CO_2.

6.2.4 SOLID PHASE MICRO-EXTRACTION

For the first time, Pawliszyn and coworkers developed solvent-free sampling technique (SPME) to measure organic compounds in groundwater

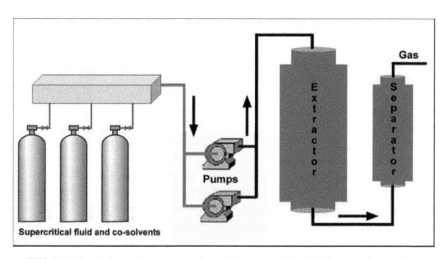

FIGURE 6.3 Schematic representation of the supercritical fluid extraction method.

(Arthur and Pawliszyn, 1990; Arthur et al., 1992; Potter and Pawliszyn, 1992). SPME can be used for analyzing volatiles, semi-volatiles, and trace amounts. SPME is a sampling technique in which polymer-coated fiber with an extracting phase was used, which adsorbs various type of analytes (Figure 6.4). It is a rapid and efficient tool for the extraction and quantification of the aroma compounds (Stashenko and Martínez, 2007). It is versatile and solvent-free technique and has integrated sampling, extraction, concentration, and sample introduction of volatile compounds into an analyzing instrument in a single step (Arthur and Pawliszyn, 1990). It is rapid, efficient, versatile, and solvent-free technique for the extraction and quantification of the aroma volatile compounds. In this method, sampling, extraction, concentration, and sample introduction of volatile compounds into an analyzing instrument occur in a single step (Arthur and Pawliszyn, 1990; Stashenko and Martínez, 2007). Because of immediate transfer of compounds into the analytical instrument, less stable compounds can be easily analyzed with minimum loss of volatile (Eisert and Pawliszyn, 1997). Several SPME fiber coatings are available in market commercially. Polarity of sample and fiber as well as the coating material SPME fiber may affect the recovery of volatiles. Mathure et al. (2011, 2014) and Bryant and McClung (2011) analyzed rice aroma volatile compounds using HS-SPME coupled with GC/GC-MS. Figures 6.4 and 6.5 depict the various parts of SPME fiber and procedure of sampling using SPME.

6.2.4.1   Type of SPME Fibers

In SPME, analytes were adsorbed from liquid and solid sample by headspace and immersion extraction using a polymer-coated fused silica fiber. Adsorbed analytes were desorbed from the fiber by exposing in the injection port of a GC or in desorption chamber of an SPME. The extraction of analytes depends on the nature and polarity of analytes, matrix composition, etc. Several commercial SPME fibers are available in the market, and based on analytes, these were applied in the research. Polydimethylsiloxane (PDMS), divinylbenzene (DVB), carboxen (CAR), polyethylene glycol (PEG), carbowax (CW), and Polyacrylate

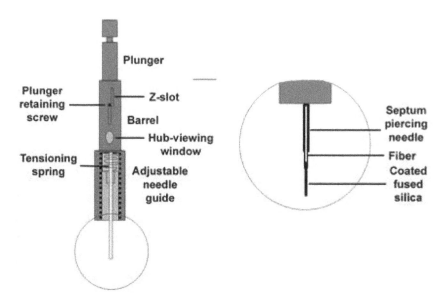

FIGURE 6.4 Schematic representation of various parts of the SPME fiber.

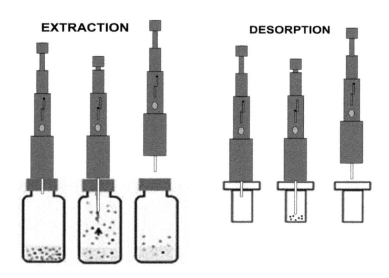

FIGURE 6.5 Schematic representation of SPME sampling.

(PA) in different combinations and thicknesses are commercially available (Fontanals et al., 2007). These fiber coatings are available in different sizes and forms. For wide analysis, these fibers are having less choice, poor selectivity, low chemical resistance, not useful for specific applications, and have limitations for polar analytes (Spietelun et al., 2010).

According to the molecular weights and polarity of the analytes, different fibers are recommended and commercially available. For low-molecular-weight or volatile compounds, 100 μm PDMS-coated fiber is used. For high-molecular-weight or semi-volatile compounds, 30 μm PDMS-coated fiber or 7 μm PDMS-coated fiber is efficient. Polar analytes can be trapped using 85 μm PA coated fiber. Highly volatile polar analytes are more efficiently released with 65 μm PDMS/DVB-coated fiber. For trace amounts of volatiles, 75 μm PDMS/CAR fiber is recommended. DVB/CAR fiber is useful for higher range of analytes (Sigma Aldrich). Apart from these fiber coating material, new types of SPME fiber coatings were developed by researchers for different analytes, viz., polypyrrole (PPY), molecularly imprinted polymers (MIP), pseudoliquids, carbonaceous, immunosorbents, sol-gel, conducting polymers, and double layer fiber coating (Spietelun et al., 2010). SPME fiber coatings and their advantage and disadvantages are depicted in Table 6.1.

6.3 IDENTIFICATION AND CONFIRMATION OF VOLATILES USING HS-SPME COUPLED WITH GC/GC-MS

6.3.1 MASS SPECTRAL DATABASES

In GC-MS, identification of compounds is achieved by comparison of query mass spectrum with reference mass spectra in a library by spectrum matching. There are many mass spectral libraries/database developed and available commercially (Table 6.2). To find the spectral match in reference library, different parameters for similarity searches were adopted by different researchers (Wei et al., 2014). Identification based on mere mass spectral similarity is not reliable. Both mass spectra and retention indices have to be used for proper identification. NIST and flavor-fragrance libraries are widely used for analysis of volatile compounds.

TABLE 6.1   SPME Fiber Coatings and Their Advantage and Disadvantages

No	Coating Type	Advantage	Disadvantage	Reference
1.	Polydimethylsi-loxane (PDMS)	Simplicity in prepara-tion, High inter-fibre reproducibility (6% RSD)	Low efficiency of extraction in case of polar compounds. equilibration times is longer	Zhou et al. (2008)
2.	Polyacryloni-trile (PAN) + C_{18}	Mechanically and chemically robust, autoclavable, excellent extraction efficiency. Wide choice of sor-bent, Good inter-fibre reproducibility (<10% RSD)	Upon autoclaving efficiency decreases upto15%	Musteata et al. (2007)
3.	Polyethylene glycol (PEG) + C_{18}	Wide choice of sor-bent, Less than 5 min equilibration time, sen-sitive than polypyrrole, less than 1% carryover.	Unstable upon au-toclaving, 15–25% inter fibre repro-ducibility	Eshaghi et al. (2007)
4.	Polypyrrole (PPY)	Autoclavable and short equilibration times	Decrease in extrac-tion efficiency due to autoclaving, low inter-fibre repro-ducibility	Wang et al. (2007)
5.	Polyacryloni-trile (PAN) as a thin membrane	Improved mechani-cally stable and have inter-fibre reproduc-ibility (<10%)	Equilibration time required is longer	Musteata et al. (2007)
6.	Supelco bio-compatible fibres	Mechanically and chemically stable. Good inter-fibre repro-ducibility (10–15%) and extraction ef-ficiency	Long equilibration times	Vuckovic et al. (2009)

6.3.2   RETENTION INDEX

Retention indices are used for conversion of retention time into system independent constant. Retention times is normalized to adjacently eluting n-alkanes. Hence, to cover the expected retention time range of all pos-sible target compounds, the range of n-alkanes are used. Retention may

TABLE 6.2 Mass Spectral Libraries/Database for the Identification of Compounds

Database	Description	Website	References
NIST Library	The library contains 242,466 general compounds spectra in which 33,782 of these spectra are registered in a sub-library.	www.nist.gov/srd/nist1a.cfm	Arthur et al. (1992).
Wiley Registry 10th (ed.). 2013	Largest among the mass spectral libraries commercially available. More than 719,000 spectra and 684,000 searchable chemical structures.	http://www.wiley.com/WileyDA/Wiley-Title/productCd-047052037X.html	Bemer and Hoff (1986)
SWGDRUG Library	For drugs and drug-related compounds analysis, has spectra for 2300 compound. It is supported by the NIST MSSEARCH program, available on-line at no charge	http://www.swg-drug.org/ms.htm	Arthur and Pawliszyn (1990)
mzCloud™	**mzCloud™** is a state of the art freely searchable collection of high resolution/accurate mass spectra, assists in identifying compounds in areas such as life sciences, metabolomics, pharmaceutical research etc. Contains 291 392 spectra.	https://www.mzcloud.org/mzcloud.org	Abdul-lah, et al. (2011)
Flavor and Fragrance Natural and Synthetic Compounds Library	Contains spectra of more than 3000 flavor and fragrance compounds.	http://www.sisweb.com/software/wiley-ffnsc.htm	Ayano and Furuhashi (1970)
MassLib	Contains spectra, structure and additional information for wide range of compounds	www.masslib.com	AMSDMZ-CLOUD (2015)
SDBS	It is integrated spectral database system for organic compounds and contains spectra for more than 30000 compounds. Access is free of charge.	http://sdbs.db.aist.go.jp	Augusto (2001)
MassBank	MassBank contains 41,092 spectra, it is first public repository for sharing mass spectral data among scientific research community..	http://www.massbank.jp/en/statistics.html	Ajarayasiri and Chai-seri (2008)

vary with the analytical instrument and parameters such as column length, thickness, diameter, carrier gas pressure, etc. whereas retention indices are pretty independent of the above experimental conditions and hence can be used for analyses of data generated from different research and analytical laboratories. Further, identification of peaks should not only depend upon on similarity of mass spectra, but it should also consider retention indices to get quality and reliable identification of compounds (Kovats, 1958; Zellner et al., 2008). Swiss chemist Ervin Kováts developed the method and equation to calculate retention indices,

For isothermal chromatography:

$$I = 100 \times \left[n + \frac{\log\left(t'_{r(unknown)}\right) - \text{lo} \, t'_{r(n)} g(\ )}{\log\left(t'_{r(N)}\right) - \left(t'_{r(n)}\right)} \right]$$

where I is the Kovats retention index, n is the number of carbon atoms in the smaller n-alkane, N is the number of carbon atoms in the larger n-alkane, and t'_r is the adjusted retention time.

For temperature programmed chromatography:

$$I = 100 \times \left[n + \frac{t'_{r(unknown)} - t'_{r(n)}}{\log\left(t'_{r(N)}\right) - \left(t'_{r(n)}\right)} \right]$$

where I is the Kovats retention index, n is the number of carbon atoms in the smaller n-alkane, N is the number of carbon atoms in the larger n-alkane and t_r is the retention time.

6.3.3 USE OF AUTHENTIC STANDARDS

Preliminary identified compounds by retention indices and mass spectral databases need to be confirmed using authentic standards. During identification using authentic standards, standards and samples should be run under the same conditions. During the identification of compounds by MS databases, minimum 5 to 6 masses and intensities were matched with unknown compounds and that having low level of confidence. Authentic standard for identification of unknown compounds is widely used by researchers and accepted as a proper identification tool using GC-FID or GC-MS. Several researchers used authentic standards for the identification of rice aroma volatiles (Buttery et al., 1988, 1999; Mathure et al., 2011, 2014).

6.4 METHODS FOR QUANTIFICATION OF VOLATILES

In SPME, calibration is very important because of only a small quantity of sample from sample matrix is used. Hence, for better characterization and more accurate information, SPME calibration is needed. Several calibration methods were developed for SPME. These methods are based on understanding of fundamental principles governing the mass transfer of analytes in multiphase systems (Ouyang and Pawliszyn, 2008). Among the several calibration methods, external standard, internal standard, standard addition, equilibrium extraction, exhaustive extraction, and diffusion-based calibration methods were used.

6.4.1 EXTERNAL STANDARD

Several standard solutions need to be prepared for calibration curve to obtain relationship between the peak responses and the target standard concentrations. For external standard calibration curve, sample extraction conditions and standard calibration curve conditions remain the same. Concentration of sample can be calculated based on calibration curve equation (Ouyang and Pawliszyn, 2008). External standard calibration methods are extensively used for quantitative analysis of environmental, food and biologic samples through SPME. Ruckriemen et al. (2015) successfully quantified principal basmati rice aroma compound 2-AP from Manuka honey. Kataoka et al. (2007) quantified isophorone in food samples by SPME coupled with GCMS through calibration curve method as an external standard method. This method commonly performed in the laboratory analysis and also for on-site sampling (Koziel et al., 2000; Ferraris et al., 2004).

6.4.2 STANDARD ADDITION

In the standard addition method, known different quantities of standard concentrations were added in the sample matrix. The responses of different concentrations of standard needs to be plotted, and the extrapolation of the plot of the response to zero defines the original concentration in the unspiked sample. The concentration of the analyte from the sample can be calculated with the help of slope of the standard plot. But this calibration method is time-consuming for a large number samples because sample preparations are required (Ouyang and Pawliszyn, 2008). For matrix samples, standard addition calibration method is generally employed by various researchers in different analyses. Mathure et al. (2011) quantified aroma volatiles from Indian scented rice cultivars by HS-SPME/GC-FID through the standard addition approach. Similarly, Wakte et al. (2011) quantified 2-AP using the calibration curve method from the flowers of *Bassia latifolia*.

6.4.3 INTERNAL STANDARD

Internal standard and standard addition methods are similar. However, in the internal standard method, the standard compound is different from the analytes, but it should be well resolved and mimic the equilibrium of the analytes. In this method, for development of calibration plot, the ratio of the peak area of the analyte to the peak area internal standard is determined. The various samples of calibration solutions that contain different concentrations of the analyte with a fixed concentration of the internal standard are analyzed, and average values are considered. In this method, we can avoid losses of analytes during sample preparation, matrix effects, and irreproducibility in parameters (Iglesias and Medina, 2008; Plutowska and Wardencki, 2008). But it is not easy to find suitable internal standards for complex samples (Ouyang and Pawliszyn, 2008). 2-AP is highly volatile, rapidly degrades, is tricky to synthesize, and is an unstable compound, thus making high purity standards unavailable. For the first time, Buttery et al. (1986) used 2,4,6-trimethylpyridine (TMP) as an internal standard because of similar physical properties. Further, Bergman et al. (2000), Wongporn-chai et al. (2003), and Grimm et al. (2011) successfully employed TMP as an internal standard. Sriseadka et al. (2006) used 2,6-dimethylpyridine (2,6-DMP) as an internal standard instead of TMP because of peak overlapping between TMP and 2 pentyl furan of matrix. Maraval et al. (2010) quantified 2-AP through isotope-labeled internal standard.

6.4.4   EQUILIBRIUM EXTRACTION

The equilibrium extraction method is a widely used for quantification using SPME. The SPME fiber coating as a small extraction phase is exposed to a sample matrix. The target analytes on the fiber adsorbed the analytes the from sample matrix and are determined by experimentation (Shurmer and Pawliszyn, 2000; Ouyang and Pawliszyn, 2008). In this method, analyte concentration can be calculated by the amount of the analytes extracted by the fiber. In this analysis, the concentration of the target analytes can be determined by distribution coefficients of the analytes between the fiber coating and the amount of the analytes on the fiber under extraction equilibrium from sample matrix. This method is generally used for on-site air and water sampling (Isetun and Nilsson, 2005; Larroque et al., 2006).

6.4.5   EXHAUSTIVE EXTRACTION

Sometimes analytes are not completely extracted using the equilibrium extraction method like SPME because distribution coefficient is very large and sample volume is small, and then, exhaustive extraction technique is employed. In this method, the concentration of the target analyte can be calculated with the volume of the sample and the amount of analyte extracted by the fiber coating (Ouyang and Pawliszyn, 2008). Here, the sample is repeatedly extracted with the fiber, and the total amount of the analyte can be extrapolated (Tena and Carrillo, 2007).

6.4.6   DIFFUSION-BASED CALIBRATION

Recently, diffusion-based calibration method has received increased attention for the on-site sampling using SPME. This method is based on non-equilibrium extraction for quantification of analytes. Recently, SPME on-site sampling devices based on non-equilibrium extraction have received increased attention. These devices are normally quantified with diffusion-based calibration methods. Several diffusion-based calibration models are developed based on Fick's first law of diffusion, interface model, cross-flow model and the kinetic processes of absorption/adsorption and desorption (Ouyang and Pawliszyn, 2008). Several researchers analyzed air and water samples by using these models (Pawliszyn, 2004; Ouyang et al., 2005). Augusto et al. (2001) and Tuduri et al. (2003) determined airborne volatile organic compounds using interface model of diffusion-based calibration methods.

6.5 ESTIMATION LIMIT OF DETECTION AND LIMIT OF QUANTIFICATION

To ensure reliable analytical procedures, quantitatively estimated aroma volatiles need to be validated under the defined conditions. The method should be validated for its accuracy, precision, specificity, linearity, range, ruggedness/ robustness, and system suitability. Method validation is based on two fundamental elements, i.e., limit of detection (LOD) and limit of quantification (LOQ), which define the limitations of an analytical method. A sufficient amount of analyte must be present to produce an analytical signal that can reliably be distinguished from "analytical noise." The lowest amount of analyte that can be detected with (stated) probability, in the sample, but unable to quantify as an exact value is called as LOD. It also called as minimum detectable concentration (MDC) of that compound. The LOD is estimation of the reliable detection limit (RDL) with high probability of producing a response significantly greater than the response at zero concentration of analyte.

The IUPAC definition states that the LOD, expressed as the concentration or quantity, is derived from the smallest measure (X_L) that can be detected with reasonable certainty for a given analytical procedure. The value of X_L is given by the equation:

$$X_L = X_{bi} + K S_{bi}$$

where X_{bi} is the mean response of the blank measures, S_{bi}, is the standard deviation of the responses of blank measures, and K is a numerical factor chosen according to the confidence level desired (McNaught and Wilkinson, 1997).

For GC-MS analysis of rice aroma volatiles, if data are acquired in the selected ion monitoring mode (SIM), the "response" represents the ion current, which registers in the specific ion channel of interest. The LOD is then defined as the mean response obtained for the blank plus 2 or 3 standard deviations, depending on whether the desired confidence level for distinguishing a positive sample from the blank is 95% or 99%, respectively (Currie, 1995).

The LOQ is the minimum concentration of an analyte in a sample that can be determined with satisfactory precision and accuracy under the specified experimental conditions. Association of Analytical Communities (AOAC) define the limit of quantitation is the lowest amount of analyte in a sample, which can be quantitatively determined with precision and accuracy appropriate to analyte and matrix considered. The LOQ is usually much higher than the concentration of LOD. LOD is an estimate of observed bias and impression at very low analyte concentration; when it meets the requirements for total error for the analyte (i.e., the assay is "fit for purpose"), then LoQ=LoD. The value of LOQ cannot be lower than that of LOD. In order to meet analytical objectives when LOD is not serving the purpose, a little higher analyte concentration must be tested to determine the LOQ. The following three methods are commonly employed for the estimation of LOD and LOQ in analytical chemistry.

6.5.1 SIGNAL-TO-NOISE (S/N) RATIO

In this method, the peak-to-peak noise at the analyte retention time is measured and the concentration of the analyte that would yield a desired value of signal-to-noise (S/N) ratio is estimated. When S/N attained 3 or >3, that concentration is accepted as LOD and S/N of value >10 gives the estimation of LOQ. The S/N ratio in rice aroma volatile analysis of analyte is measured in matrix.

6.5.2   BLANK DETERMINATION

This method is applied when there is result in blank analysis with a non-zero standard deviation (Eurachem, 1998). Then using blank determination, LOD is estimated as value of blank sample signal or analyte concentration plus three times of standard deviation. LOQ is the analyte concentration corresponding to the sample blank value plus ten times of standard deviations as shown in the following equations:

$$LOD \cong \bar{X}_{bi} + 3S_{bi}$$

$$LOQ \cong \bar{X} + 10S_{bi}$$

where $\bar{X}_{bi}$ is the mean concentration of the blank and S_{bi} is the standard deviation of the blank.

6.5.3   LINEAR REGRESSION

It is assumed that the instrument response (y) is linearly related to the standard concentration(x) for a limited range of concentration in linear calibration curve. It can be expressed in a model such as:

$$y = a + bx$$

This model is used to compute the sensitivity b and the LOD and LOQ (ICHQ2B, 1996). Therefore, the LOD and LOQ can be expressed as:

$$LOD \cong 3S_a/b$$

$$LOQ \cong 10S_a/b$$

where S_a is the standard deviation of the response and b is the slope of the calibration curve. The standard deviation of the response can be obtained from the standard deviation of either y-residuals, or y-intercepts, of regression lines. This method is applicable in all cases, and it is most suitable when the analysis method does not involve background noise. It uses a range of low values close to zero for calibration curve, and with a more homogeneous distribution, it will result in a more relevant assessment. Among all these three methods, signal to noise (S/N) is more suitable for the estimation of LOD and LOQ of 2-AP and other rice aroma volatiles analysis using the HS-SPME-GC-MS or FID detector. Maraval et al. (2010) estimated 0.1 ng/g of LOD and 0.4 ng/g of LOQ as the lowest amount of 2-AP giving a response that corresponded at least to a signal-to-noise ratio (S/N) of 3 and 10, respectively.

6.6   METHODS USED FOR RICE AROMA ANALYSIS

There are more than 400 different volatile compounds reported in the rice grains. Due to increasing demand for safety, solvent free, lower time consuming, reliable, less labor consuming, more accurate, and higher repeatability analytical procedures. Researchers developed sensitive and automated instruments that allow detecting trace quantities analytes. In the last few decades, sample preparation for the GC analysis of rice aroma volatiles has been mainly carried out by means of distillation, solvent extraction, dynamic headspace extraction, supercritical extraction, and solid phase extraction. Amongst these, solid phase micro-extraction is very rapid, simple, and sensitive method for the analyzed flavor and fragrance compounds (Talon and Montel, 1999). Hence, SPME was mostly used as the sampling method for the GC and GC-MS analysis of volatile compounds. Furthermore, SPME-GC-MS is an appropriate method for the quantitative analysis of volatile aroma compounds of aromatic rice varieties. From the HS-SPME-GCMS chromatograms of aromatic and non-aromatic rice varieties, aroma grams can be developed that can be evaluated by multivariate data analysis with the aim of recognizing a rice variety.

6.6.1   RICE AROMA ANALYSIS USING TRADITIONAL METHODS

Identification of rice aroma volatile was initiated in the early 1970s (Mitsuda et al., 1968; Ayano and Furuhashi, 1970; Chikubu, 1970), and till now, more than 400 volatiles have been identified. In 1977, Bullard and Holguin isolates 73 rice flavor compounds from unprocessed rice. They isolated volatiles by rotary evaporator at 50°C, and the liquid nitrogen-trapped volatiles were further identified using GC-MS. The identified compounds were from alcohols, aldehydes, alkyl aromatics, furans, ketones, terpenes, and naphthalenes groups. Yajima et al. (1978) steam distilled 80% polished Koshihikari rice variety, and the extract was further separated into acidic, weak acidic, basic, and neutral fractions. The compounds from the natural extract were identified using GC-MS. A total of 100 compounds belonging to hydrocarbons (13), alcohols (13), aldehydes (16), ketones (14), acids (14), esters (8), phenols (5), pyridines (3), pyrazines (6), and other compounds (8) were identified. In their further studies (Yajima et al., 1979), they distillated scented *japonica* rice cultivar for 20 min, and vapor-containing volatiles were trapped with reduced pressure and condensed with ethyl ether. They were further separated into different fractions, and compounds from the natural extract identified using GC-MS. They identified 114 volatile compounds, and α-pyrrolidinone was identified as a key odorant in Kaorimai. Tsugita et al. (1978) isolated volatile compounds from rice bran of Koshihikari variety using steam distillation method and identified by comparing their GC-MS data and GC retention times of authentic compounds. Some compounds were tentatively identified by comparing their GC-MS data with MS data in the literature. They have reported more than 250 volatiles belonging to alkanes, alkene, aromatic, alcohols, aldehydes, ketones, esters, lactones, acetal, furans, pyridines, pyrazines, quinolines, thiazoles, thiophenes, phenols, and acids.

In 1983, Buttery et al., analyzed 10 rice varieties and 2-acetyl-1-pyrroline was evaluated and reported to have popcorn-like odor. Based on the odor quality evaluations, they confirmed this compound as the aroma in more aromatic rice varieties. Tava and Bocchi (1999) collected volatiles by steam distillation from lines of aromatic rice American cv. A301 and commercial basmati rice. In both varieties, they identified and quantified volatile compounds from hydrocarbons, aldehydes, alcohols, ketones, heterocyclic, terpenes, disulfides, and phenols classes. Widjaja et al. (1996) analyzed volatile components in aromatic and nonaromatic rice using Likens-Nickerson simultaneous distillation-extraction method. The extracted volatile compounds were separated by a GC with an olfactory port and GC-MS. Total 70 volatile compounds were identified in which alkanals, alk-2-enals, alka-2,4-dienals, 2-pentylfuran, 2-acetyl-l-pyrroline, and 2-phenylethanol are most important aroma compounds. According to them, nonfragrant rice showed n-hexanal, (E)-Zheptenal, l-octen-3-ol, n-nonanal, (E)-2-octenal, (E)-,2, 4-decadienal, 2-pentylfuran, 4-vinyl-guaiacol, and 4-vinylphenol were more in concentration than in scented rice varieties. Bhattacharjee et al. (2003) extracted aroma volatiles from a commercial brand of Basmati rice by using supercritical carbon dioxide and Likens–Nickerson extraction methods. These extracted volatiles were further characterized by GC-MS and identified by matching mass spectra with the Wiley/NIST spectral library. They observed that the supercritical carbon dioxide extraction method was better than the Likens–Nickerson extraction method because the former yielded 18 more volatile compounds than the latter (17). Ajarayasiri and Chaiseri (2008) comparatively analyzed volatile compounds from black and white glutinous rice varieties and identified aroma active compounds. Volatile compounds were extracted from cooked rice samples using stem distillation method and identified by comparing mass spectra with the mass spectrum library, and confirmed through retention index and authentic standards. A total of 24 volatile compounds were identified, in which based on odor thresholds and odor active values, 10 compounds (1,2-propanediol, 2,3-butane-diol, 2-butanol, phenol, 1-propanol, butanoic acid, indole, benzaldehyde, 4-vinyl-2-methoxyphenol, and 1-tetradecene) were confirmed as aroma active compounds.

The purge and trap method was used to isolate volatiles from cooked Koshihikari milled rice by the Tenax GC trap, and 40 volatiles were analyzed and identified through GC and GC-MS (Tsugita et al., 1980). Buttery et al. (1988) used Tenax trap to study volatiles from cooked California long grain rice, and they found heptanal, hexanal, (E)-α-heptenal, 2-pentylfuran, 2-acetyl-1-pyrroline, octanal, hexanol, nonanal, (E)-2-nonenal, benzaldehyde, decanal, nonanol, (E)-2-decenal, (E,E)-2,4-decadienal, 2-phenylethanol, 4-vinylguaiacol, and 4-vinylphenol are major aroma volatiles compounds in cooked California long grain rice. Yang et al. (2008) also isolates volatiles by the Tenax trapping method and further analyzed them by GC-MS in cooked black rice. A total of 35 volatile compounds were identified based on comparison of their mass spectra and relative abundances with NIST 02 and Wiley 7 spectral libraries. The identified compounds were confirmed by comparison of the Kovats retention index (RI) and mass spectra with authentic standards. The identified compounds belonged to aromatic (10), nitrogen-containing (4), alcohol (6), aldehyde (10), ketone (3), and terpenoids (2); among these, 25 volatile compounds belonging to aromatics (8), aldehydes (10), alcohols (3), ketones (2), and nitrogen-containing (2) were recorded as odor active compounds.

Jezussek et al. (2002) characterized aroma compounds in cooked brown rice varieties like Improved Malagkit Sungsong, Basmati 370, and Khashkani. Cooked rice was further analyzed by GC-FID, GC-Olfactometry, and GC-MS. They observed 40 volatiles in improved Malagkit Sungsong and 48 volatiles in Basmati 370 and Khashkani. They confirmed that 2-AP, (E,E)-deca-2,4-dienal, 4,5-epoxy-(E)-dec-2-enal, bis-(2-methyl-3-furyl)-disulfide, 3-hydroxy-4,5-dimethyl-2(5H)-furanone, 2-methoxy-4-vinylphenol, 2-amino acetophenone, phenylacetic acid, and vanillin were the important odorants. Yang et al. (2008) analyzed cooked grains of basmati, jasmine, two Korean japonica cultivars, black and nonaromatic rice types by using a dynamic headspace system with Tenax trap, GC-MS, and GC-olfactometry. Out of 36 volatile compounds, 25 were considered to be major odor-active compounds. Maraval et al. (2008) identified 33 volatile compounds in cooked rice of Aychade, Fidji, Giano and commercial Thai (aromatic) and Ruille (non-aromatic) cultivars by GC-O and GC-MS. All compounds were identified by comparing mass spectra of available standards, mass spectra libraries (Wiley, NIST, INRA database), and retention indices. Identified compounds were quantified by the internal standard method.

6.6.2 OPTIMIZATION OF SPME CONDITIONS FOR RICE

6.6.2.1 Optimization of Sample Form and Quantity

Sample weight and volume also play an important role in efficiency of SPME analyses; hence, they need to be optimized for better results. For HS-SPME optimization, sample size depends upon the volume of vial to be used so that sufficient headspace volume should be available. Mathure et al. (2011) used 1.5 to 0.5 g rice grain sample in 4 mL screw cap vial and found that 1 g sample gave higher peak area and can be used for further SPME analyses. Further, 7–8 g powdered grain in 27 mL vial was found to be optimum for isolation of 2-AP (Wongpornchai et al., 2004;

Tananuwong and Lertsiri, 2010). Similarly, 2 g of rice sample was used in a 10 mL vial by Ghiasvand et al. (2007). Bryant and McClung (2011) used only 20 mg sample of whole milled rice kernels in a 10 mL vial.

6.6.2.2 Water Quantity

Addition of water or solvent into rice sample allows to release maximum quantity of rice aroma volatiles. The quantity of water depends on the sample quantity for optimum extraction volatiles. Mathure et al. (2011) recorded significant increase in 2-AP area with addition of water up to 300 µl in 1 g of sample. Ghiasvand et al. (2007) added 200 mL ultrapure water to 2 g rice sample, and the samples were then agitated for 10 min, which showed increase in peak area. This may be due to starch gelatinization with water addition that might have helped to release aroma compounds. Addition of water and its partition within matrix and headspace for releasing maximum aroma and optimization of addition of water quantity in the sample play a critical role. Several researchers have recorded increased 2-AP recovery during extraction over a narrow range of water addition (Grimm et al., 2001; Laguerre et al., 2007).

6.6.2.3 Extraction Temperature

Variations in equilibrium of volatile or sensitivity of volatile compounds are known to attribute to temperature-dependent variation. Therefore, optimization of temperature for SPME is one of the important steps in analysis. The optimized equilibrium and extraction temperatures will give accurate quantification of rice aroma volatiles. Mahattanatawee and Rouseff (2014) reported that among the 30, 40, 50 and 60°C equilibrium temperatures, 30°C is optimum for the analysis of cooked rice sample. Tananuwong and Lertsiri (2010) performed pre-incubation in a silicone oil bath at 120°C for 15 min. Grimm et al. (2011) reported that the adsorption rates of onto the fiber are not equal. According to them, 4 h adsorption time was maximum for recovery of 2-AP and TMP. Extraction temperature used by researchers varied from 70°C to 100°C to determine optimum temperature. Relative area of 2-AP increased 1.4 times with increase in temperature from 70°C to 80°C, and 80°C was found to be the optimum temperature of extraction (Mathure et al., 2011). Similar observations were recorded by Grimm et al. (2001), Wongpornchai et al. (2004), Lin et al. (2010) and Bryant and McClung (2011). Ghiasvand et al. (2007) extracted volatile compounds at 60°C.

6.6.2.4 Extraction Time

Grimm et al. (2001) reported that 10–15 min of a pre-incubation time is sufficient for optimal extraction of most volatile components. Mathure et al. (2011) varied extraction period for the maximum release of 2-AP. They observed that 20 min adsorption time considerably affected 2-AP recovery. Zeng et al. (2009) observed variation in composition of volatiles: the longer extraction period reduces the adsorption of 2-AP over other volatiles, resulting in decreased 2-AP peak area during cooking. Wongpornchai et al. (2004) exposed fiber for 30 min to headspace for extraction of volatile. Bryant and McClung (2011) allowed adsorption for 18 min, and Mahattanatawee and Rouseff (2014) and Tananuwong and Lertsiri (2010) performed extraction for 15 min. Thus, 15–20 min extraction time was reported as ideal for the maximum release of 2-AP.

6.6.3 RICE AROMA ANALYSIS USING HS-SPME COUPLED WITH GC-MS

Grimm et al. (2001) confirmed SPME is a rapid and sensitive analytical technique for the screening of 2-AP and other volatile compounds in rice. They screened 21 rice varieties by SPME and solvent extraction method and found that aromatic rice varieties showed several nanogram of 2-AP, whereas non-aromatic rice showed trace amounts of 2-AP in 0.75 g of sample. According to them, the recovery of 2-AP from SPME headspace was 0.3% of the total 2-AP content in the rice sample. Further, Monsoor and Proctor (2004) analyzed commercially milled head and broken rice of mixed long grain varieties (Wells, Cocodrie, and Drew) for their volatile components. Each volatile was extracted by SPME using carboxen/polydimethylsiloxane (CAR/PDMS) fiber coupled with GC-MS and 28 volatile compounds were identified by a combination of NIST mass spectral and GC retention times of the standard compounds. Among 28 identified volatiles, 10 volatiles (pentanal, pentanol, hexanal, 2-heptanone, heptanal, heptanol, 2-pentyl furan, octanal, 2-octenal, and nonanal) were quantified based on an internal standard method. Laguerre et al. (2007) successfully employed SPME coupled with GC-MS for discrimination of scented rice from non-scented rice cultivars. They formed fingerprints of 61 rice samples. Zeng et al. (2009) identified 96 volatile compounds in three glutinous rice cultivars during cooking by a modified headspace SPME with GC–MS. Amongst them, 15 (hexanal, 2-pentylfuran, (E)-2-heptenal, 1-hexanol, nonanal, 1-octen-3-ol, (E)-2-nonenal, (E,Z)-2,4-decadienal, (E,E)-2,4-decadienal, g-nonalactone, 2-pentadecanone, 2-methoxy-4-vinylphenol, 4-vinylphenol, indole, and vanillin) were identified by comparing their mass spectra (Wiley and NIST/ EPA/NIH mass libraries) and confirmed by RI values of authentic compounds. The remaining compounds were identified by their corresponding mass spectra and RI values on the DB Wax capillary column. A total of 27 (2-methyl-3-octanone, 2-hexylfuran, 7-methyl-2-decene, 2-heptylfuran, (E)-4-nonenal, (E)-3-nonen-2-one, 2-methyl-5-decanone, (Z)-2-octen-1-ol, 6-dodecanone, 2,3-octanedione, 6,10-dimethyl-2-undecanone, dodecanal, a-cadinene, 1-decanol, d-cadinene, 1-undecanol, tridecanal, (1S,Z)-calamenene, 3-hydroxy-2,4,4-trimethylpentyl isobutanoate, 2,4,4-trimethylpentan-1,3-diol di-iso-butanoate, hexadecanal, cubenol, d-cadinol, 5-iso-propyl-5H-furan-2-one, di-isobutyl adipate, a-cadinol, and glycerin) were reported as a new compounds.

Mathure et al. (2011) identified and quantified aroma volatiles of Indian scented rice cultivars by HS-SPME coupled with GC-FID. Each volatile was identified by comparing their mass spectra with NIST library. They identified 8 compounds (2-AP, hexanal, nonanal, decanal, benzyl alcohol, vanillin, guaiacol, and indole) that were separated and quantified by the standard addition approach using GC coupled with FID from 33 scented and 2 non-scented rice cultivars. In a further study (2014), they analyzed 23 volatiles (pentanal, hexanal, heptanal, octanal, 2-acetyl-1-pyrroline, 1-hexanol, nonanal, trans-3-octen-2-one, trans-2-octenal, 1-tetradecene, 1-octen-3-ol, decanal, trans-2-nonenal, (e,e)-nona-2,4-dienal, guaiacol, benzyl alcohol, 2-phenylethanol, 4 vinyl guaicol, nonanoic acid, 2 amino acetophenone, 4 vinyl phenol, indole, and vanillin) from 91 Indian rice cultivars belonging to non-basmati scented (77), basmati (9), and non-scented (5). All the volatiles were identified on the basis of their mass spectra by comparing the spectra with the records of the NIST library and with authentic standards. Each volatile was quantified by standard addition approach and limit of detection (LOD) were calculated.

Bryant and McClung (2011) analyzed 7 aromatic (Aromatic se2, Dellmati, Dellrose, IAC 600, Jasmine 85, JES, and Sierra) and 2 non-aromatic (Cocodrie and Wells) cultivars for their volatile composition using SPME with GC/GCMS. They identified 93 volatile compounds belonging to aldehydes (11), alkanes (15), alkenes (10), ketones (10), alcohols (19), amines (4), acids (5), and 19 miscellaneous. Every compound was identified by the presence of selected ions and their ratio and by comparing the mass spectra to the reference spectra in the NIST mass spectral database. Recently, by SPME with GC/GCMS, Mahattanatawee and Rouseff (2014) reported 30 aroma active volatile compounds (dimethyl sulfide, hexanal, 3-methyl-2-butene-1-thiol, octanal, 1-octen-3-one, 2-methyl-3-furanthiol, 2-acetyl-1-pyrroline, hexanol, dimethyl trisulfide, nonanal, (e)-2-octenal, methional, decanal, (e)-2-nonenal, 1-octanol, (e,z)-2,6-nonadienal, (e)-2-decenal, (e,e)-2,4-nonadienal, dodecanal, 2-acetyl-2-thiazoline, geranyl acetate, (e,e)-2,4-decadienal, b-damascone, b-damascenone, a-ionone, 2-phenylethanol, and b-ionone) in Jasmine, Basmati, and Jasmati rice.

6.7 FACTORS AFFECTING SPME QUANTIFICATION IN RICE VOLATILES

Several factors are responsible for the quantification of rice aroma volatile compounds. Quantification of volatiles from rice grains can be done simply by external standard, internal standard, and standard addition methods. Because of the heterogeneous nature, quantification of their volatiles by headspace sampling can be challenging. Solid rice grains can be ground or powdered for increasing the sample surface area. This enables realistic times for equilibration of standard solutions of volatiles with the sample matrix and for equilibration between the sample and the fiber. While optimizing sampling conditions aroma profiles may alter because of metabolic perturbation or damage to the sample. The following factors influence the quantification of rice aroma volatiles.

6.7.1 pH AND SALT CONCENTRATION

Adsorption of volatile on SPME fiber can be affected by the pH of the sample. pH can play an important role in case of analysis of weakly acidic or basic analytes (Yang et al., 2002). Furthe,r fiber coatings are sensitive to the strong pH environments. In many instances, salt addition to sample may decrease the partition coefficients of analyte, and therefore, more amount of target compound enter the headspace phase. It has been observed that the volatiles like 2-AP are better released under alkaline conditions.

6.7.2 SAMPLE AND HEADSPACE VOLUMES

Concentration of a compound in the headspace is directly proportional to its original concentration in the sample and inversely proportional to partition coefficient. In case of compound with low partition coefficient, change in sample volume significantly affect the amount of compound in headspace (Spietelun et al., 2013). Care should be taken during sample amount and volume adjustment so that there should be enough headspace volume available for volatile; otherwise, decreased headspace will negatively affect the SPME efficiency in case of aromatic rice grain (Lin et al., 2010; Mathure et al., 2011).

6.7.3 TEMPERATURE

Extraction temperature is one of the most important factors in SPME analyses. The partition coefficient of a volatile is inversely proportional to its vapor pressure, and vapor pressure increases with temperature; therefore, with decrease in partition coefficient, more amount of analyte will travel to the headspace phase (Kolb et al., 1992). Analyte passage from the sample to the headspace can be accelerated by increasing the temperature; hence, there is a need to choose the optimum extraction and preincubation temperature to enhance the volatility of target volatile compounds in the headspace (Panavaite et al., 2006). Lin et al. (2010) and Mathure et al. (2011) optimized extraction temperature for volatile analysis of aromatic rice and found that temperature of 80°C is optimum for rice grain.

6.7.4 EQUILIBRATION AND EXTRACTION TIME

Equilibration and extraction time are another important factor that determines the efficiency of SPME analysis and is dependent on other parameters such as sample temperature, partition coefficient of the analyte, etc. Generally, equilibration time is longer and most of the time equilibration time is actually more than the GC analysis time. By manipulating the temperature, sample size, and agitating the sample, we can reduce or optimize the time parameters. Optimized conditions for equilibrium temperature, extraction temperature and sample weight were used, and it was found that 60 min for equilibration and 50 min for extraction were optimum for aromatic rice (Lin et al., 2010).

6.7.5 FIBER SELECTIVITY

Fiber coating is very important parameter that decides the efficiency of SPME analyses. Polar, nonpolar, or combinations of both polar-nonpolar fiber coatings are available. Generally, combinations of polar/nonpolar sorbents (PDMS/DVB, PDMS/CAR, or CW/DVB) provide extraction coatings that give good recoveries for majority of polar, organic volatiles (Spietelun et al., 2013). Mahattanatawee and Rouseff (2014) evaluated PDMS, CAR/DVB, PDMS/DVB, CAR/PDMS, and DVB/CAR/PDMS fiber for their ability and found that DVB/CAR/PDMS was better for extraction and concentration of the headspace volatiles of cooked rice.

6.8 RECENT DEVELOPMENTS IN SPME FOR RICE AROMA ANALYSIS

New cost-effective, portable, electronic nose prototype with inbuilt data processing ability for aromatic rice classification has been developed by Abdullah et al. (2011). This e-nose uses Hierarchical Cluster Analysis (HCA) and Principal Component Analysis (PCA) for data analysis. It was found that this e-nose is able to successfully classify the aromatic rice with high accuracy. Further, Jana et al. (2015) developed electronic nose instrument to identify aromatic varieties. It is made up of different modules:

an odor handling module that delivers odor to the sensor array; a sniffing module for a sensor array; a water bath module for sample preparation; and a computing module that quantify the aroma data acquired by sensors. It is also used for cluster analysis of data acquired by sensor array. It also exploits various algorithms for identification of different rice varieties such as probabilistic neural network (PNN), back-propagation multilayer perceptron (BPMLP), and linear discriminant analysis (LDA). Cold-Fiber Solid-Phase Micro-Extraction (CF–SPME) is a newly developed device, useful for collection and concentration of volatile compounds, coupled to a gas chromatography time-of-flight mass spectrometry (GC–TOF–MS). It is used for volatile profiling of Iranian aromatic rice. The system described uses an internally cooled and soft extraction phase. A large number of compounds can be identified in the headspace above various food samples using this system (Ghiasvand et al., 2007).

6.9   CONCLUSIONS

Among the micro-extraction methods used for rice aroma analysis, SPME takes a leading position due to their simplicity and the possibility of automation. SPME is a very useful technique for analyzing volatiles and non-volatiles compounds occurring in very low concentrations in various foods and environmental applications. HS-SPME coupled with GC-MS proved to be a good tool to study the rice aroma volatile compounds. This advanced technique is easy to handle, accurate, solvent free, and low cost. By using these methods, trace compounds have also been analyzed. The accurate quantitative analysis of rice aroma volatiles through HS-SPME is depend on method used for quantification. Quantitative aroma grams generated by SPME-GCMS are very useful for characterizing and authenticating aromatic rice cultivars.

6.10   SUMMARY

Aroma is a complex combination of volatile compounds that which affect the aroma quality of rice. Several instrumental analysis tools have been developed for the characterization of rice aroma. Among the different techniques, headspace–solid phase micro extraction (HS-SPME) is a very simple, rapid, efficient, and, most importantly, solvent-free method. Hence, SPME has been widely used in combination with gas chromatography (GC), high-performance liquid chromatography (HPLC), and mass spectrometry (MS). SPME coupled with GC-FID or GCMS is ideally suited for analyzing aroma compounds, combining a simple and efficient sample preparation with versatile and sensitive detection. For SPME-GC-MS-based characterization of rice aroma volatiles, it is necessary to understand several steps as well as analytical practices followed. In this review, all details of identification and quantification of rice aroma volatiles using HS-SPME-GC-MS/FID are discussed.

ACKNOWLEDGMENTS

Altafhusain Nadaf acknowledge the financial assistance from the Department of Atomic Energy, Board of Research in Nuclear Sciences, Mumbai, India (Sanction No. 2006/37/45/BRNS), Rahul Zanan acknowledge to Science and Engineering Research Board (SERB), Department of Science and Technology, India (Sanction No. SR/FT/LS-350/2012) and Vidya Hinge to WOS-A scheme of Department of Science and Technology, India (Sanction No. SR/WOS-A/LS433/2011(G)) for financial assistance. Kiran Khandagale acknowledge Council of Scientific and Industrial Research (CSIR), India (Sanction No. 09/137/(0541)/2012-EMR-1) for the award of SRF.

KEYWORDS

- 2-acetyl-1-pyrroline (2AP)
- gas chromatography-flam ionized detector (GC-FID)
- gas chromatography-mass spectroscopy (GC-MS)
- head space–solid phase micro extraction (HS-SPME)
- optimization of HS-SPME
- quantification of rice volatiles
- rice aroma volatiles

REFERENCES

Abdullah, A. H., Adom, A. H., Shakaff, A. Y., Ahmad, M. N., Zakaria, A., Fikri, N. A., & Omar, O. (2011). An electronic nose system for aromatic rice classification. *Sensor Letter, 9*(2), 850–855.

Ajarayasiri, J., & Chaiseri, S. (2008). Comparative study on aroma-active compounds in Thai, Black and White glutinous rice varieties. *Kasetsart Journal (Natural Science), 42*, 715–722.

AMSDMZCLOUD (2015). Advanced mass spectral database, mzCloud. URL: https://www.mzcloud.org/mzcloud.org. Accessed 15.11.15.

Arthur, C. L., & Pawliszyn, J. (1990). Solid phase microextraction with thermal desorption using fused silica optical fibers. *Analytical Chemistry, 62*, 2145–2148.

Arthur, C. L., Pratt, K., Motlagh, S., Pawliszyn, J., & Belardi, R. P. (1992). Environmental analysis of organic compounds in water using solid-phase microextraction. *Journal of High Resolution Chromatography, 5*(11), 741–744.

Augusto, F., Koziel, J., & Pawliszyn, J. (2001). Design and validation of portable SPME devices for rapid field air sampling and diffusion based calibration. *Analytical Chemistry, 73*(3), 481–486.

Ayano, Y., & Furuhashi, T. (1970). Volatiles from cooked kaorimai rice. *Chiaba Diagaku Engeigakubu Gakujutsu Hokoku, 18*, 53.

Bemer, D. K., & Hoff, B. J. (1986). Inheritance of scent in American long grain rice. *Food Science and Technology, 26*, 876–878.

Bergman, C. J., Delgado, J. T., Bryant, R., Grimm, C., Cadwallader, K. R., & Webb, B. D. (2000). Rapid gas chromatographic technique for quantifying 2-acetyl-l-pyrroline and hexanal in rice (*Oryza sativa*, L.). *Cereal Chemistry, 77*, 454–458.

Bhattacharjee, P., Ranganathan, T. V., Singhal, R. S., & Kulkarni, P. R. (2003). Comparative aroma profiles using supercritical carbon dioxide and Likens-Nickerson extraction from a commercial brand of Basmati rice. *Journal of Science, Food and Agriculture, 83*, 880–883.

Blank, I., Fay, L. B., Lakner, F. J., & Schlosser, M. (1997). Determination of 4- hydroxy2-5-dimethyl-3(2H)-furanone and 2(or 5)-ethyl-4-hydroxy-5(or 2)-methyl-*3*(2H)-furanone in pentose sugar-based Maillard model systems by isotope dilution assays. *Journal of Agriculture and Food Chemistry, 45*, 2642–2648.

Bryant, R. J., & McClung, A. M. (2011). Volatile profiles of aromatic and non-aromatic rice cultivars using SPME/GC–MS. *Food Chemistry, 124*, 501–513.

Bullard, R. W., & Holguin, G. (1977). Volatile components of unprocessed rice (*Oryza sativa*). *Journal of Agriculture and Food Chemistry, 31*, 99.

Buttery, R. G., Ling, L. C., & Mon, T. R. (1986). Quantitative analysis of 2-acetyl-1 pyrroline in rice. *Journal of Agriculture and Food Chemistry, 34*, 112–114.

Buttery, R. G., Ling, L. C., Juliano, B. O., & Turnbaugh, J. G. (1983). Cooked rice aroma and 2-acetyl-1-pyrroline. *Journal of Agriculture and Food Chemistry, 31*, 823–826.

Buttery, R. G., Orts, W. J., Takeoka, G. R., & Nam, Y. (1999). Volatile flavor components of rice cakes. *Journal of Agriculture and Food Chemistry, 47*(10), 4353–4356.

Buttery, R., Turnbaugh, J., & Ling, L. (1988). Contributions of volatiles to rice aroma. *Journal of Agriculture and Food Chemistry, 36*, 1006–1009.

Champagne, E. T. (2008). Rice aroma and flavor: A literature review. *Cereal Chemistry,* *85*(4), 445–454.

Chikubu, S. (1970). Stale flavor of stored rice. *Japan Agricultural Research Quarterly, 5,* 63–68.

Currie, L. A. (1995). Nomenclature in evaluation of analytical methods including detection and quantification capabilities (IUPAC Recommendations 1995). *Pure and Applied Chemistry, 67,* 1699–1723.

Dhulappanavar, C. V. (1976). Inheritance of scent in rice. Euphytica, *25*(1), 659–662.

Eisert, R., & Pawliszyn, J. (1997). Automated in-tube solid-phase microextraction coupled to high-performance liquid chromatography. *Analytical Chemistry, 69,* 3140–3147.

Eshaghi, A., Zhang, X., Musteata, F. M., Bagheri, H., & Pawliszyn, J. (2007). Evaluation of bio-compatible poly (ethylene glycol) based solid-phase microextraction fiber for in vivo pharmacokinetic studies of diazepam in dogs. *Analyst., 132*(7), 672–678.

Eurachem, (1998). The fitness for purpose of analytical methods. *A Laboratory Guide to Method Validation and Related Topics.* LGC, Queens Rd, Teddington.

Ferraris, D., Ko, Y. S., Calvin, D., Chiou, T., Lautar, S., Thomas, B., Wozniak, K., Rojas, C., Kalish, V., & Belyakov, S. (2004). Ketopyrrolidines and ketoazetidines as potent dipeptidyl peptidase IV (DPP IV) inhibitors. *Bioorg. Med. Chern. Lett., 14,* 5579–5583.

Fitzgerald, M. A., Hamilton, N. R. S., Calingacion, M. N., Verhoeven, H. A., & Butardo, V. M. (2008). Is there a second gene for fragrance in rice? *Plant Biotechnol. J., 6,* 416–423.

Fontanals, N., Marce, R. M., & Borrull, F. (2007). New materials in sorptive extraction techniques for polar compounds. *J. Chromatogr. A., 1152,* 14–31.

Ghiasvand, A. R., Setkova, L., & Pawliszyn, J. (2007). Determination of flavor profile in Iranian fragrant rice samples using cold-fibre SPME–GC–TOF–MS. *Flavor Fragr. J., 22,* 377–391.

Grimm Grimm, C. C., Champagne, E. T., Lloyd, S. W., Easson, M., Condon, B., & McClung, A. (2011). Analysis of 2-acetyl-l-pyrroline in rice by HSSE/GCIMS. *Cereal Chern., 88*(3), 271–277.

Grimm, C. C., Bergman, C., Delgado, J. T., & Bryant, R. (2001). Screening for 2-acetyl-1-pyrroline in the headspace of rice using SPME/GC-MS. *J. Agric. Food Chem., 49*(1), 245–249.

Grosch, W. (2001). Evaluation of the key odorants of foods by dilution experiments, aroma models and omission. *Chem. Senses, 26*(5), 533–545.

ICHQ2B (1996). *Validation of Analytical Procedure: Methodology,* international conference on harmonisation of technical requirements for registration of pharmaceuticals for human use. URL: http://www.fda.gov/downloads/drugs/guidancecomplianceregulatoryinformation/guidances/ucm073384.pdf. Accessed 15.11.15.

Iglesias, J., & Medina, I. (2008). Solid-phase microextraction method for the determination of volatile compounds associated to oxidation of fish muscle. *J. Chromatogr., 1192*(1), 9–16.

Isetun, S., & Nilsson, U. (2005). Dynamic field sampling of airborne organophosphate triesters using solid-phase microextraction under equilibrium and non-equilibrium conditions. *Analyst, 130,* 94–98.

Jana, A., Bhattacharyya, N., Bandyopadhyay, R., Tudu, B., Mukherjee, S., Ghosh, D., & Roy, J. K. (2015). Fragrance measurement of scented rice using electronic nose. *Int. J. Smart Sensing Intell. Syst.*, *8*(3), 1730–1747.

Jezussek, M., Juliano, B. O., & Schieberle, P. (2002). Comparison of key aroma compounds in cooked brown rice varieties based on aroma extract dilution analyses. *J. Agric. Food Chem.*, *50*, 1101–1105.

Kataoka, H., Terada, Y., Inoue, R., & Mitani, K. (2007). Determination of isophorone in food samples by solid-phase microextraction coupled with gas chromatography–mass spectrometry. *J. Chromatogr. A.*, *1155*(1), 100–104.

Kolb, B., Welter, C., & Bichler, C. (1992). Determination of partition coefficients by automatic equilibrium headspace gas chromatography by vapor phase calibration. *Chromatographia*, *34*, 235–240.

Kovats, E. (1958). Gas chromatographic characterization of organic compounds. Part 1: Retention indices of aliphatic halides, alcohols, aldehydes and ketones. *Helv. Chim. Acta.*, *41*(7), 1915–1932.

Koziel, J., Jia, M., & Pawliszyn, J. (2000). Air Sampling with Porous Solid-Phase Microextraction Fibers. *Anal. Chem.*, *72*, 5178–5186.

Laguerre, M., Mestres, C., Davrieux, F., Ringuet, J., & Boulanger, R. (2007). Rapid discrimination of fragrant rice by solid-phase microextraction, mass spectrometry, and multivariate analysis used as a mass sensor. *J. Agric. Food Chem.*, *55*, 1077–1083.

Larroque, V., Desauziers, V., & Mocho, P. (2006). Development of a solid phase microextraction (SPME) method for the sampling of VOC traces in indoor air. *J. Environ. Monit.*, *8*, 106–111.

Lin, C., Hsieh, T., & Hoff, B. (1990). Identification and quantification of the "popcorn"-like aroma in Louisiana aromatic Della rice (*Oryza sativa* L.). *J. Food Sci.*, *55*, 1466–1467.

Lin, J. Y., Fan, W., Gao, Y. N., Wu, S. F., & Wang, S. X. (2010). Study on volatile compounds in rice by HS-SPME and GC-MS. 10th International working conference on stored product protection, *Julius-Kühn-Archiv.*, *425*.

Mahatheeranont, S., Keawsa-ard, S., & Dumri, K. (2001). Quantification of the rice aroma compound, 2-acetyl-1-pyrroline, in uncooked Khao Dawk Mali 105 brown rice. *J. Agric. Food Chem.*, *49*, 773–779.

Mahattanatawee, K., & Rouseff, R. L. (2014). Comparison of aroma active and sulfur volatiles in three fragrant rice cultivars using GC–Olfactometry and GC–PFPD. *Food Chem.*, *154*, 1–6.

Maraval, I., Mestres, C., Pernin, K., Ribeyre, F., Boulanger, R., Guichard, E., & Gunata, Z. (2008). Odor-active compounds in cooked rice cultivars from Camargue (France) analyzed by GC-O and GC-MS. *J. Agric. Food Chem.*, *56*, 5291–5298.

Maraval, I., Sen, K., Agrebi, A., Menut, C., Morere, A., Boulanger, R., Gay, F., Mestres, C., & Gunata, Z. (2010). Quantification of 2-acetyl-1-pyrroline in rice by stable isotope dilution assay through headspace solid-phase microextraction coupled to gas chromatography-tandem mass spectrometry. *Anal. Chim. Acta.*, *675*(2), 148–155.

MassBank, (2015). High quality mass spectral database, Mass Bank, National Bioscience Database Center, Science and Technology Agency and the mass spectorometry society, Japan. URL: http://www.massbank.jp/en/statistics.html. Accessed 15.11.15.

MassLib (2015). Spectra databases, MassLib. URL: www.masslib.com. Accessed 15.11.15.

Mathure, S. V., Jawali, N., Thengane, R. J., & Nadaf, A. B. (2014). Comparative quantitative analysis of headspace volatiles and their association with *BADH2* marker in non-basmati scented, basmati and non-scented rice (*Oryza sativa* L.) cultivars of India. *Food Chem., 142*, 383–391.

Mathure, S. V., Wakte, K. V., Jawali, N., & Nadaf, A. B. (2011). Quantification of 2-Acetyl-1-pyrroline and other rice aroma volatiles among Indian scented rice cultivars by HS-SPME/GC-FID. *Food Anal. Method, 4*, 326–333.

Mathure, S., Shaikh, A., Renuka, N., Wakte, K., Jawali, N., Thengane, R., & Nadaf, A. (2012). Characterization of aromatic rice (*Oryza sativa* L.) germplasm and correlation between their agronomic and quality traits. *Euphytica, 179*, 237–246.

McNaught, A. D., & Wilkinson, A. (1997). *IUPAC compendium of chemical terminology.* 2nd ed., Blackwell Science.

Mitsuda, H., Yasumoto, K., & Iwami, K. (1968). Analysis of volatile components in rice bran. *Agric. Biol. Chem., 32*(4), 453–458.

Monsoor, M. A., & Proctor, A. (2004). Volatile component analysis of commercially milled head and broken rice. *J. Food Sci., 69*(8), 662–636.

Musteata, M. L., Musteata, F. M., & Pawliszyn, J. (2007). Biocompatible solid-phase microextraction coatings based on polyacrylonitrile and solid-phase extraction phases. *Anal. Chem., 79*(18), 6903–6911.

NISTMSRIL (2015). NIST mass spectral and retention index libraries, National Institute of Standards and Technology, Gaithersburg, MD. URL: www.nist.gov/srd/nist1a.cfm. Accessed 15.11.15.

Ouyang, G., & Pawliszyn, J. (2008). A critical review in calibration methods for solid-phase microextraction. *Anal. Chim. Acta., 627*(2), 184–197.

Ouyang, G., Chen, Y., Setkova, L., & Pawliszyn, J. (2005). Calibration of solid-phase micro-extraction for quantitative analysis by gas chromatography. *J. Chromatogr. A., 1097*(1–2), 9–16.

Panavaite, D., Padarauskas, A., & Vickackaite, V. (2006). Silicone glue coated stainless steel wire for solid phase microextraction. *Anal. Chim. Acta., 571*(1), 45–50.

Pawliszyn, J. (2004). *Hyphenated Techniques in Speciation Analysis* (Szpunar, J., & Lobinski, R., eds.) *Analytical Chemistry, 76*(13), 248.

Plutowska, B., & Wardencki, W. (2008). Aromagrams–aromatic profiles in the appreciation of food quality. *Food Chem., 101*, 845–872.

Potter, D. W., & Pawliszyn, J. (1992). Detection of substituted benzenes in water at the PGML level using solid-phase microextraction and gas chromatography-ion trap mass spectrometry, *J. Chromatogr., 625* 247–255.

Reineccius, G. (2006). *Flavor Chemistry and Technology.* Boca Raton, F. L., Taylor & Francis. pp. 520.

Reinke, R. F., Welsh, L. A., Reece, J. E., Lewin, L. G., & Blakeney, A. B. (1991). Procedures for quality selection of aromatic rice varieties. *Int. Rice Res. Newslett., 16*, 10–11.

Ruckriemen, J., Schwarzenbolz, U., Adam, S., & Henle, T. (2015). Identification and quantitation of 2-Acetyl-1-pyrroline in Manuka Honey (*Leptospermum scoparium*). *J. Agric. Food Chem., 63*, 8488−8492.

Sakthivel, K., Sundaram, R. M., Shobha Rani, N., Balachandran, S. M., & Neeraja, C. N. (2009). Genetic and molecular basis of fragrance in rice. *Biotechnol. Adv., 27*, 468–473.

SDOCSDBS (2015). Spectral Database for Organic Compounds, SDBS, National Institute of Advanced Industrial Science and Technology (AIST), Japan. URL: http://sdbs. db.aist.go.jp. Accessed 15.11.15.

Shurmer, B., & Pawliszyn, J. (2000). Determination of distribution constants between a liquid polymeric coating and water by a solid-phase microextraction technique with a flow-through standard water system. *Anal Chem., 72*(15), 3660–3664.

Sood, B. C., & Sidiq, E. A. (1978). A rapid technique for scent determination in rice. *Indian J. Genetic Plant Breed., 38*, 268–271.

Spietelun, A., Kloskowski, A., Chrzanowski, W., & Namiesnik, J. (2013). Understanding solid-phase microextraction: key factors influencing the extraction process and trends in improving the technique. *Chem. Rev., 113,* 1667–1685.

Spietelun, A., Pilarczyk, M., Kloskowski, A., & Namiesnik, J. (2010). Current trends in solid-phase microextraction (SPME) fibre coatings. *Chem. Soc. Rev., 39*, 4524–4537.

Sriseadka, T., Wongpornchai, S., & Kitsawatpaiboon, P. (2006). Rapid method for quantitative analysis of the aroma impact compound, 2-acetyl-1-pyrroline, in fragrant rice using automated headspace gas chromatography. *J. Agric. Food Chem., 54*, 8183–8189.

Stashenko, E. E., & Martinez, J. R. (2007). Sampling volatile compounds from natural products with headspace/solid-phase micro-extraction. *J. Biochem. Biophy. Methods., 70*, 235–242.

SWGDRUGMSL (2015). SWGDRUG Mass Spectral Library. URL: http://www.swgdrug. org/ms.htm. Accessed 15.11.15.

Talon, R., & Montel, M. (1999). Applied SPME, *Royal Society of Chemistry,* Cambridge, UK Coden: 67TUA8, 364–371.

Tananuwong, K., & Lertsiri, S. (2010). Changes in volatile aroma compounds of organic fragrant rice during storage under different conditions. *J. Sci. Food Agric., 90*(10), 1590–1596.

Tanchotikul, U., & Hsieh, T. C. Y. (1991). An improved method for quantification of 2-acetyl-1-pyrroline, a "popcorn"- like aroma, in aromatic rice by high-resolution gas chromatography/mass spectrometry/ selected ion monitoring. *J. Agric. Food Chem., 39*, 944–947.

Tava, A., & Bocchi, S. (1999). Aroma of cooked rice (*Oryza sativa*): Comparison between commercial Basmati and Italian line 135–3. *Cereal Chem., 76*, 526–529.

Tena, M. T. Y., & Carrillo, J. D. (2007). Multiple solid phase microextraction. Theory and applications. *Trends Anal. Chem., 26*, 206–214.

Theodoridis, G., Koster, E. H. M., & De Jong, G. J. (2000). Solid-phase microextraction for the analysis of biological samples. *J. Chromatogr. B., 745*, 49–82.

Tsugita, T., Kurata, T., & Fujimaki, M. (1978). Volatile components in steam distillate of rice bran: identification of neutral and basic compounds. *Agric. Biol. Chem., 42,* 643–651.

Tsugita, T., Kurata, T., & Kato, H. (1980). Volatile components after cooking rice milled to different degrees. *Agric. Biol. Chem., 44*, 835–840.

Tuduri, L., Desauziers, V., & Fanlo, J. L. (2003). A simple calibration procedure for volatile organic compounds sampling in air with adsorptive solid-phase microextraction fibres. *Analyst, 128*(8), 1028–1032.

Verma, D. K., & Srivastav, P. P. (2016). Extraction technology for rice volatile aroma compounds. In: *Food Engineering Emerging Issues, Modeling, and Applications* (eds. Meghwal, M., & Goyal, M. R.). In book series on Innovations in Agricultural and Biological Engineering, Apple Academic Press, USA. pp. 245–291.

Verma, D. K., Mohan, M., & Asthir, B. (2013). Physicochemical and cooking characteristics of some promising basmati genotypes. *Asian Journal of Food Agro-Industry*, *6*(2), 94–99.

Verma, D. K., Mohan, M., Prabhakar, P. K., & Srivastav, P. P. (2015). Physico-chemical and cooking characteristics of Azad basmati. *International Food Research Journal*, *22*(4), 1380–1389.

Verma, D. K., Mohan, M., Yadav, V. K., Asthir, B., & Soni, S. K. (2012). Inquisition of some physico-chemical characteristics of newly evolved basmati rice. *Environment and Ecology*, *30*(1), 114–117.

Vuckovic, D., Shirey, R., Chen, Y., Sidisky, L., Aurand, C., Stenerson, K., & Pawliszyn, J. (2009). In vitro evaluation of new biocompatible coatings for solid-phase micro-extraction: implications for drug analysis and in vivo sampling applications. *Anal. Chim. Acta.*, *638*(2), 175–185.

Wakte, K. V., Kad, T. D., Zanan, R. L., & Nadaf, A. B. (2011). Mechanism of 2-acetyl-1-pyrroline biosynthesis in *Bassia latifolia* Roxb. flowers. *Physiol. Mol. Biol. Plants*, *17*(3), 231–237.

Wang, Y., & Ha, J. (2012). Determination of hexanal in rice using an automated dynamic headspace sampler coupled to a gas chromatograph–mass spectrometer. *J. Chromatogr. Sci.,* doi: 10.1093/chromsci/bms161.

Wang, Y., Nacson, S., & Pawliszyn, J. (2007). The coupling of solid-phase microextraction/surface enhanced laser desorption/ionization to ion mobility spectrometry for drug analysis. *Anal. Chim. Acta.*, *582*(1), 50–54.

Weber, D. J., Rohilla, R., & Singh, U. S. (2000). Chemistry and bio- chemistry of aroma in aromatic rice. In: *Aromatic Rices* (Singh, R. K., Singh, U. S., & Khush, G. S., eds.) Oxford & IBH Publ. Co. Pvt. Ltd., New Delhi, India. pp. 29–46.

Wei, X., Koo, I., Kim, S., & Zhang, X. (2014). Compound identification in GC-MS by simultaneously evaluating the mass spectrum and retention index. *Analyst.*, *139*, 2507–2514.

Widjaja, R., Craske, J. D., & Wootton, M. (1996). Changes in volatile components of paddy, brown, and white fragrant rice during storage. *J. Sci. Food Agric.*, *71*, 218–224.

WLFFNSC, (2015). Wiley library FFNSC – Mass spectra of flavors and fragrances of natural and synthetic compounds. URL: http://www.sisweb.com/software/wiley-ffnsc.htm. Accessed 15.11.15.

WMSL, (2015). Wiley mass spectral library. URL: http://www.wiley.com/WileyDA/WileyTitle/productCd-047052037X.html. Accessed 15.11.15.

Wongpornchai, S., Dumri, K., Jongkaewwattana, S., & Siri, B. (2004). Effects of drying methods and storage time on the aroma and milling quality of rice (*Oryza sativa* L.) cv. Khao Dawk Mali 105. *Food Chem.*, 87, 407–414.

Wongpornchai, S., Sriseadka, T., & Choonvisase, S. (2003). Identification and quantitation of the rice aroma compound, 2-acetyl-1- pyrroline, in bread flowers (*Vallaris glabra* Ktze). *J. Agric. Food Chem.*, *51*, 457–462.

Yajima, I., Yanai, T., Nakamura, M., Sakakibara, H., & Habu, T. (1978). Volatile flavor components of cooked rice. *Agric. Biol. Chem., 42*, 1229–1233.

Yang, D. S., Shewfelt, R. L., Lee, K. S., & Keys, S. J. (2008). Comparison of odor-active compounds from six distinctly different rice flavor types. *J. Agric. Food Chem., 56*, 2780–2787.

Yang, S. S., Huang, C. B., & Smetena, I. (2002). Optimization of headspace sampling using solid-phase microextraction for volatile components in tobacco. *J. Chromatogr., 942*(1), 33–39.

Yi, M., New, K. T., Vanavichit, A., Chai-arree, W., & Toojinda, T. (2009). Marker assisted backcross breeding to improve cooking quality traits in Myanmar rice cultivar Manawthukha. *Field Crops Res., 113*, 178–186.

Zellner, B., Bicchi, C., Dugo, P., Rubiolo, P., Dugo, G., & Mondello, L. (2008). Linear retention indices in gas chromatographic analysis: a review. *Flavor Fragr. J., 23*, 297–314.

Zeng, Z., Zhang, H., Zhang, T., Tamogami, S., & Chen, J. Y. (2009). Analysis of flavor volatiles of glutinous rice during cooking by combined gas chromatography-mass spectrometry with modified headspace solid-phase micro-extraction method. *J. Food Comp. Anal., 22*(4), 347–353.

Zhou, S. N., Oakes, K. D., Servos, M. R., & Pawliszyn, J. (2008). Application of solid-phase microextraction for in vivo laboratory and field sampling of pharmaceuticals in fish. *Environ. Sci. Technol., 42*(16), 6073–6079.

CHAPTER 7

SUPERCRITICAL FLUID EXTRACTION (SCFE) FOR RICE AROMA CHEMICALS: RECENT AND ADVANCED EXTRACTION METHOD

DEEPAK KUMAR VERMA,[1] JYOTI P. DHAKANE,[2] DIPENDRA KUMAR MAHATO,[3] SUDHANSHI BILLORIA,[4] PARAMITA BHATTACHARJEE,[5] and PREM PRAKASH SRIVASTAV[6]

[1]Research Scholar, Department of Agricultural and Food Engineering, Indian Institute of Technology, Kharagpur–721 302, West Bengal, INDIA, Mobile: +91-7407170260, +91-9335993005, Telephone: +91-3222-281673, Fax: +91-3222-282224, E-mail: deepak.verma@agfe.iitkgp.ernet.in, rajadkv@rediffmail.com

[2]Research Scholar, Indian Agricultural Research Institute, Pusa Campus, New Delhi–110012, INDIA, Mobile: +91-8745000441, E-mail: jyotip.dhakane@gmail.com

[3]Senior Research Fellow, Indian Agricultural Research Institute, Pusa Campus, New Delhi–110012, INDIA, Mobile: +91-9911891494, +91-9958921936, E-mail: kumar.dipendra2@gmail.com

[4]Research Scholar, Department of Agricultural and Food Engineering, Indian Institute of Technology, Kharagpur–721302, West Bengal, India, Mobile: +91-8768126479, E-mail: sudharihant@gmail.com

[5]Reader, Department of Food Technology & Biochemical Engineering Jadavpur University, Kolkata-700 032, India, Mobile: +91-9874704488 E-mail: pb@ftbe.jdvu.ac.in, yellowdaffodils07@gmail.com

[6]Associate Professor, Department of Agricultural and Food Engineering, Indian Institute of Technology, Kharagpur–721302, West

Bengal, India, Mobile: +91-9434043426,
Tel.: +91-3222-283134, Fax: +91-3222-282224,
E-mail: pps@agfe.iitkgp.ernet.in

CONTENTS

7.1   INTRODUCTION

Supercritical fluid extraction (SCFE) is one of the most innovative and reliable methods for separation and extraction process of volatile aroma compounds by use of supercritical fluids (SCFs) as the extraction solvent (Figure 7.1) (McHugh and Krukonis, 1986; Hawthorne et al., 1988; Sinha et al., 1992; Anklam et al., 1998; King, 2002). SCF could be easily removed from the extracted material, almost leaving no trace by decreasing pressure and therefore is eco-friendly (Stashenko et al., 1996, 2004). Therefore, the resulting treatment is simple when SCF is used for the extraction of volatile aroma compounds.

SCFs are a kind of clean solvent in the view of "Green Chemistry" and nontoxic compared to organic solvents (de Melo et al., 2014). Although costlier, SCFE is preferred over organic solvents because of higher extraction efficiency and residue-free extract. Besides, it allows working at moderate temperatures, which invariably reduces thermal degradation, and due to the absence of light and oxygen, oxidation

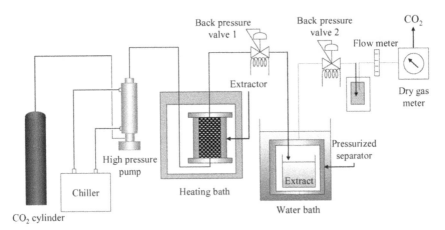

FIGURE 7.1 Schematic diagram of supercritical fluid extraction (SCFE) set-up. (Reprinted from Machmudah, S., Sulaswatty, A., Sasaki, M., Goto, M., & Hirose, T. (2006). Supercritical CO_2 extraction of nutmeg oil: experiments and modeling. *Journal of Supercritical Fluids, 39*, 30–39. © 2006 Elsevier. With permission.)

reactions are prevented. Owing to these advantages, SCFE is used as an alternative to liquid extraction, and this is proved to be an efficient sample preparation technique (Stashenko et al., 2004) and is simple to use for the extraction of volatile/aromatic compounds (Lou and Chen, 1996).

Supercritical state is obtained when a fluid is forced to a pressure and temperature above its critical point where there is no distinction between its liquid and gaseous phases (Herrero et al., 2006). It behaves both like gas and liquid. It effuses through solid surfaces as well as acts as a solvent for solutes. The density of a SCF matches to that of a liquid and its viscosity to a gas, while the diffusivity lies between the two states (Table 7.1). Though there are a number of solvents that have been used for SCFE, the most widely used SCF is carbon dioxide (CO_2), because of its lower critical temperature (T_c) and pressure (P_c) and higher critical density (Table 7.2). CO_2 is sometimes modified by addition of water, ethanol, or methanol (Sun and Temelli, 2006; Shi et al., 2009; Zhang and Li, 2010) as co-solvents. These are highly polar compounds that significantly changes the solvent properties of supercritical CO_2 (SC-CO_2) as and when added (Kislik, 2012).

TABLE 7.1 Physico-Chemical Properties of gases, SCFs, and Liquids

Properties	Gas	Supercritical fluid	Liquid
	$P=1$ atm; $T=21°C$	$P=1$ atm; $T=15–30°C$	$P=P_c$; $T=T_c$
Density (g/cm³)	10^{-3}	0.3–0.8	1.0
Viscosity (g/cm.s)	10^{-4}	$10^{-4}–10^{-3}$	10^{-2}
Diffusivity (cm²/s)	0.1	$10^{-3}–10^{-4}$	$<10^{-5}$

Source: Herrero et al. (2006).

TABLE 7.2 Chemical Solvents for SCFE Techniques

Solvents	T_c (K)	P_c (Mps)	Critical Density kg/m³
Methane	192	4.60	162
Ethylene	283	5.03	218
Carbon dioxide	304	7.38	468
Ethane	305	4.88	203
Propylene	365	4.62	233
Propane	370	4.24	217
Ammonia	406	11.3	235
Water	647	22.0	322

Source: Supercritical Fluid Extraction (SFE). URL: http://www.separationprocesses.com/Extraction/SE_Chp06.htm. Accessed on 25-12-2015.

SCFEs have already been used for a number of years for the extraction of food constituents. SCFE has recently become an alternative to conventional extraction procedures that avoids some of the drawbacks linked to previous extraction technology. There are numerous articles that deal with the advantages, applications, and possibilities of SCFE in flavor analysis (Obaya-Valdivia and Guerrero-Barajas, 1993; Polesello et al., 1993; Kallio and Kerrola, 1995; Taylor and Larick, 1995; Diaz et al., 1997; Gere et al., 1997; Valcarcel and Tena, 1997; Anklam et al., 1998).

7.2 PRINCIPLE AND INSTRUMENTATION

7.2.1 PRINCIPLE

Supercritical fluid extraction (SCFE) is the process of separation of one matrix (i.e., the extract) from another matrix (i.e., the sample) using super critical fluids (SCFs), mostly CO_2, as the extracting solvent. CO_2, a well-established extraction solvent for botanicals, is considered as the king among the solvents available in market. SCFs are highly compressed gases, which possesses combined properties of liquids and gases in an interesting manner (Sapkale et al., 2010). These fluids are superior to the conventional solvents with respect to their excellent reactivity with the matrices, which are impossible to achieve with the conventional solvents (Sapkale et al., 2010). It is a fast process compared to traditional solvent extraction and can be easily recovered by releasing the pressure without leaving any traces behind.

The methodology of SCFE is based upon the principle that a fluid (gas) above its critical point exhibits the solution properties of a liquid solvent. At critical temperature (T_c), the gas is converted to liquid by increasing the pressure, while at critical pressure (P_c), the liquid is converted to a gas by increasing temperature. This is the main feature of the critical point (CP) on the pressure-temperature curve as shown in Figure

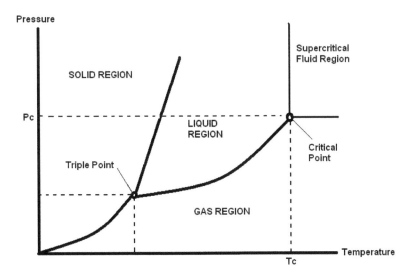

FIGURE 7.2 Pressure-temperature phase diagram of carbon dioxide (Reprinted from Hizir, M. S., & Barron, A. R. (2016). Basic principles of supercritical fluid chromatography and supercritical fluid extraction. URL: http://cnx.org/contents/wmqmYmjc@2/Basic-Principles-of-Supercriti. https://creativecommons.org/licenses/by/4.0/).

7.2 (Cavalcanti and Meireles, 2012). In the supercritical region ($T > T_c$ and $P > P_c$), the compound is present as a single-phase fluid that is non-condensing; important changes in properties such as polarity and solubility of SCFs are observed with small changes in pressure at constant temperature (Brunner, 1994). These characteristics make SCFs very attractive as small changes in pressure of process solvents affect recovery and type of compounds to be extracted.

The combination of low surface tension and viscosity and high solubility is principally important to increase extraction from raw material where the target compounds are deep inside the pores of solid matrices (Azmir et al., 2013). CO_2 is considered as a most suitable supercritical solvent because it has moderate critical parameters, viz. 7.38 MPa (74 bar) pressure and 31.04°C temperature (Otto and Thomas, 1967); thus, the sample matrix of biological materials can be processed at temperatures around 35°C (Sapkale et al., 2010). Moreover, CO_2 is non-toxic, non-flammable, odorless, tasteless, and inert gas (Das and Panda, 2015).

7.2.2 INSTRUMENTATION

Presently, SCFE is a well-known method for separation and extraction because of its design and easy operating method (Li et al., 2010). The wide ranges of commercial instruments from laboratory scale to industrial scale are available in market to carry out SCFEs. The basic instrumentation needed to build an SCFE instrument will slightly vary depending on the application and solid or liquid extraction. Schematic diagram of an SCFE experimental apparatus (AKICO Co., Japan) with its basic components is depicted in Figure 7.1 (Machmudah et al., 2006). Typical SCFE apparatus is a semi-continuous flow extractor. The complete arrangement consists of reservoir of mobile phase usually liquid CO_2, extraction vessel with thermostatic bath, high pressure pumps to pressurize the gas, co-solvent vessel and pump, back pressure regulators (BPRs), and collection and separation vessel. Different types of measurement instruments like flow meter, dry/wet gas meter, and pressure indicators could be attached to system.

7.2.2.1 CO_2 Cylinder

Carbon dioxide (CO_2) is stored in liquid form in a dip tube cylinder that is connected to a high pressure metered pump through pipeline. The usual initial pressure in CO_2 cylinders should be nearly 50 bar or above to develop necessary operating pressure in the extraction vessel. The liquid CO_2 is usually supplied below 5°C to the extraction vessel (Sapkale et al., 2010).

7.2.2.2 Extraction Vessel

The extraction vessel (sample holder) is equipped with means of heating such as oven for small vessel or coil or electrically heated jacket for larger vessels because the adiabatic expansion of CO_2 results in significant cooling. Extraction vessels are available in wide range of sizes useful for laboratory or pilot scale extraction. The typical capacity of extraction vessel for a laboratory scale model is as small as 5 mL to 300 mL (Sun and Temelli, 2006; Nobre et al., 2009). A typical extractor of 500 mL capacity has dimension of 20 cm height and 7 cm inside diameter (Machmudah et al., 2006). The pressure requirement for extraction varies from at least 74 bar to 350 bar or more. The rubber seals used in the vessels can swell by absorbing CO_2. Hence, care must be taken to avoid rupture of the rubber seal during depressurization of the vessel (Sapkale et al., 2010).

7.2.2.3 High Pressure Pump

In the SCFE system, the CO_2 pump sets high pressures required for SCF work in the extraction vessel. The reciprocating CO_2 pumps or syringe pumps are recommended for use in small-scale extractions while for larger scale extractions, diaphragm pumps are mostly preferred (Sapkale et al., 2010). Additionally, modifier (co-solvent) pump is also connected to the extraction vessel wherever necessary. The modern systems are equipped with rupture disc assembly in order to prevent accidents during over-temperature or overpressure conditions. In the case of liquid samples, the extraction column is frequently equipped with different ports for the introduction of the sample at variable heights, where another additional pump is needed to introduce the sample into the system (Brunner, 1995).

7.2.2.4   Back Pressure Regulators (BPR)

The pressure set by CO_2 pump inside the extraction chamber, to keep SC-CO_2 always under the desired conditions, is maintained using a variable restrictor or a BPR (Sun and Temelli, 2006). The smaller systems are provided either with a capillary tube or a needle valve in order to maintain pressure at different flow rates. On the other hand, larger systems utilize the back pressure regulator that maintains the upstream pressure by the use of a spring, compressed air, or electronically driven valve (Sapkale et al., 2010). The surge tank connected to a pipe carrying liquid CO_2 is intended to maintain sudden changes of pressure in the flow by filling when the pressure increases and emptying when it drops. There are two back pressure regulators provided with the extraction system to control the pressure. BPR 1 and BPR 2 are used for controlling the pressure in the extractor and the separato,r respectively, as shown in Figure 7.1.

7.2.2.5   Thermostatic Baths

Thermostatic baths are provided in the SCFE system to achieve the temperature of the extraction vessel at operating conditions ($T > T_c$) and another at the separation vessel (Figure 7.1) (Machmudah et al., 2006; Sun and Temelli, 2006). Additional thermostatic bath or pre-heater may be provided to bring co-solvent temperature equal to operating temperature of solvent where co-solvent module is provided. The desired extraction temperature is monitored by a thermocouple and regulated by a controller (Sun and Temelli, 2006)

7.2.2.6   Separation and Collection Vessel

The extract separated from the sample matrix is carried into the collection vessel or several fractionation vessels in order to perform cascade depressurization along with SC-CO$_2$ and co-solvent. The separation of the extract from SC-CO$_2$ is carried out under reduced pressure than the extraction vessel because the solubility in the lower density CO$_2$ is much lower, and the material precipitates for collection from where CO$_2$ can be recycled or depressurized to atmospheric pressure and vented through the pipeline and pure or co-solvent containing extract is taken out. The desired soluble materials in the solvent can be fractioned using a series of vessels at reduced pressure (Brunner, 1995; Sapkale et al., 2010).

7.2.2.7   Gas Flow and Pressure Measuring Devices

To check the flow of SC-CO$_2$ in the system after the collection vessel, the flow meter and dry gas meter are fixed in the pipeline. Pressure indicators are fitted on the pipeline at the extraction vessel as well as at the separation and collection vessel in order to assure required gas pressure built up inside these vessels. To control pressure and temperature of the system, proportional–integral–derivative controllers (PID controllers) are commonly used in models available in the market. Integrated programmable logic controller monitors are used for maintaining a desired set point during the extraction process.

7.3   EXTRACTION PROCESS

Initially, the extraction vessel is charged with a fixed amount of sample placed in the sample basket. Liquid CO$_2$ from a cylinder with siphon attachment is passed through a chiller to decrease the temperature of CO$_2$ down to −11°C and operating pressure (P) is maintained higher than the critical pressure (P_c) using a high-pressure pump. On the other hand, compressed CO$_2$ is flowed into the extractor that is kept in the heating bath in order to maintain its operating temperature (T) greater than the critical temperature (T_c). The fluid coming out of the extractor is expanded to a pressurized separator at 2.5 MPa and 273 K by BPR 1 followed by expansion to ambient pressure by BPR 2. The flow of CO$_2$ is measured by the flow meter and the dry gas meter. The extracted material is finally collected from the pressurized separator and used for further analysis (Machmudah et al., 2006).

7.4 MERITS AND DEMERITS

7.4.1 MERITS

SCFE is an environment friendly and very advantageous extraction technology in the course of analysis of volatile aroma compounds from rice sample than the other conventional extraction methods. Xu et al. (2011) reviewed the period from 2005 to 2011 which covers the recent advances and developments of SCFE in the extraction from the plant materials. The merits and demerits of SCFE over conventional extraction methods can be summarized as follows (Lang and Wai, 2001; Ahuja and Diehl, 2006; Handa, 2008):

1. Possibility of extraction and purification of volatile aroma compounds present in solid or liquid sample matrix. At low temperature, damage from heat could be avoided, which is ideal for extraction of thermolabile compounds.
2. There is no solvents residue (i.e., reduces the risk of storage), better diffusivity, lower viscosity and surface tension and hence fast extraction (since the complete extraction step is performed in about 20 min) over the other extraction technology.
3. The SCF is preferred over liquid solvent as its solvation power can be tuned either by changing temperature and/or pressure.
4. Useful for highly volatile compounds as can be connected on-line to a chromatograph immediately after extraction for analysis.
5. The SCF can be used several times for complete extraction.
6. Use of SCF for separation of solute from solvent takes less time than the conventional extraction process.
7. The amount of sample required in SCFE extraction is very low.
8. SCFE is considered as an eco-friendly method as very little amount of organic solvent is utilized.
9. The SCF can be recycled and used again.
10. SCFE is useful for both laboratory and industrial purpose.

7.4.2 DEMERITS

1. SCF penetration takes little time and is rapid into the interior of a solid compare to solute diffusion from solid to exterior and take prolonged time.

2. Scale in SCFE is not possible because of absence of fundamental and molecular-based model of solutes.
3. SCFE technology is very expensive compared to others and consistency and reproducibility may vary in continuous production.
4. The low polarity of CO_2 makes it unsuitable for most pharmaceuticals and drug samples without the use of a chemical modifier.

7.5 CASE STUDY ON SCFE METHOD FOR RICE AROMA CHEMICALS

A comparative evaluation of the extraction of the aroma constituents of a popular commercial brand of Basmati rice using Likens–Nickerson extraction and SCFE with CO_2 was carried out. SCFE at 50°C and 120 bar for 2 h provided appreciable extraction of the volatile constituents of the rice when compared with Likens–Nickerson extraction. The advantages of smaller sample size, shorter time of extraction, and negligible possibility of artifacts with the SCFE technique merit its use for the recovery of aroma volatiles from Basmati rice (Bhattacharjee et al., 2003).

Yahya (2011) experimented pandan leaf for the extraction of aroma compound to use the compound that enhance rice flavor using supercritical carbon dioxide (SC-CO_2) and Soxhlet extraction and hexane as solvent to extract 2-acetyl-1-pyrroline (2-AP) from pandan leaves (*Pandanus amaryllifolius* Roxb). A comparative study on the extraction of 2-AP from scented leaves of *P. amaryllifolius* was done by Bhattacharjee et al. (2005) using SCFE with CO_2 extraction in comparison to the Likens–Nickerson extraction method. The experiment was performed SCFE at 450 bar pressure for 3 h at 60°C with a constant flow rate of 0.1 l min^{-1} of CO_2. Under this instrumental condition, Bhattacharjee et al. (2005) reported 7.16 ppm yield of 2-AP. This 2-AP yield was found higher than the 2-AP yield found in the Likens–Nickerson extraction method. From the study of Bhattacharjee et al. (2005), a process parameter for quantitative analysis of 2-AP from extracts was optimized for SCFE as a new densitometric method, and it was concluded that a novel application of extracted 2-AP could be used in rice flavoring industry. The effects of different extraction pretreatments such as particle size and drying on the extraction yield and concentration of 2-AP were investigated. The identification and quantification of 2-AP were carried out by GC-MS and gas chromatography-flame ionization detection (GC-FID), respectively. This work aimed to provide an understanding of the phenomenon that occurs during cooking and storage, typically by the changes of 2-AP absorption when cooking rice grains with pandan leaves. The parameters were investigated at optimal water conditions. Although 2-AP concentration obtained in the SC-CO$_2$ extracts was low, the extracts were pure without any contamination. Grinding and freeze-drying methods proved to be good pre-treatments for SCFE. The absorption of 2-AP during the cooking of rice grains did not increase with time. This unexpected result indicated that the phenomena occurring during cooking are quite complex. This work also quantified the potential of pandan leaves to enhance the flavor of cooked rice, particularly under excess water conditions. Storage for 15 min at 24.0±1.0°C is considered as the optimum time for obtaining cooked rice with a high quality of flavor. In the experiment, the volatile components of rice were isolated by extraction with SC-CO$_2$ under pressure followed by atmospheric steam distillation and enrichment of steam volatiles on Porapak Q.

Although SCFE can recover the majority of the aroma compounds, it is also more selective, and therefore less effective, than solvent extraction, i.e., fewer compounds are extracted with SCFE than with solvent extraction (Polesello et al., 1993). At present, SCFE has been applied for the extraction of volatile aroma compounds from rice (Table 7.3) and other

TABLE 7.3   List of Extracted Major Rice Volatile Aroma Compounds Using SCFE Techniques

Aroma Compounds		Aroma Compounds	
1.	2-decenal	2.	Hexanal
3.	2-heptanone	4.	Hexanol
5.	2-octanone	6.	Nonanal
7.	Butan-2-one-3-Me	8.	Octanal
9.	Decanol	10.	Octanol
11.	Dodecane	12.	Pentanal
13.	Heptadecane	14.	Tetradecane
15.	Heptanol	16.	Undecane
17.	Hexadecane	18.	dfgvbhyt

Sources: Bhattacharjee, et al. (2003).

plant samples such as fruits, vegetables, spices, and so on (Schultz and Randall, 1970; Temelli et al., 1988; Tuan and Ilangantileke, 1997; Morales et al., 1998; Díaz-Maroto et al., 2002; Duarte et al., 2004).

7.6 SUMMARY AND CONCLUSION

In the recent decades, the use of SCFE technology has received special attention that provides promising extraction and fractionation of food (Xu et al., 2011). There are many literatures about the extraction using SCFE (Salgin et al., 2006; Salgin, 2007; Fiori, 2007, 2009; Talansier et al., 2008; Passos et al., 2009; Comim et al., 2010; Corso et al., 2010; Yilmaz et al., 2011), but very few research works were found in literature on the application of SCFE in rice aroma compound extraction (Bhattacharjee et al., 2003; Yahya, 2011). The present chapter addressed and discussed in detail the used of SCFE as modern and novel extraction technology for rice aroma compounds, principle and instrumentation behind the operation of SCFE extraction method, factors influencing method performance, case study on research progress, and strength and weakness of rice aroma research. There are number of references that exist on this recent and advance extraction technology in food, but very few works are existing on the topic entitled "Supercritical Fluid Extraction (SCFE) for Rice Aroma Chemicals: A Recent and Advance Extraction Method." Study on this technology is necessary for the further application of SCFE for aroma chemical extraction of the rice in future, which can provide different perspectives.

ACKNOWLEDGMENT

Deepak Kumar Verma and Prem Prakash Srivastav are indebted to the Department of Science and Technology, Ministry of Science and Technology, Government of India for an individual research fellowship (INSPIRE Fellowship Code No.: IF120725; Sanction Order No. DST/INSPIRE Fellowship/2012/686; Date: 25/02/2013).

KEYWORDS

- 2-acetyl-1-pyrroline (2-AP)
- aroma chemicals
- basmati
- extraction
- Likens–Nickerson
- pandan leaves
- rice
- supercritical fluid extraction
- volatile aroma

REFERENCES

Ahuja, S., & Diehl, D. (2006). Sampling and sample prepration. In: *Comprehensive Analytical Chemistry,* (Ahuja, S., & Jespersen N., eds.). vol. 47, Oxford, UK, Elsevier (Wilson & Wilson), pp. 15–40.

Anklam, E., Berg, H., Mathiasson, L., Sharman, M., & Ulberth, F. (1998). Supercritical fluid extraction (SFE) in food analysis: a review. *Food Add. Cont., 15*(6), 729–750.

Bhattacharjee, P., Kshirsagar, A., & Singhal, R. S. (2005). Supercritical carbon dioxide extraction of 2-acetyl-1-pyrroline from *Pandanus amaryllifolius* Roxb. *Food Chemi., 91*, 255–259.

Bhattacharjee, P., Ranganathan, T. V., Singhal, R. S., & Kulkarni, P. R. (2003). Comparative aroma profiles using supercritical carbon dioxide and Likens-Nickerson extraction from a commercial brand of Basmati rice. *J. Sci. Food Agric., 83*, 880–883.

Brunner, G. (1995). *Gas Extraction*: An introduction to fundamentals of supercritical fluids and the application to separation processes. Springer-Verlag, New York, LLC: New York, pp. 63.

Cavalcanti, R. N., & Meireles, M. A. A. (2012). Fundamentals of Supercritical fluid extraction. In: *Comprehensive Sampling and Sample Preparation: Analytical Techniques for Scientists* (Pawliszyn, J., ed.). vol. *2, Theory of Extraction Techniques.* Elsevier Inc., pp. 117–133.

Comim, S. R. R., Madella, K., Oliveira, J. V., & Ferreira, S. R. S. (2010). Supercritical fluid extraction from dried banana peel (*Musa* spp., genomic group AAB): Extraction yield, mathematical modeling, economical analysis and phase equilibria. *J. Supercrit. Fluid, 54*, 30–37.

Corso, M. P., Fagundes-Klen, M. R., Silva, E. A., Cardozo Filho, L., Santos, J. N., Freitas, L. S., & Dariva, C. (2010). Extraction of sesame seed (*Sesamun indicum* L.) oil using compressed propane and supercritical carbon dioxide. *J. Supercrit. Fluid, 52*, 56–61.

Das, D., & Panda, A. K. (2015). Supercritical CO_2 extraction of active components of fenugreek (*Trigonella foenum-graecum* L.) seed. *Inter. J. Pharma Res. & Rev., 4*(7), 1–9.

De Melo, M. M. R., Silvestre, A. J. D., & Silva, C. M. (2014). Supercritical fluid extraction of vegetable matrices: Applications, trends and future perspectives of a convincing green technology. *J. Supercrit. Fluid, 92,* 115–176.

Diaz, O., Cobos, A., De la Hoz, L., & Ordonez, J. A. (1997). Supercritical carbon dioxide in the production of food from plants. *Aliment. Equipos Tecnol., 16*(8), 55.

Diaz-Maroto, M. C., Perez-Coello, M. S., & Cabezudo, M. D. (2002). Supercritical carbon dioxide extraction of volatiles from spices: comparison with simultaneous distillation-extraction. *J. Chromato., 947,* 23–29.

Duarte, C., Moldao-Martins, M., Gouveia, A. F., Costa, S. B., Leitao, A. E., & Bernardo-Gil, M. G. (2004). Supercritical fluid extraction of red pepper (*Capsicum frutescens* L.). *J. Supercrit. Fluid, 30,* 155–161.

Fiori, L. (2007). Grape seed oil supercritical extraction kinetic and solubility data: Critical approach and modeling. *J. Supercrit. Fluid, 43,* 43–54.

Fiori, L. (2009). Supercritical extraction of sunflower seed oil: Experimental data and model validation. *J. Supercrit. Fluid, 50,* 218–224.

Gere, D. R., Randall, L. G., & Callahan, D. (1997). Supercritical fluid extraction: principles and applications. In: *Instrumental Methods in Food Analysis*, (Pare, J. R. J., & Belanger J. M. R., eds.). vol. *18, Techniques and Instrumentation in Analytical Chemistry*, Pub. Elsevier, pp. 421–484.

Handa, S. S. (2008). An overview of extraction techniques for medicinal and aromatic plants. In: *Extraction Technologies for Medicinal and Aromatic Plants* (Handa, S. S., Khanuja, S. P. S., Longo, G., Rakesh, D. D., eds.). International Centre for Science and High Technology ICS-UNIDO, Trieste, Italy. pp. 21–52.

Hawthorne, S. B., Krieger, M. S., & Miller, D. J. (1988). Analysis of flavor and fragrance compounds using supercritical fluid extraction coupled with gas chromatography. *Anal. Chem., 50,* 472–477.

Herrero, M., Cifuentes, A., & Ibanez, E. (2006). Sub- and supercritical fluid extraction of functional ingredients from different natural *Sources:* plants, food-by-products, algae and microalgae a review. *Food Chem., 98,* 136–148.

Hizir, M. S., & Barron, A. R. (2016). Basic principles of supercritical fluid chromatography and supercritical fluid extraction. URL: http://cnx.org/contents/wmqmYmjc@2/Basic-Principles-of-Supercriti Accessed on 25–06–2016.

Kallio, H., & Kerrola, K. (1995). Dense gas extraction as a preparation method in food analysis. *Curr. Status Future Trends Anal. Food Chem., Proc. Eur. Conf. Food Chem., 8th,* (Sontag, G., & Pfannhauser, W., eds.). vol. *1,* Austrian Chemical Society, Vienna, pp. 30.

King, J. W. (2002). Supercritical fluid extraction: present status and prospects. *Grasas y Aceites, 53*(1), 8–21.

Kislik, V. S. (2012). Advances in development of solvents for liquid–liquid extraction. In: *Solvent Extraction Classical and Novel Approaches*. Elsevier, B. V, pp. 451–481.

Lang, Q., & Wai, C. M. (2001). Supercritical fluid extraction in herbal and natural product studies-a practical review. *Talanta, 53*(4), 771–782.

Li, H., Wu, J., Rempel, C. B., & Thiyam, U. (2010). Effect of operating peramaters on oil and phenolic extraction using supercritical CO_2. *J. Am. Oil Chem. Soc., 87,* 1081–1089.

Lou, M. T., & Chen, R. (1996). Advances of application techniques of supercritical fluid extraction. *Analy. Instrum., 4*, 5–9.

Machmudah, S., Sulaswatty, A., Sasaki, M., Goto, M., & Hirose, T. (2006). Supercritical CO_2 extraction of nutmeg oil: experiments and modelling. *Journal of Supercritical Fluids, 39*, 30–39.

McHugh, M., & Krukonis, V. (1986). *Supercritical Fluid Extraction, Principles and Practice*. Butterworth, USA.

Morales, M. T., Berry, A. J., McIntyre, P. S., & Aparicio, R. (1998). Tentative analysis of virgin olive oil aroma by supercritical fluid extraction-high-resolution gas chromatography-mass spectrometry. *J. Chromato., 819*, 267–275.

Nobre, B. P., Palavra, A. F., Pessoa, F. L. P., & Mendes, R. L. (2009). Supercritical CO_2 extraction of trans-lycopene from Portuguese tomato industrial waste. *Food Chem., 116*, 680–685.

Obaya-Valdivia, A., & Guerrero-Barajas, C. (1993). Supercritical fluid extraction applied to food processing industries. *Tecnol. Aliment., 28*, 22.

Passos, C. P., Silva, R. M., Da Silva, F. A., Coimbra, M. A., & Silva, C. M. (2009). Enhancement of the supercritical fluid extraction of grape seed oil by using enzymatically pre-treated seed. *J. Supercrit. Fluid, 48*, 225–229.

Polesello, S., Lovati, F., Rizzolo, A., & Rovida, C. (1993). Supercritical fluid extraction as a preparative tool for strawberry aroma analysis. *J. High Res. Chromatog., 16*, 555–559.

Salgin, U. (2007). Extraction of jojoba seed oil using supercritical CO_2+ethanol mixture in green and high-tech separation process. *J. Supercrit. Fluid, 39*, 330–337.

Salgin, U., Doker, O., & Calimli, A. (2006). Extraction of sunflower oil with supercritical CO2, Experiments and modelling. *J. Supercrit. Fluid, 38*, 326–331.

Sapkale, G. N., Patil, S. M., Surwase, U. S., & Bhatbhage, P. K. (2010). A Review Supercritical Fluid Extraction. *Int. J. Chem. Sci., 8*(2), 729–743.

Schultz, W. G., & Randall, J. M. (1970). Liquid carbon dioxide for selective aroma extraction. *Food Techno., 24*, 1282–1286.

Schultz, W. G., & Randall, J. M. (1970). Liquid carbon dioxide for selective aroma extraction. *Food Techno., 24*, 1282–1286.

Shi, J., Yi, C., Xue, S. J., Jiang, Y., Mac, Y., & Li, D. (2009). Effects of modifiers on the profile of lycopene extracted from tomato skins by supercritical CO_2. *J. Food Engg, 93*, 431–436.

Sinha, N. K., Guyer, D. E., Gage, D. A., & Lira, C. T. (1992). Supercritical carbon dioxide extraction of onion flavors and their analysis by gas chromatography–mass spectrometry. *J. Agric. Food Chem., 40*, 842–845.

Stashenko, E. E., Jaramillo, B. E., & Martinez, J. R. (2004). Analysis of volatile secondary metabolites from Colombian *Xylopia aromatica* (Lamarck) by different extraction and headspace methods and gas chromatography. *J. Chromato., 1025*, 105–113.

Stashenko, E. E., Puertas, M. A., & Combariza, M. Y. (1996). Volatile secondary metabolites from *Spilanthes americana* obtained by simultaneous steam distillation-solvent extraction and supercritical fluid extraction. *J. Chromato., 752*, 223–232.

Sun, M., & Temelli, F. (2006). Supercritical carbon dioxide extraction of carotenoids from carrot using canola oil as a continuous co-solvent. *J. Supercrit. Fluid, 37*, 397–408.

Supercritical Fluid Extraction (SFE). URL: http://www.separationprocesses.com/Extraction/SE_Chp06.htm. Accessed on 25–12–2015.

Talansier, E., Braga, M., Rosa, P., Paoluccijeanjean, D., & Meireles, M. (2008). Supercritical fluid extraction of vetiver roots: A study of SFE kinetics. *J. Supercrit. Fluid, 47,* 200–208.

Taylor, D. L., & Larick, D. K. (1995). Volatile content and sensory attributes of supercritical carbon dioxide extracts of cooked chicken fat. *J. Food Sci., 60*(6), 1197–1200.

Temelli, F., Chen, C. S., & Braddock, R. J. (1988). Supercritical fluid extraction in citrus oil processing. *Food Techn., 42,* 145–150.

Tuan, D. Q., & Ilangantileke, S. G. (1997). Liquid CO_2 extraction of essential oil from Star anise fruits (*Illicium verum*H.). *J. Food Engg., 31,* 47–57.

Valcarcel, M., & Tena, M. T. (1997). Applications of supercritical fluid extraction in food analysis. *Fresenius J. Anal. Chem., 358*(5), 561–573.

Xu, L., Zhan, X., Zeng, Z., Chen, R., Li, H., Xie, T., & Wang, S. (2011). Recent advances on supercritical fluid extraction of essential oils. *Afri. J. Pharm. Pharmaco., 5*(9), 1196–1211.

Yahya, F. B. (2011). Extraction of aroma compound from pandan leaf and use of the compound to enhance rice flavor. *PhD Thesis*, School of Chemical Engineering, The University of Birmingham, Birmingham, UK.

Yilmaz, E. E., Ozvural, E., B., & Vural, H. (2011). Extraction and identification of proanthocyanidins from grape seed (*Vitis vinifera*) using supercritical carbon dioxide. *J. Supercrit. Fluid, 55*(3), 924–928.

Zhang, Z., & Li, G. (2010). A review of advances and new developments in the analysis of biological volatile organic compounds. *Microchem. J.,* 95, 127–139.

PART IV

GENETIC, BIOCHEMICAL, AND BIOTECHNOLOGICAL DEVELOPMENT AND FUTURE PERSPECTIVES

CHAPTER 8

AROMA VOLATILES AS BIOMARKERS FOR CHARACTERIZING RICE (*ORYZA SATIVA* L.) FLAVOR TYPES AND THEIR BIOSYNTHESIS

VIDYA HINGE,[1] RAHUL ZANAN,[2] DEO RASHMI,[3] and ALTAFHUSAIN NADAF[4]

[1]*Women Scientist (WOS-A), Department of Botany, Savitribai Phule Pune University, Pune–411007, India, Mobile: +91-7588611028, Tel.: +91-20-25601439, Fax: +91-20-25601439, E-mail: vidyahinge17@gmail.com*

[2]*Young Scientist (SERB), Department of Botany, Savitribai Phule Pune University, Pune–411007, India, Mobile: +91-9689934273, Tel.: +91-20-25601439, Fax: +91-20-25601439, E-mail: rahulzanan@gmail.com*

[3]*Research Scholar, Department of Botany, Savitribai Phule Pune University, Pune–411007, India, Mobile: +91-9960844746, Tel.: +91-20-25601439, Fax: +91-20-25601439, E-mail: deo.rashmi9@gmail.com*

[4]*Associate Professor, Department of Botany, Savitribai Phule Pune University, Pune–411007, India, Mobile: +91-7588269987, Tel.: +91-20-25601439, Fax: +91-20-25601439, E-mail: abnadaf@unipune.ac.in*

CONTENTS

8.1   INTRODUCTION

Rice (*Oryza sativa* L.) is the staple food for more than one half of world population, and an estimated 3.5 billion people worldwide consume rice (Verma and Srivastav, 2017). Ninety percent of the world rice crop is grown and consumed in Asia. Rice has a pivotal role in shaping the cultures, diets, and economies of thousands of millions of people in Asia. World rice production in the year 2013-14 was 476 million tons with consumption of 481 million MT and trading of 41 million MT in world market (FAO, 2015).

Around the world, many rice cultivars are grown, and they exhibit a diverse array of properties. *O. sativa* (Asian rice, grown worldwide) and *O. glaberrima* (African rice, grown in parts of West Africa) are two important species of rice grown as cereals for human nutrition. *O. sativa* consists of two major varietal groups, *indica* (Hsien) and *japonica* (Keng) (Chou, 1948; Ting, 1957). Amongst these, *indica is the* major rice type grown in India, Pakistan, Sri Lanka, Indonesia, central and southern China, Philippines, and in some African countries of the tropical and subtropical region. *Japonica* is a group of rice varieties cultivated extensively in northern and eastern China and in some areas of the world.

Based on the isozyme markers (Glaszmann, 1987), SSRs (Garris et al., 2005; Agrama et al., 2010), SNP (Caicedo et al., 2007), two major varietal groups were distinguished into five genetically distinct subpopulations like *indica* (Southern Asia, South China, Southeast Asia, South Pacific, Central Africa, and around the equator), *aus* (Bangladesh and India), *temperate japonica* (Europe and the North Pacific), *tropical japonica* (America and part of the South Pacific and Oceania), and *aromatic* (Bangladesh and India). Based on phylogenetic relationship, *japonica* varietal group comprises *temperate japonica*, *tropical japonica*, and *aromatic* whereas *indica* and *aus* subpopulations have a distinct ancestry and are recognized as members of the *Indica* varietal group (Kovach et al., 2009) (Figure 8.1). More than 40,000 varieties of rice are cultivated and differ with respect to size, texture, glutinous nature, aroma, and cooking quality (Vachhani et al., 1962). Among these five subpopulations, fragrant accessions were identified in *aromatic* (Basmati and Sadri), *indica* (Jasmine), and *tropical japonica*.

Worldwide consumer preference for rice is based on the appearance and aroma (Meullenet et al., 2001). Even small variations in sensory properties, particularly aroma, can increase consumer rejection of rice and its products. Therefore, aroma is one of the most important grain quality traits and a key factor in determining market price and is related to

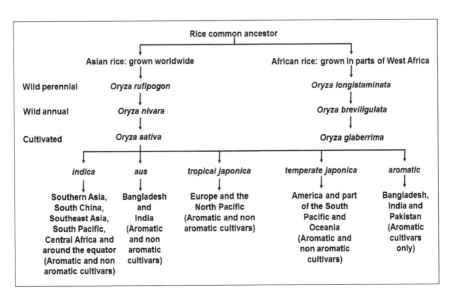

FIGURE 8.1 Evolutionary pathway and classification of the two cultivated rice cultivars (Khush, 2000; Kovach et al., 2009).

both local and global identity (Bhattacharjee et al., 2002; Fitzgerald et al., 2009; Verma et al., 2012, 2013, 2015). Aromatic rice contributes 15–18% of the rice trade and procures highest prices in the world market. The price of aromatic rice in the world trade market is USD $ 1,150/MT or more, whereas coarse rice is USD $440-580/MT (FAO, 2015).

Basmati and Jasmine are premium long grain fragrant rice cultivars preferred in the world market. Their characteristic fragrance in both the raw and cooked states is the major cause of their high value. Further associated quality attributes are the distinctive shape of the grain and highest elongation after cooking (Singh et al., 2000). Basmati rice has been cultivated traditionally since centuries in the north and north western part of the Indian sub-continent, whereas Jasmine rice is cultivated in north and north eastern Thailand.

There are several local short grain fragrant *indica* rice cultivars that are preferred in worldwide domestic market and fetch highest market price than non-fragrant ones. The fragrant rice varieties include Kalijira, Lousiana, red rice, black rice, Carolina rice, Jasmine, Arboria, Della rice, Texmati, and Wehani. Kalijira is a non-glutinous miniature scented rice from Bangladesh. Lousiana is a long grain rice having pecans, nutty, and rich aroma, grown only in the bayou country of southern Louisiana. Jasmine rice is an excellent white rice, which has a jasmine-type aroma after cooking and cooks to nice firm rice. It is just slightly sticky when compared with Basmati. Black rice has nutty flavor and is grown in Indonesia and the Philippines. Della is another variety of rice that was developed in America to mimic the Basmati grain. It is good rice, but kernels after cooking are not as long and slender as Basmati grains, because it swells both lengthwise and widthwise, like regular long grain rice. Similar to this, Texmati rice was created in the United States to mimic Basmati Rice. It is a good rice but has failed to have all the characteristics of Basmati. Wehani is long grained rice that is aromatic with a musky smoky nutty flavor. China Black rice is a non-glutinous rice having excellent texture, color and flavor. It has a nutty taste, soft texture, and beautiful rich deep purple color (Sarkarung et al., 2000).

The pleasant odor of different rice cultivars was controlled by a blend of various volatiles (Weber et al., 2000). More than 300 volatiles have been reported (Maga, 1984; Tsugita, 1985; Champagne, 2008), but to date, all volatiles from different rice cultivars are not collectively represented. Most of the volatiles are similar in aromatic and non-aromatic rice with their differed proportion. Among these volatiles, only a few/selected compounds are responsible, contributing in rice aroma and those are called as aroma active compounds (AACs). The AACs are varied on the basis of their contribution in particular rice flavor; hence, AACs may vary in different rice cultivars. Among all the fragrant rice cultivars 2-acetyl-1-pyrroline (2-AP) is correlated and confirmed as the major compound of rice aroma (Buttery et al., 1983), and it was found that a single recessive gene is responsible for its synthesis (Bradbury et al., 2005). The biosynthesis of other volatiles leading to aroma in rice remains unknown.

Researchers grouped rice cultivars into 5 types using SSR markers (Agrama et al., 2010). But, the relationship among the rice type with the aroma volatiles is not established. In the present investigation, all reported volatile compounds are listed and based on AACs, and the relationship among the rice type is established. The biosynthetic pathways of AACs and possible metabolic engineering for aroma improvement are discussed.

8.2 AROMA VOLATILES CHARACTERIZATION IN RICE CULTIVARS

In the early 1970s, researchers initiated to identify volatile compounds in rice (Mitsuda et al., 1968; Ayano and Furuhashi, 1970; Chikubu, 1970; Tanaka et al., 1978). This was followed by Bullard and Holguin (1977) who reported 73 volatiles in unprocessed rice of which 54 were newly reported in raw rice. Further, Yajima et al. (1978) reported 100 compounds (13 hydrocarbons, 13 alcohols, 16 aldehydes, 14 ketones, 14 acids, 8 esters, 5 phenols, 3 pyridines, 6 pyrazines, and 8 other compounds) in cooked rice. Subsequently, in 1979, they detected 114 compounds in cooked Kaorimai (Scented rice, *O. sativa japonica*) and α-pyrrolidinone was identified as a key odorant in Kaorimai.

In 1983, Buttery et al., reported that 2-AP is a key compound responsible for the rice aroma. 2-AP is mainly associated with the aroma flavor and present in all parts of the rice plant (stems, leaves, and grains) except roots (Buttery et al., 1983; Lorieux et al., 1996). Yoshihashi et al.

(2002) quantified 2-AP in various parts of the plant and under different conditions of aromatic rice. They confirmed that 2-AP is not synthesized during post-harvest or cooking processes; it is synthesized during development stage at paddy fields in aerial parts of plants. Buttery et al. (1986) analyzed different aromatic rice varieties for 2-AP content and confirmed the very high flavor contribution to all aromatic rice varieties. Paule and Powers (1989) and Ishitani and Fushimi (1994) found a positive correlation of 2-AP with the rice aroma. Further based on sensory evaluation, Nagaraju et al. (1991) classified aromatic rice varieties into traditional basmati (Basmati 370 and Pakistan Basmati), evolved basmati (Pusa basmati), and small-grained non-basmati (Badshabhog, Krishnabhog, Haldigudhi, and Marasingbhog).

Maga (1984) reviewed the reported volatile compounds in wild rice, scented rice, raw rice, cooked rice, rice bran, Soong-Neung, and rice cake. Hussain et al. (1987) reported 83 volatiles belonging to hydrocarbon (13), aldehydes (16), ketones (14), alcohols (13), esters (8), acids (14), and phenols (5) and compared volatile profile of aromatic Basmati with that of non-aromatic rice. Pentadecan-2-one, hexanol, and 2-pentylfuran were found in Basmati rice. More than 64 volatile compounds were identified in cooked California long grain rice. Based on odor activity values, 2-AP, (E,E)- deca-2,4-dienal, nonanal, hexanal, (E)-non-2-enal, octanal, decanal, **2-methoxy-4-vinylphenol**, and 4-vinylphenol are identified as a major aroma contributors (Buttery et al., 1988). Hernandez et al. (1989) identified 28 volatiles from rice seedlings (Mars and PI346833) in which hexanal, (E)-2-hexenal, (Z,Z)- and (E,E)-2,4-heptadienal were the major constituents in both the varieties (50% of total volatiles).

According to Mahatheeranont et al. (1995) 2-AP, butyl acetate, diethyl carbonate, butyl cyclopropane, 1,4-dimethylbenzene, isocyanato-methyl-benzene, hexanal, nonanal, 7-octen-4-ol, 2-(2-propoxyethoxy)ethanol and 2,6,-bis(1,1,-dimethyl-ethyl)-4-methylphenol are the important volatile compounds of uncooked KDML 105 brown rice. Petrov et al. (1996) identified 9 volatile compounds to distinguish fragrant and non-fragrant rice. In a comparative study, *n*-hexanal, (E)-2-heptenal, 1-octen-3-ol, *n*-nonanal, (E)-2-octenal, (E)-2, (E)-4-decadienal, 2-pentylfuran, 4-vinylguaiacol, and 4-vinylphenol were reported to be in higher amount in non-fragrant rice than in the four fragrant rice cultivars (Widjaja et al., 1996).

Buttery et al. (1999) obtained more than 60 volatiles from rice cakes, and amongst them, 1-hydroxy-2-propanone, 2, 5-dimethylpyrazine, pyrazine, 2-methylpyrazine, furfuryl alcohol, hexanal, furfural, pentanol, 3-hydroxy-2-butanone and ethyl-3,6-dimethylpyrazine are the major volatile compounds. 3-Methylbutanal, dimethyl trisulfide, 2-ethyl-3,5-dimethylpyrazine, 4-vinylguaiacol, hexanal, 2-methylbutanal, (E,E)-2,4-decadienal, 2-acetyl-1-pyrroline, 1-octen-3-one and 1-octen-3-ol were found to contribute to the aroma and flavor. Similarly, Tava and Bocchi (1999) identified and quantified several compounds in Basmati and B-53 rice samples belongs to hydrocarbons, aldehydes, alcohols, ketones, heterocyclic, terpenes, disulfides, and phenols. Mahatheeranont et al. (2001) identified more than 140 volatile compounds from uncooked Khao Dawk Mali 105 brown rice. Jezussek et al. (2002) identified 41 odor active aroma compounds in cooked brown rice of the Improved Malagkit Sungsong, Basmati 370, and Khaskhani by aroma extract dilution analyses (AEDA). Hien et al. (2006) analyzed aroma volatiles in several aromatic and non-aromatic rice varieties and reported that non-aromatic rice varieties did not show 2-AP content. Similarly, Nadaf et al. (2006) reported 2-AP in marketed basmati, Pusa basmati, and local scented cultivar Ghansal, whereas non-scented Krishnahansa and Pusa basmati lacked the presence of 2-AP.

Laguerre et al. (2007) developed "fingerprint" for 61 rice samples based on extracted volatile compounds detected using HS-SPME with GC-MS and soft independent modeling of class analogy (SIMCA) procedure. They also observed good discrimination between scented and non-scented rice samples. A total of 38 aroma volatiles belonging to aldehydes, ketones, alcohols, furan derivative, fatty acids, ester, phenolic compounds, hydrocarbons, etc. were identified during cooking in the Akitakomachi rice cultivar (Zeng et al., 2007). Ajarayasiri and Chaiseri (2008) observed 12 aroma-active compounds in the cooked black glutinous rice and 14 in white glutinous rice. Maraval et al. (2008) identified 60 compounds of cooked rice from scented (Aychade, Fidji) and nonscented (Ruille) cultivars grown in the Camargue area in France. Yang et al. (2008) analyzed aroma volatiles in six different rice types and indicated that guaicol, indole, p-xylene, and 2-AP influenced the difference in aroma among the different rice types. A total of 96 flavor volatile compounds of glutinous rice during cooking were observed in Tatsukomochi (86), Kinunohada (90), and

Miyakoganemochi (94) rice cultivars, belonging to aldehydes, alcohols, ketones, heterocyclic compounds, phenolic compounds, fatty acids, esters, hydrocarbons, etc. (Zeng et al., 2009). Sukhonthara et al. (2009) character-ized aroma volatile compounds from red and black rice bran and identified 129 volatile compounds in red and black rice bran. Nonanal, myristic acid, 6,10,14-trimethyl-2-pentadecanone, and (E)-β-ocimene were important compounds in red rice bran, whereas nonanal, myristic acid, pentadeca-nal, caproic acid, and pelargonic acid were major compounds in black rice bran. Mathure et al. (2011) identified and quantified rice aroma volatiles among 35 rice samples (33 aromatic and 2 non-aromatic) and reported highest amount of 2-AP in Indrayani Brand 2 (0.552 ppm), followed by Kamod (0.418 ppm), and Basmati Brand 5 (0.411 ppm). Along with 2-AP, nonanal, hexanal, decanal, vanillin, benzyl alcohol, guaiacol, and indole content were significantly varied between the rice types Basmati, Ambe-mohar, Kolam, and Indrayani. Similarly, Bryant and McClung (2011) ana-lyzed seven aromatic and two non-aromatic cultivars for their volatiles at before and after storage. Among 93 identified volatile compounds, 64 were first time identified in rice. Maximum compounds were observed in freshly harvested rice, and few new compounds being identified after storage. Subsequently, Piyachaiseth et al. (2011) identified 188 volatile compounds of flash-fried rice, 154 from stir-fried rice (SFR), and 57 of steamed rice (SR). Khorheh et al. (2011) identified 75, 55, and 66 com-pounds in Hashemi, HD5, and HD6 rice respectively, in which 58 unique compounds were detected for the first time. Cho (2012) analyzed vola-tiles from wild rice (*Zizania palustris* L.) and African rice species (*Oryza sativa* L., *Oryza glaberrima* Steud., and interspecific hybrids). A total of 73 volatile compounds were identified and quantified in cooked wild rice, in which 58 were newly reported and 34 are AACs. Mathure et al. (2014) quantitatively analyzed 23 volatiles from non-basmati scented (77), bas-mati (9), and non-scented (5) rice cultivars. 2-AP content was the maximum in non-basmati scented cultivars (15) compared to basmati cultivars, and a positive correlation of 2-AP with 1-tetradecene and indole was observed. However, 2-AP was correlated negatively with benzyl alcohol. Nonanal, octanal, decanal, and 1-octen-3-ol separates basmati cultivars from non-scented cultivars. Benzyl alcohol, 2-phenylethanol, 2-amino acetophe-none, indole, 1-hexanol, and nonanoic acid exhibited significant variation among rice categories. Mahattanatawee and Rouseff (2014) comparatively analyzed volatile flavors in cooked rice samples of Jasmine, Basmati, and Jasmati. A total of 26 aroma active volatiles in Jasmine, 23 in Basmati, and 22 in Jasmati were observed. Liyanaarachchi et al. (2014) observed aldehydes (hexanal, benzaldehyde, and octanal) and alcohols (2-butoxy ethanol, benzyl alcohol, octanol, phenol, 1-pentanol, alpha terpineol, and hexanol) in traditional aromatic rice varieties in Sri Lanka.

Rice aroma volatiles are very important for assuring whole grain or grain products quality. Because of importance of volatile composition in the rice flavor, a number of studies have been done for identifying the volatile compounds from different rice cultivars. All reported volatile compounds in different rice cultivars with their chemical class and odor descriptions are depicted in Table 8.1. The volatile analysis from reported aromatic and non-aromatic rice cultivars showed a total of 484 volatile compounds, which included 44 (9%) N-containing aromatic, 57 (12%) aliphatic hydrocarbons, 46 (10%) aromatic hydrocarbon, 24 (5%) non-aromatic cyclic hydrocarbon, 64 (13%) ketone, 49 (10%) aliphatic aldehyde, 12 (3%) aromatic aldehyde, 61 (13%) alcohol, 31 (6%) ester, 21 (4%) phenol containing, 13 (3%) sulfur containing, 6 (1%) Cl/N containing, 25 (5%) furans, and 31 (6%) carboxylic acid (Figure 8.2). Hydrocarbon-containing compounds (aliphatic, aromatic, and non-aromatic cyclic hydrocarbon) are the maximum contributors (127 compounds, 26%) of total volatile compounds, followed by ketone, alcohol, and aldehyde-containing compounds (13% each). Among the total aldehyde compounds (13%), aliphatic aldehyde (10%) are more in number than aromatic aldehyde (3%). Cl/N-containing compounds showed least number of compounds (1%). About 66% of volatile compounds (313 compounds) are belonging to the hydrocarbon, ketone, alcohol, and aldehyde groups.

8.3 AROMA ACTIVE VOLATILE COMPOUNDS CONTRIBUTING TO RICE FLAVOR

Though more than 480 volatile compounds have been identified in aromatic and non-aromatic rice, only 2-AP is characterized as a principle aroma compound in all aromatic rice cultivars (Buttery et al., 1983; Yoshihashi

TABLE 8.1 Aroma Volatiles Compounds in Rice Cultivars

Sr. No.	Compound name	Odor description	Reference
N-containing aromatic			
1	2-Acetyl-1-pyrroline	Cooked jasmine rice	a
2	2-Acetylpyrrole	Musty and nutty	b
3	N-FurfurylPyrrole	Earthy-green aroma	c
4	Pyrrole	Nutty type odor	d
5	Pyrolo[3,2-d]pyrimidin-2,4(1H,3H)-dione	Mild chlorine odor	c
6	5-Amino-3,6-dihydro-3-imino-1(2H) pyrazine acetonitrile	-	c
7	Pyridine	Sour, putrid	d
8	2-Methylpyridine	Sweat, astringent	d
9	4-Methylpyridine	Sweetish odor	e
10	3-Methylpyridine	Strong odor	e
11	3-Vinylpyridine	Pungent, unpleasant	e
12	1H-Indole	Floral, sweet, burnt	e
13	Pyrazine	Musty, nutty, peanut odor	e
14	Ethylpyrazine	Nutty type odor	d
15	Ethenylpyrazine	Nutty, green flavor	d
16	2-Methylpyrazine	Nutty, roasted,	e
17	3-ethyl-2,5-dimethylpyrazine	Cooked rice, nutty	e
18	2,3-dimethyl-5-ethylpyrazine	Nutty, pleasant	e
19	2-ethyl-3,5-dimethylpyrazine	Nutty, roasted	e
20	2-ethyl-6-methylpyrazine	Almond	e
21	2-ethyl-5-methylpyrazine	Coffee type odo	e
22	2-isoamyl-6-methylpyrazine	Grassy alcoholic, sweet	e
23	2-ethyl-3-methylpyrazine	Green, nutty, cocoa, musty	e
24	2,3-Dimethylpyrazine	Nutty, almond	e
25	2,5-Dimethylpyrazine	Nutty, roasted	e
26	2,3,5-Trimethylpyrazine	Nutty, musty type flavor	e
27	α-pyrrolidinone	Mild, fishy odor	e
28	β-Quinoline	Medicinal, earthy	e

TABLE 8.1   (Continued)

Sr. No.	Compound name	Odor description	Reference
29	Nicotine	Fishy odor	e
30	N,N-dimethyl chloestan-7-amine	-	c
31	2-Butyl-1,2-azaborolidine	-	c
32	7-Chloro-4-hydroxyquinoline	Phenolic odor	c
33	2-Acetopyridine	Popcorn type odor	f
34	2 amino acetophenone	Fruity, grape-like	g
35	Pyrolo[3,2-d]pyrimidin-2,4(1H,3H)-dione	Nutty	c
36	1-(1Hpyrrol-2-yl)-ethanone	Earthy	c
37	2-Acetyl-2-thiazoline	Cooked jasmine rice	h
38	Benzothiazole	Meaty type	i
39	Benzonitril	Sweet almond	l
40	2-isobutyl-3-methoxypyrazine	Earthy,bell pepper	g
41	3-methylindole	Mothball-like	j
42	2,6-Dimethylpyrazine	Chocolate, nuty,	j
43	Isocyanato-methylbenzene	Sharp fruity odor	k
44	2-methoxy-3,5-dimethylpyrazine	Musty, moldy aroma	g
Alkane			
1	Nonane	Gasoline-like odor	e
2	Decane	Gasoline odor	e
3	Dodecane	Gasoline like	m
4	Tetradecane	Gasoline-like	e
5	Tridecane	Gasoline-like to odorless	n
6	Pentadecane	Mild odor	e
7	Heptadecane	-	e
8	Hexadecane	Gasoline-like	e
9	Nonadecane	Sweet, deep-rosy,	e
10	Octadecane	White,	e
11	Geranyl acetate	Green waxy oily fatty	o
12	Ethyl acetate	Fruity odor	e
13	Butyl acetate	Ethereal type odor	e
14	2,6,10,14-tetramethyl-Hexadecane	Odorless	p
15	6-Methyl-octadecane	-	c

TABLE 8.1 (Continued)

Sr. No.	Compound name	Odor description	Reference
16	2-Methyl decane	Pungent acrid odor	c
17	3-Methyl-Tetradecane	-	q
18	2- methyl- Tridecane	-	p
19	3-Methyl-Pentadecane	-	q
20	Tritetracontane	Odorles	c
21	3,5,24-Trimethyl tetracontane	-	c
22	2,6,10-Trimethylpentadecane	Paraffinic	q
23	9- Methyl- Nonadecane	-	p
24	4-Methyldecane	Pungent, acrid odor	r
25	2-Methylheptane	Odorless	s
26	4-Acetoxypentadecane	-	c
27	Pentacosane	Strong odor	e
28	Tetracosane	Oligosulfide odor	t
29	Docosane	Waxy type odor	p
30	Dotriacontane	Odorless	p
31	Tricosane	Waxy type odor	p
32	Heneicosane	Waxy type odor	p
33	n-Eicosane	Waxy type odor	e
Alkene			
1	1-Butene	Slightly aromatic	u
2	1-Tetradecene	Mild pleasant odor	u
3	2-tetradecene	-	u
4	5-Octadecene	Mild hydrocarbon odor	u
5	7-Tetradecene	-	c
6	3-ethyl-1,5-octadiene	-	f
7	3-Tridecene	Mild pleasant odor	c
8	Dodecene	Pleasant odor	v
9	thujopsene	-	f
10	(E)-5-Methyl-4-decene	-	r
11	((Z)-3-Undecene	Hydrocarbon odor	r
12	1-Hexacosene	-	c
13	Ethylene	Sweet and musky	w
14	3,5-Dimethyl-1-hexene	Slightly sweet odor	c

TABLE 8.1 (Continued)

Sr. No.	Compound name	Odor description	Reference
15	Nonene	Gasoilne	c
16	17-Pentatricontene	Pineapple odor	c
17	1-Docosene	-	c
18	Butyl cyclopropane	Pleasant or desired	k
19	7-Methyl-2-decene	-	x
20	α-cadinene	Woody type odor	x
21	δ-Cadinene	Woody type odor	x
Alkyne			
1	4Methyl-2-pentyne	Pleasant odor	c
2	1-Tetradecyne	-	c
3	Octadecyne	-	c
Aliphatic Aldehyde			
1	ethanal	Ethereal, and an pungent	y
2	Formaldehyde	Pungent, suffocating odor	y
3	Butanal	Pungent, aldehyde odor	y
4	Propanal	Fruity odour	y
5	2-Methylpropanal	Wet cereal or straw	y
6	Isovaleraldehyde	Fruity, Almond, Toasted	z
7	2-Methylbutanal	Chocolate type odor	y
8	2-Methylpentanal	Ethereal,.fruity.odor	y
9	Hexanal	Green	y
10	2-Hexenal	Sweet, fruity, fresh	y
11	2,4-Hexadienal	Green type	y
12	Octanal	Citrusy	y
13	Heptanal	Grass, fresh	y
14	2-Heptenal	Fatty green	y
15	2,4-Heptadienal	Citrus, grass	y
16	Pentanal	Strong, Acrid	y
17	Nonanal	Grassy, citrus, floral	y
18	2-Pentenal	Green type odor	w

TABLE 8.1 (Continued)

Sr. No.	Compound name	Odor description	Reference
19	2-methyl-2-pentenal	Fruity, cooked rice,	y
20	(E)-2-Octenal	Green, nutty	c
21	2- Butyl-2- Octenal	Fatty type odor	c
22	2-Butyl-1-octenal	Fatty type odor	c
23	2,4-Octadienal	Fatty type odor	c
24	2-undecenal	-	p
25	Citral	Lemon odor	e
26	β-Cyclocitral	Minty type odor	r
27	β-homocyclocitral	Camphoreous odor	r
28	Decanal	Fatty, citrusy	y
29	(E,Z)-2,6-Nonadienal	Green, metallic	g
30	(Z,E)-2,4-heptadienal	Fatty type odor	y
31	(E)-2-Decenal	Green, herbal	y
32	Undecanal	Soapy, floral, citrus	e
33	(E)-2-Undecenal	Fatty, sweet	q
34	2-Dodecenal	Aldehydic type odor	y
35	Dodecenal	Minty, soapy	x
36	2-Ethylhexanal	Mild odor	s
37	(E)-2-Nonenal	Metallic	y
38	(E,E)-2,4-Nonadienal	Fatty, metallic	y
39	Tetradecanal	Sweet rose odor	p
40	Pentadecanal	Waxy type odor	c
41	(E,E)-2,4-Decadienal	Fatty, metallic, citrus	y
42	(E,Z)-deca-2,4-dienal	Fatty green	g
43	2-Methyl-hexadecanal	-	c
44	4,5-epoxy-(E)-dec-2-enal	Metallic	g
44	(4E)-4-Nonenal	-	x
45	(Z)-hex-3-enal	Green leaf like	g
46	(Z)-non-2-enal	Fatty, green	g
47	(Z)-dec-2-enal	Fatty green	g
48	Dodecanal	Citrus type odor	x
49	Tridecanal	Fatty-waxy, Citrus-like	x

TABLE 8.1   (Continued)

Sr. No.	Compound name	Odor description	Reference
Aromatic Aldehyde			
1	m-Tolualdehyde	Sweet aromatic, almond	s
2	Benzaldehyde	Nutty,sweet	y
3	Phenylacetaldehyde	Herbal, floral	y
4	Safranal	Herbal, Woody type odor	aa
5	2-Ethylbenzaldehyde	Roasted, nutty	y
6	2-hydroxybenzaldehyde	Medicinal and an spicy	f
7	p -methylbenzaldehyde	Cherry-like scent	y
8	Cinnamaldehyde	Spicy type odor	d
9	para-Methylbenzaldehyde	Cherry-like scent	y
10	α-Hexylcinnamic aldehyde	-	y
11	3,5-Di-tert-butyl-4-hydroxybenzaldehyde	-	y
12	Myrtenal	Spicy type odor	f
Ketone			
1	Acetone	Fruity, nail polish remover	y
2	Geranyl acetone	Magnolia, green	p
3	Neryl acetone	Fatty type odor	p
4	2-Butanone	Pleasant, Pungent	y
5	Diphenylketone	Rose-like odor	s
6	2-Pentadecanone	Floral, Fatty type	z
7	3-Methyl-3-buten-2-one	Herbal type odor	z
8	Nonadecanone	-	z
9	2-Heptadecanone	Floral type odor	z
10	6,10,14-Trimethyl-2-pentadecanone	-	ab
11	4-Cyclopentylidene-2-butanone	Woody and fruity	r
12	5-Ethyl-6-methyl-3-hepten-2-one	Citrus type odor	i
13	Pentan-2-one	Fruity, sweet banana-like	y
14	3-Penten-2-one	Sharp, fishy	y
15	2-Hexanone	Acetone like odor	y

TABLE 8.1   (Continued)

Sr. No.	Compound name	Odor description	Reference
16	Cyclohexanone	Peppermint or acetone-like	e
17	2-Octanone	Dairy, fruity, cooked meat	y
18	2-Nonanone	Green, weedy, herbal	y
19	3-nonanone	Spicy, sweet	f
20	6,10-Dimethyl-5,9 undecadien-2-one	Floral type flavor	c
21	2,2,6-Trimethylcyclohexanone	Citrus, honey, pungent	r
22	Carvone	Caraway or dill	e
23	2-Tridécanone	-	e
24	2-Decanone	Mild, orange blossom odor	y
25	Undecalactone	-	e
26	2,6,6-Trimethyl-2-cyclohexene-1,4-dione	Musty, and an citrus type	c
27	6,10-Dimethyl-2-undecanone	-	x
28	2-Undecanone	Floral and fatty pineapple	y
29	2-Dodecanone	Citrus type, fatty typ	y
30	Methylheptenone	Pungent, Citrus, herbal	e
31	1-Octen-3-one	Mushroom	h
32	β-Damascone	Sweet honey	o
33	Farnesyl Acetone	Weak, Floral, Creamy,	e
34	3,5-octadien-2-one	Fruity	ac
35	3-Nonen-2-one	Fruity type odor	x
36	7,9-Di-tert-butyl-1-oxaspiro-(4,5) deca-6,9-diene-2,8-dione	-	c
37	1-(2-methylphenyl)-ethanone	Fruity type odor	f
38	α/ β -Ionone	Raspberry , floral	g
39	2-Heptanone	Fruit,spicy	y
40	p-Menthan-3-one	-	aa
41	6-Methyl-2-heptanone	Camphoreous	e
42	6-Methyl-5-hepten-2-one	Herby, green	y

TABLE 8.1 (Continued)

Sr. No.	Compound name	Odor description	Reference
43	b-Damascenone	Sweet, honey	o
44	(3E)-3-Octen-2-one	Citrus, herbal, floral	aa
45	(3E)-6-Methyl-3,5-heptadien-2-one	Spicy, green type	j
46	Isophorone	Peppermint-like odor	z
47	5-Methy-3-hepten-2-one	Citrus, green type	c
48	6,10-Dimethyl-5,9 undecadien-2-one	Floral type flavor	c
49	2,6-Bis(1,1-dimethylethyl)-2,5-cyclo-hexadiene-1,4-dione	-	c
50	5-Methyl-2-hexanone	Pleasant fruity odor	c
51	1-hydroxy-2-propanone	Caramellic, sweety	h
52	3-hydroxy-2-butanone	Buttery-oily, creamy-fatty	h
53	butan-2,3-dione	Buttery	g
54	Nonalactone	Coconut-like	g
55	2-Tridecanone	Oily, nutty	x
56	3-Octanone	Herbal, mushroom	j
57	γ-Octalactone	-	g
58	non-1-en-3-one	Mushroom-like	g
59	2-Methyl-3-octanone	-	g
60	(E)-3-Nonen-2-one	Powdery, mushroom	x
61	2-Methyl-5-decanone	Acidic, fruity	x
62	6-Dodecanone	Citrus, Fatty	x
63	2,3-Octanedione	Buttery flavor	x
64	3,5,5-Trimethyl-2-cyclopenten-1-one	Sweet type odor	f
Sulfur-containing			
1	Hydrogen sulfide	Foul odor of rotten eggs	j
2	Dimethyl sulfide	Cooked, sulphury	i
3	2-Methyl-3-furanthiol	Meaty	g
4	Dimethyl trisulfide	Sulphury, cabbage	h
5	Methional	Cooked potato	g
6	Prenylthiol	Nutty, sulfury	o
7	Dimethyl disulfide	Sulfurous type odor	i
8	1-Propene-1-thiol	Sulphurous type odor	s

TABLE 8.1   (Continued)

Sr. No.	Compound name	Odor description	Reference
9	Butanethiol	-	o
10	3-Dimethyl-2-(4-chlorphenyl)-thioac-rylamide	-	c
11	Methanethiol	Rotten cabbage or eggs	o
12	[(1-Methylethyl)thio] cyclohexane	-	c
13	3-methyl-2-butene-1-thiol	Thiol nutty, sulphury	o
Chlorine containing			
1	1-Chloro-3-methyl butane	-	c
2	1-Chloro-nonadecane	-	c
3	2-Chloro-3-methyl-1-phenyl-1-buta-none	-	c
Nitrogen containing			
1	1-Nitrohexane	-	q
2	Propiolonitrile	ether-like odor	c
3	O-decylhydroxamine	-	c
Alcohol			
1	Ethanol	Pleasant smell	y
2	1-Nonanol	Rose or fruity odor	l
3	1-Butanol	Banana-like, sweet	z
4	2-Butanol	Fruity type	ad
5	2-Ethylhexanol	Citrus, fatty type	e
6	Isobutanal	Sweet, musty	y
7	1-Propanol	Mild, alcohol-like	y
8	Methanol	Repulsive pungent odor	l
9	Dodecanol	Waxy type odor	e
10	Pentanol	Moderately strong odor	y
11	Isopentanol	Mild, choking alcohol odor	y
12	2-Ethyl-3-buten-1-ol	-	c
13	(E)-2-Pentenol	Fermented type odor	c
14	Hexanol	Green	ad
15	1-Heptanol	Green, solvent	ab

TABLE 8.1   (Continued)

Sr. No.	Compound name	Odor description	Reference
16	1-Hexanol, 2-ethyl	Heavy, earthy	e
17	2 Nonen-1-ol	Green type odor	c
18	Carveol	Spicy type odor	r
19	2,3-Butanediol	Buttery, creamy	h
20	α- Terpineol	Tarry, cold pipe like	e
21	2-Penten-1-ol	Earthy type odor	c
22	3,4-Dimethylcyclohexanol	-	r
23	1-Hexadecanol	-	ab
24	2-Hexadecanol	-	c
25	1,2-Propanediol	-	u
26	2-Decen-1-ol	Waxy type odor	c
27	Undecanol	Floral citrus like odor,	x
28	2,4-Hexadien-1-ol	Green type	y
29	3,7-Dimethyl-1-octanol	Fatty type flavor	r
30	1-Hepten-3-ol	Green type odor	t
31	2-Hexyl-1-octanol	-	c
32	2-Butyl-1-octanol	Sweet odor	c
33	2-Methyl-1-hexadecanol	-	c
34	2-Heptanol	Citrus, fruity type	e
35	1-octen-3-ol	Straw, earthy, mushroom	ae
36	1-Octanol	Fatty, metallic	l
37	(E)-2-HEXENOL	Green, Green	e
38	2-Butoxyethanol	Fruity odor	t
39	n-Nonadecanol	-	c
40	2-Hexyl-1-octanol	Waxy,	c
41	(3Z)-3-Hexen-1-ol	Green type odor	c
42	2-Ethyl-4-methyl-1-pentanol	-	c
43	3, 4-Dimethylcyclohexanol	-	r
44	Z-10-pentadecen-1-ol	Waxy type odor	c
45	Linalool	Sweet, floral, petitgrain-like	e
46	trans-linalool oxide	Floral, Green flavor	e

TABLE 8.1 (Continued)

Sr. No.	Compound name	Odor description	Reference
47	cis-linalool oxide	Sweet, floral, creamy	e
48	Menthol	Peppermint odor	e
49	cis-2-tert-butyl-Cyclohexanolacetate	-	p
50	Anizole	Phenolic, Gasoline	p
51	Cyclodecanol	Camphor-like odor	c
52	2-Hexyl-1-decanol	Odor free	c
53	2-Hexyl-decanol	Mild sweet odor	c
54	(2Z)-2-Octen-1-ol	Green type odor	x
55	7-Octen-4-ol	-	k
56	2-(2-Propoxyethoxy)ethanol	Mild pleasant odor	k
57	2-Butoxyéthanol	Fruity odor	af
58	(E)-2-Octen-1-ol	Green type odor	x
59	Cubenol	Spicy type odor	x
60	δ-Cadinol	Balsamic type odor	x
61	L-α-terpineol	Citrus-woody profile	f
Phenol-containing			
1	Phenol	Sweet and tarry	z
2	Biphenyl	Pleasant	d
3	4-Vinylphenol	Clove, Phenolic,	e
4	2-methylphenol	Medicinal odor	z
5	2,6-Dimethoxy-phenol	Smoky type odor	p
6	4 vinylguaicol	Spicy,fruity	z
7	Guaiacol	Smoky	z
8	Vanillin	Vanilla-like	g
9	2,6-Dimethylaniline	-	p
10	Toluene	Sweet, Pungent, benzene-	y
11	2-Phenylethanol	Honey-like	z
12	2-Phenoxyethanol	Pleasant odor	r
13	Benzyl alcohol	Floral	e
14	5-Ethyl-4-methyl-2-phenyl-1,3-dioxane	Powerful, floral, indole-like	c
15	acetophenone	Sweet, almond odor	z
16	2,2-Dihydroxy-1-phenyl-ethanone	Green flavor	c

TABLE 8.1 (Continued)

Sr. No.	Compound name	Odor description	Reference
17	2,2-Dihydroxy-1-phenylethanone	-	c
18	Isoeugenol	Spicy type odor	q
19	p-Cresol	Phenolic	j
20	m-Cresol	Phenolic	g
21	Butylatedhydroxytoluene	Slight, phenolic	x
Furans			
1	2,5-Dimethylfuran	Meaty type odor	y
2	2-Butylfuran	Chocolate odor	y
3	2-N-Pentylfuran	Licorice, beany	y
4	2-Acetylfuran	Pleasant aromatic	y
5	2-Methyl-5-propionylfuran	Almond, Caramel,	h
6	2,5-dimethyl-3(2H)- furanone	Caramellic type odor	h
7	1-Benzofuran	Gray, Phenolic Sweet	l
8	5-methyl-2-furancarboxaldehyde	Caramellic type odor	v
9	2-Ethylfuran	Rubbery, Pungent, Acid,	l
10	2-Propylfuran	-	s
11	furfural	Almond-like	j
12	5-Methyl-2-furaldehyde	Caramellic, Sweet	d
13	γ-Nonalactone	Sweet coconut-like	e
14	(±)-γ-Decanolactone	Fatty, coconut	e
15	5-methyl-2-furanmethanol	-	f
16	furfuryl alcohol	Floral	h
17	3-hydroxy-4,5-dimethyl-2(5H)-fura-none	Seasoning-like	g
18	2-Hexylfuran	-	x
19	4-hydroxy-2,5-dimethyl-3(2H)-fura-none	Caramel like	g
20	bis-(2-methyl-3-furyl)-disulfide	Meaty	g
21	3a,4,5,7a-tetrahydro-3,6-dimethyl-2(3H)-benzofuranone	Sweat Spicy	g
22	5-Ethyl-3-hydroxy-4-methyl-2(5H)-furanone	Seasoning-like	g
23	2-Methyl-3-furanthiol	Sulfurous type odor	g
24	2-Heptylfuran	Green type odor	x

TABLE 8.1 (Continued)

Sr. No.	Compound name	Odor description	Reference
25	5-iso-Propyl-5H-furan-2-one	Cooling, minty	x
Non-aromatic Cyclic Hydrocarbon			
1	4-Cyclohexyl-dodecane	Detergent-like odor	c
2	Heptylcyclohexane	Pungent fruity green	c
3	n-Heptadecylcyclohexane	-	c
4	n-Octyl-Cyclohexane	-	p
5	Cyclosativene	-	q
6	Germacrene-D	Woody type odor	u
7	Azulene	-	v
Monoterpenes			
8	p-Cymene	Terpenic type odor	y
9	α-Pinene	Herbal type odor	y
10	d-limonene	Lemon	z
11	L- Limonene	Fresh, sweet	z
12	Camphene	Woody camphoreous	p
13	α-Phellandrene	Terpenic type odor	s
14	Menthone	Minty type odor	e
15	Eucalyptol	Fresh, camphora-ceous	f
Sesquiterpene			
16	Valencen	-	r
17	α-Farnesene	Woody type odor	u
18	B-Elemene	Fruity, Dry	ag
19	Isolongifolene	Fresh woody	i
20	Longifolene	vegetal/flowery	ah
21	B-Caryophyllene	Spicy type odor	u
22	Aromadendrene	Woody type odor	r
23	Turmerone	Spicy, earthy, warm-woody	q
24	(E,E)-Farnesol	Floral type odor	ag
Aromatic Hydrocarbon		-	
1	Indene	-	v
2	O-Xylene	Strong, Sweetish like	y
3	m-Xylene	-	y

TABLE 8.1 (Continued)

Sr. No.	Compound name	Odor description	Reference
4	p-Xylene	-	y
5	Naphthalene	Naphthalene	y
6	1-Methoxy-naphthalene	-	p
7	1,4-Dimethylnaphthalene	-	p
8	2,6-Dimethylnaphthalene	-	q
9	1- Ethyl- Naphthalene	Gasoline, Sweet aroma	p
10	2,3,5-Trimethylnaphthalene	-	p
11	2-acetylnaphthalene	Berry; blossom;	p
12	2-Methylnaphthalene	Floral type odor	y
13	1-Methylnaphtalene	Naphthyl type odor	y
14	β-Bisabolene	Balsamic odor	p
15	Ethylbenzene	Gasoline-like odor	y
16	2,6-Bis(1,1-dimethylethyl)-2,5-cyclo-hexadiene-1,4-dione	-	y
17	4-Ethyltoluene	Pleasant odor	y
18	Benzene	Aromatic, gasoline	y
19	Decyl benzene	weak oily odor	c
20	Carene	-	y
21	n-Propylbenzene	-	y
22	1,2,4-Trimethylbenzene	Aromatic odor	y
23	1,3,5-Trimethylbenzene	-	y
24	O-Diethylbenzene	Aromatic odor	y
25	P-Diethylbenzene	Aromatic odor	y
26	m-Diethylbenzene	Aromatic odor	y
27	1,4-Diethylbenzene	Aromatic-smelling	y
28	1,4-dimethyl-2-ethylbenzene	-	y
29	1,3-Dimethyl-4-ethylbenzene	-	y
30	1,3-Dimethoxybenzene	-	q
31	1,3-dimethyl-5-ethylbenzene	-	y
32	1-Ethyl-3-methylbenzene	-	s
33	1,2-Dimethyl-3-ethylbenzene	-	y
34	1,2,3-Trimethylbenzene	Aromatic odor	y
35	1,2,4-Trimethyl benzene	-	p

TABLE 8.1   (Continued)

Sr. No.	Compound name	Odor description	Reference
36	1,2,3,4-Tetramethylbenzene	Sweet odor	y
37	1,2,3,5-tetramethylbenzene	Camphor-like odor.	y
38	1,2,4,5-Tetramethylbenzene	Camphor-like odor	y
39	1-methylnaphtalene	Metallic odor	p
40	2-Ethylnaphthalene	-	y
41	1-Chloro-3,5-bis(1,1-dimethylethyl)2-(2-propenyloxy)benzene	-	y
42	2-Acetyl-naphthalene	-	i
43	2,3,6-Trimethylnaphthalene	Licorice type odor	p
45	(1S,Z)-Calamenene	Herbal type odor	x
46	Styrene	-	f
Ester			
1	Methyl 2-furoate	Fungal type odor	p
2	Octylformate	Fruity	p
3	Hexadecanoic acid, methyl ester	Waxy type odor	z
4	Benzenebutyrate	Fruity type odor	z
5	2-hydroxy, 2-methyl- Benzoic acid	Acidic, Fruity	p
6	Methylsalicylate	Minty type odor	ai
7	Isopropyldodecanoate	Green type odor	s
8	Ethyl Hexadecanoate	Waxy type odor	i
9	Ethyl hexanoate	Fruity, Green, Pine-apple	e
10	Methyl benzoate	Pleasant smell	e
11	ethyl benzoate	Sweet, fruity, me-dicinal	z
12	Benzyl acetate	Flowery	aj
13	Ethyl linoleate	-	z
14	Ethyl oleate	Fatty type odor	z
15	Methyl linoleate	Waxy type odor	e
16	Methyl Oleate	-	e
17	Isobutyl nonyl ester oxalic acid	-	c
18	Methyl stearate	Waxy type odor	z
19	Octadecanoic acid, ethyl ester	-	c
20	Isobutyl hexadecyl ester oxalic acid	-	c

TABLE 8.1 (Continued)

Sr. No.	Compound name	Odor description	Reference
21	Hexylpentadecyl ester-sulphurous acid	-	c
22	Dibutyl phthalate	Bland type odor	e
23	Diethyl phthalate	Slight aromatic	z
24	Ethyl palmitate	Waxy type odor	e
25	(Z)-9-Octadecenoic Acid Ethyl Ester	Mild hydrocarbon	c
26	Methyl palmitate	Waxy type odor	z
27	Isobutyl salicylate	Sweet, green, floral,	p
28	Diethyl carbonate	Mild pleasant odor	k
29	2,4,4-Trimethylpentan-1,3-dioldi-iso-butanoate	-	x
30	3-Hydroxy-2,4,4-trimethylpentyliso-butanoate	-	x
31	Diisobutyladipate	Estery type odor	x
Carboxylic acid			
1	Icosanoic acid	-	e
2	Nonadecylic acid	-	e
3	Linoleic acid	-	z
4	Oleic acid	-	v
5	2,2,4-Trimethyl-3-carboxyisopropyl, isobutyl ester pentanoic acid	-	c
6	Hexadecyl ester, 2,6-difluro-3-methyl benzoic acid	-	c
7	Acetic acid	Sour	y
8	Stearic acid	Pungent odor	z
9	Tridecanoic acid	Waxy type odor	z
10	Lauric acid	Fatty type odor	e
11	Pentadecanoic acid	Waxy type odor	e
12	(2E)-2-Hexadecenoic acid	Waxy type flavor	i
13	Succinic acid	Odorless	r
14	1,2- Benzenedicarboxylicacid	Slight aromatic,	p
15	Decanoic acid	Strong, rancid	e
16	Furoic acid	Distinct odor	e
17	Nonanoic acid	Waxy type odor	e
18	n-heptanoic acid	Cheesy type odor	e
19	1-Hexanoic acid	Fatty type odor	j

TABLE 8.1 (Continued)

Sr. No.	Compound name	Odor description	Reference
20	Myristic acid	Waxy type odor	z
21	Benzoic acid	Faint, pleasant odor	z
22	Ethyl myristate	Waxy type odor	z
24	Caprylic acid	Fatty type odor	j
25	Hexadecanoic acid	Waxy type odor	e
26	2-Ethylcaproic acid	Mild odor	e
27	Butanoic acid	Sweaty, rancid	g
28	Pentanoic acid	Sweaty	g
29	3-Methylbutanoic acid	Cheese-like, sweaty	g
30	2-Methylbutanoic acid	Cheese-like, sweaty	g
31	Phenylacetic acid	Honey-like	g

Sources: a: Buttery, et al. (1983), b: Laguerre, et al. (2007), c: Bryant and McClung, (2011), d: Withycombe, et al. (1978), e: Yajima, et al. (1979), f: Cho, (2012), g: Jezussek, et al. (2002), h: Buttery, et al. (1999), i: Grimm, et al. (2011), j: Buttery, et al. (1988), k: Mahatheeranont, et al. (1995), l: Tsugita, (1985), m: Legendre, et al. (1978), n: Suvarnalatha, et al. (1994), o: Mahattanatawee and Rouseff, (2014), p: Givianrad, (2012), q: Khorheh, et al. (2011), r: Hinge, et al. (2015), s: Calingacion, et al. (2015), t: Mahatheeranont, et al. (2001), u: Ajarayasiri and Chaiseri, (2008), v: Piyachaiseth, et al. (2011), w: Tsuzuki, et al. (1977), x: Zeng, et al. (2009), y: Bullard and Holguin, (1977), z: Yajima, et al. (1978), aa: Yang, et al. (2008), ab: Petrov, et al. (1996), ac: Sukhonthara, et al. (2009), ad: Mitsuda, et al. (1968), ae: Widjaja, et al. (1996), af: Liyanaarachchi, et al. (2014), ag: Cheng, et al. (2007), ah: Maraval, et al. (2008), ai: Wilkie and Wootton, (2004), aj: Endo, et al. (2007).

et al., 2002). However, only a few volatiles have been contributed to the characteristic of rice aroma (Yajima et al., 1978; Buttery et al., 1982, 1988; Widjaja et al., 1996; Jezussek et al., 2002). The volatiles contributing to the flavor type have been characterized by their odor units. Charm analysis and aroma extract dilution analysis (AEDA) with gas chromatography-olfactometry (GC-O), OSME, surface of nasal impact frequency (SNIF), and odor active value (OAV) are used for determining the contribution of volatile compound in the flavor (Acree, 1997).

Charm analysis and AEDA involved smelling a serial dilution of the aroma sample and recording the presence or absence of individual odorants (Blank, 1997). In charm analysis, dilution value over the entire time of elution of compound from the column was measured, whereas in AEDA, the maximum dilution value was detected, and data are expressed as a

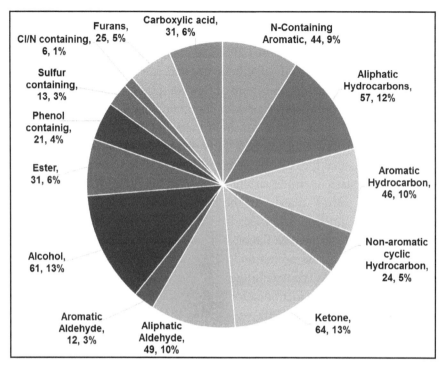

FIGURE 8.2 Classification volatile compounds from reported rice cultivars.

flavor dilution factor (FD factor). The FD factor is calculated as the ratio of odorant concentration in the initial extract to its concentration in the most diluted extract in which the odor was detected by GC-O. AEDA is simple and requires extraction into solvent and dilution of the sample. It is time-consuming, and only one panelist can smell the sample at a time; multiple assessments of each odorant assure relatively precise results (Van Ruth, 2001). In some analyses, non-diluted isolates were assessed by sniffing and based on relative sensory intensity, and the importance of odorant was measured as the overall aroma. Several panelists need to evaluate the odor intensity (Grosch, 2001). In the SNIF method, those odorants were taken that were detected by the maximum number of panelists. Eight to ten untrained panelists are required for evaluation of this method (Reineccius, 2006).

Flavor is complex mixture of compounds; hence, there are limitations of explaining flavor based on OAV. An odor unit or odor active value is the ratio of concentration of the individual compound to its odor threshold. Contribution of individual volatile compound can be determined for the characteristic aroma odor threshold of that compound. The odor threshold is the minimum concentration of odor compound that is perceivable by the human sense of smell. Odor threshold value (OTV) or aroma threshold value (ATV) or flavor threshold (FT) is defined as the lowest concentration of a substance that can be detected by a human nose and expressed as a concentration in water or air (Rothe, 1976). Marsili (2007) evaluated different sample matrix for the determination of sensory threshold. The volatile compound that exceeds OAV than 1 are said to be OAC and contributing into aroma or flavor. Those compounds having high OAV contribute more to the aroma and are important in defining flavor. The identification of OAC contributing rice flavor based on OAV is a relatively easier method for the identification of aroma contributing volatile using HS-SPME-GCMS or GC-FID or HS-GC-O. Odor thresholds values for reported rice aroma active volatile are listed in Table 8.2. Odor threshold values for major rice aroma volatiles in water were reported by Buttery et al. (1988, 1999) and in air by Yang et al. (2008).

Buttery et al. (1988) determined odor thresholds of 64 known rice volatiles in water. Based on OAC from cooked California long grain rice (variety L-202), nine aroma volatiles (2-acetyl-1-pyrroline, (E,E)-2,4-decadienal, nonanal, hexanal, (E)-2-nonenal, octanal, decanal, 4-vinylguaiacol, and 4-vinylphenol) were reported as probable major contributors to the flavor. In a further study, they identified and quantified more than 60 volatiles from rice cakes, and based on concentration/threshold ratios, 10 compounds (3-methylbutanal, 2-ethyl-3,5-dimethylpyrazine, dimethyl trisulfide, 4-vinylguaiacol, hexanal, 2-methylbutanal, (E,E)-2,4-decadienal, 2-acetyl-1-pyrroline, 1-octen-3-one, and 1-octen-3-ol) were found to contribute to the aroma (Buttery et al., 1999).

Similarly, Jezussek et al. (2002) identified 41 odor active compounds (OACs) from Improved Malagkit Sungsong, Basmati 370, Khaskhani, and indica variety using the FD factor. A total of 12 and 14 volatile compounds were identified from cooked black and white glutinous rice, respectively, in which butanoic acid, indole, 4-vinyl-2-methoxyphenol, and 1-tetradecene

TABLE 8.2 Odor Active Compounds from Different Rice Types

S. N.	Name of Compound	Odor type	OTV[c]	Rice type
N-containing aromatic				
1	2-Acetyl-1-pyrroline	Popcorn-like	0.01 ppb[a]	*ind*[2,6,7] *aro*[6,7 8], *trop japo*[1,3,4], *temp japo*[5]
2	2-methoxy-3,5-dimethylpyrazine	Earthy	0.4 ng/L[a]	*aro*[2], *trop japo*[2], *ind*[2]
3	2-Amino acetophenone	Musty, wet popcorn	0.2 ng/L[a]	*aro*[2], *trop japo*[2], *ind*[2]
4	Indole	Floral, sweet, burnt	140 ppb[a]	*aro*[2], *trop japo*[1,2], *ind*[2]
5	3-Methylindole	Mothball-like	140 ppb[a]	*aro*[2], *trop japo*[2], *ind*[2]
6	2-Isobutyl-3-methoxypyrazine	Earthy, green bell pepper	0.002 ppb[a]	*aro*[2], *trop japo*[2], *ind*[2]
7	2-Acetyl-2-thiazoline	Cooked jasmine rice	1.3 ppb[a]	*aro*[7]
Alkane				
1	Geranyl acetate	Floral		*aro*[7]
Alkene				
1	1-Tetradecene		60 ppb[a]	*trop japo*[3]
Aliphatic Aldehydes				
1	Pentanal	Nutty, sweet	12 ppb[a]	*ind*[6,7], *aro*[2,4,6,7 8], *trop japo*[1]
2	Hexanal	Green tomato, green	4.5 ppb[b]	*ind*[2,6,7,8], *aro*[2,6,7, 8], *trop japo*[1,4], *temp japo*[5]
3	Heptanal	Floral	3 ppb[a]	*ind*[2,6,7,8], *aro*[2,6,7,8], *trop japo*[1,4]
4	Octanal	Citrus, fruity, floral, and fatty	0.7 ppb[a]	*ind*[2,6,7,8], *aro*[6,7,8], *trop japo*[1,4], *temp japo*[5]
5	(E)-2-octenal	Nutty, green, and fatty	3 ppb[a]	*ind*[2,6,7,8], *aro*[2,6,7,8], *trop japo*[4]
6	Nonanal	Strong fruity or floral odor	1 ppb[a]	*ind*[2,6,7], *aro*[2,6,7 8], *trop japo*[1,3,4], *temp japo*[5]
7	(3Z)-3-Hexenal	Green, leaf-like	90 ppb[a]	*aro*[2]
8	Decanal	Sweet, waxy, floral	2 ppb[a]	*ind*[2,6,7], *aro*[6,7 8], *trop japo*[1,3,4], *temp japo*[5]

TABLE 8.2   (Continued)

S. N.	Name of Compound	Odor type	OTV[c]	Rice type
9	(2E,4E)-2,4-Deca-dienal	Fatty, metallic, citrus	0.07 ppb[a]	*ind*[2,7], *aro*[7,8], *trop japo*[1]
10	(E,Z)-deca-2,4-dienal	Fatty green	0.07 ppb[a]	*ind*[2,7], *aro*[2,7,8], *trop japo*[1]
11	(2Z)-2-Nonenal	Fatty, green	0.08 ppb[a]	*ind*[2], *aro*[2], *trop japo*[2]
12	(2E)-2-Nonenal	Fatty, green	0.08 ppb[a]	*ind*[2,7], *aro*[2,7,8], *trop japo*[1,2]
13	(2E,6Z)-2,6-Nona-dienal	Green, metal-lic	0.01 ng/g[a]	*ind*[2,7], *aro*[2,7], *trop japo*[2]
14	(2E)-2-Decenal	Green herbal geranium	15 ng/l[a]	*ind*[2,7], *aro*[2,4,7], *trop japo*[1,2,4]
15	(2Z)-2-Decenal	Fatty green	2.7 ng/l[b]	*ind*[2,7], *aro*[2,4,7], *trop japo*[1,2,4]
16	(2E,4E)-2,4-Nona-dienal	Nutty, fatty	0.2 ng/l[b]	*ind*[2,6,7], *aro*[2,4,6,7], *trop japo*[1,2,4]
17	Dodecanal	Minty, soapy	0.1 ppb[a]	*aro*[7]
18	3-(3-Pentyl-2-oxira-nyl)acrylaldehyde	Metallic	0.12 ng/l[a]	*aro*[2], *ind*[2], *trop japo*[2]
Aromatic Aldehydes				
1	Phenylacetaldehyde	Herbal, floral	4 ppb[a]	*trop japo*[1],*aro*[2,7,8]
2	Vanillin	Vanilla like	58 ppb[a]	*aro*[6,7], *ind*[2,6], *trop japo*[2], *temp japo*[5]
Aliphatic Alcohol				
1	2,3-Butanediol	Buttery, creamy	>10⁵ ppb[a]	*trop japo*[3]
2	1-octen-3-ol	Straw, earthy, raw mush-room	1 ppb[a]	*aro*[4,6,8], *ind*[4,6,8], *trop japo*[4]
3	1-Octanol	Fatty, metallic	110 ppb[a]	*aro*[7,8], *ind*[7], *trop japo*[1]
4	Hexanol	Green	2.5 ppm[a]	*aro*[7,8], *ind*[7], *temp japo*[5]
Aromatic Alcohol				
1	2-Methoxy-4-vinyl-phenol	Spicy,fruity	3 ppb[a]	*ind*[2,6,7], *aro*[6,7,8], *trop japo*[1,3,4], *temp japo*[5]
2	2-Methoxyphenol	Smoky	3 ppb[a]	*ind*[2,6], *aro*[6], *trop japo*[1,3,4]

TABLE 8.2 (Continued)

S. N.	Name of Compound	Odor type	OTV[c]	Rice type
3	4-Vinylphenol	Clove, Phenolic, Smokey	10 ppb[a]	*ind*[2,6,7], *aro*[6,7,8], *trop japo*[1,3,4], *temp japo*[5]
4	2-Phenylethanol	Honey-like	1100 ppb[a]	*aro*[7]
5	p-Cresol	Phenolic	55 ppb[a]	*aro*[2], *ind*[2], *trop japo*[2]
6	m-Cresol	Phenolic	55 ppb[a]	*aro*[2], *ind*[2], *trop japo*[2]
Ketone				
1	1-Octen-3-one	Mushroom	0.05 ppb[a]	*aro*[6,7,8], *trop japo*[1], *ind*[6], *temp japo*[5]
2	β-Ionone	Violet-like	1.0 ppb[a]	*aro*[2,7], *ind*[2,7], *trop japo*[2]
3	a-Ionone	Floral	0.5 ppb[a]	*aro*[7], *ind*[7]
4	1-Nonen-3-one	Mushroom-like	8 x 10^{-6} ppb[a]	*aro*[2], *ind*[2], *trop japo*[2]
5	Butan-2,3-dione	Buttery	5.40x10^{-3} ppm[a]	*aro*[2], *ind*[2], *trop japo*[2]
6	p-menthan-3-one	Mint	4.7 ng/l[b]	*aro*[4]
7	6-Methyl-5-hepten-2-one	Banana-like	50 ppb[a]	*aro*[8]
8	(E)-3-Octen-2-one	Citrus, herbal, floral	6.7 ng/l[b]	*aro*[4,8], *ind*[8]
9	b-Damascone	Sweet, honey	1 ppb[a]	*ind*[7]
10	b-Damascenone	Sweet, honey	0.002 ppb[a]	*aro*[7], *ind*[7]
Furan Containing				
1	4-hydroxy-2,5-dimethyl-3(2H)-furanone	Caramel like	25 ppb[a]	*aro*[2], *ind*[2], *trop japo*[2]
2	bis-(2-methyl-3-furyl)-disulfide	Meaty	2 x 10^{-5}ppb[a]	*aro*[2], *ind*[2], *trop japo*[2]
3	3-hydroxy-4,5-dimethyl-2(5H)-furanone	Curry, walnut	0.08 ppb[a]	*aro*[2], *ind*[2], *trop japo*[2]
4	3a,4,5,7a-tetra-hydro-3,6-dimethyl-2(3H)-benzofuranone	Sweet, spicy	1- 4 x 10^{-5}ng/l[b]	*aro*[2], *ind*[2], *trop japo*[2]
5	5-ethyl-3-hydroxy-4-methyl-2(5H)-furanone	Seasoning-like	5.0 ppb[a]	*aro*[2], *ind*[2], *trop japo*[2]

TABLE 8.2   (Continued)

S. N.	Name of Compound	Odor type	OTV[c]	Rice type
6	γ-Nonalactone	Coconut-like	30-65 ppb[a]	*aro[2], ind[2]*
7	γ-Octalactone	Coconut-like	400 ppb[a]	*aro[2], ind[2], trop japo[2]*
8	2-Pentylfuran	Floral, fruit, nutty	5[a]	*aro[2,8], ind[8]*
9	2-Methyl-3-furan-thiol	Meaty	0.007 ppb[a]	*aro[7], ind[7]*
Carboxylic acid				
1	Acetic acid	Sour	50000 ng/g[a]	*aro[2], ind[2], trop japo[2]*
2	Butanoic acid	Sweaty, rancid	240 ng/g[a]	*aro[2], ind[2], trop japo[2]*
3	2-and 3-methylbuta-noic acid	Cheese-like, sweaty	0.1 ppm[a]	*aro[2], ind[2], trop japo[2]*
4	Pentanoic acid	Sweaty	2100 ng/g[a]	*aro[2], ind[2], trop japo[2]*
5	Phenylacetic acid	Honey-like	10000 ng/g[a]	*aro[2], ind[2], trop japo[2]*
6	Hexanoic acid	Sweaty	3000 ng/g[a]	*aro[2], ind[2], trop japo[2]*
Sulphur Containing				
1	Methional	Cooked potato	0.2 ng/g[a]	*aro[2,7], trop japo[2], ind[2,7]*
2	Dimethyl sulphide	Cooked, sulfury	0.1 ppb[a]	*ind[2,7]*
3	3-Methyl-2-butene-1-thiol	Nutty, sulfury	0.0035 ppb[a]	*ind[2,7]*
4	Dimethyl trisulfide	Sulfury, cabbage-like	0.07 ppm[a]	*aro[7], ind[2,7]*

a: odor threshold value detected in water; b: odor threshold value detected in air; c: complied from Buttery, et al. (1988, 1999), Rychlik, et al. (1998), Yang, et al. (2008). 1: Buttery, et al. (1988); 2: Jezussek, et al. (2002); 3: Ajarayasiri and Chaiseri, (2008); 4: Yang, et al. (2008); 5: Maraval, et al. (2008); 6: Mathure, et al. (2014); 7: Mahattanatawee and Rouseff, (2014); 8: Hinge, et al. (2015).

contributed to aroma, showed odor-active values exceed to 1. 2,3-Butane-diol, 2-Acetyl-1-pyrroline, 1-Octen-3-ol, Nonanal, 1 Ethyl octanoate, (E,E)-Nona-2,4-dienal, Germacrene-D and Methyl dodecanoate showed 1 or more than 1 average FD factor, indicating contribution toward the aroma contribution in black and white glutinous rice (Ajarayasiri and Chaiseri, 2008). Yang et al. (2008) analyzed aroma compounds of six different rice flavor types (jasmine, basmati, black rice, 2 Korean japonica cultivars, and a nonaromatic rice) by using the dynamic headspace system with Tenax trap, GC-MS, and GC-O analysis. Out of 36 volatiles from cooked samples, 25 were found to be major OACs, in which 13 OACs (2-AP; hexanal; (E)-2-nonenal; octanal; heptanal; nonanal; 1-octen-3-ol; (E)-2-octenal; (E,E)-2,4-nonadienal; 2-heptanone; (E,E)-2,4-decadienal; decanal; and guaiacol) separated individual rice flavor types.

Maraval et al. (2008) characterized volatile compounds in cooked rice from scented (Aychade, Fidji, and Thai) and nonscented (Ruille) cultivars. Out of 40 volatile compounds, 11 were found odor active and being specific to some of the rice. Further, Mathure et al. (2014) quantitatively analyzed 23 headspace volatiles from 91 (aromatic and non-aromatic) Indian rice cultivars. They reported that 15 compounds (2-AP, nonanal, hexanal, octanal, (E,E)-nona-2,4-dienal, trans-2-nonenal, decanal, pentanal, heptanal, trans-2-octenal, 4 vinyl guaiacol, 4 vinyl phenol, guaiacol, vanillin, and 1-octen-3-ol) exceeded the odor active value of 1 and contribute to rice aroma. Mahattanatawee and Rouseff (2014) characterized aroma volatiles from three cooked fragrant (Jasmine, Basmati, and Jasmati) rice type. They observed 26 aroma active volatiles in Jasmine, 23 in Basmati, and 22 in Jasmati. Recently, Hinge et al. (2015) characterized aroma volatiles at vegetative and mature stages in Ambemohar 157 and Basmati 370 rice cultivars. Among the 26 volatiles, 2-AP, 1-octanol, 1-octen-3-ol, (E)-3-octen-2-one, 2-pentylfuran, phenylacetaldehyde and aliphatic aldehydes pentanal octanal, (E)-2-nonenal, nonanal, heptanal, hexanal, decanal, and (E)-2-octenal exceeded odor active value of 1 and were identified as major contributors in the aroma of scented rice.

In review of abovementioned researchers, 68 OACs were reported from different rice type (*temperate japonica, tropical japonica, aromatic,* and *indica*). All the reported OACs are depicted in Table 8.2. These reported compound belong to N-containing aromatic (7), aliphatic hydrocarbons (2), aliphatic aldehyde (18), aromatic aldehyde (2), aliphatic alcohol (4), aromatic alcohol (6), ketone (10), furan containing (9), carboxylic acid (6), and sulfur containing (4). According to previous reports (Table 8.1), the maximum number of volatiles are from hydrocarbon group (27% compounds), but the least number (2) of volatiles are AACs. Similarly, the second largest group is ketone, alcohol, and aldehyde (13% compounds from each group), in which aldehyde having more number of AACs (20 compounds) followed by ketone and alcohol (10 compounds each). N-containing aromatic group contributed 9% (44 compounds) of total compounds; however, 7 compounds are AACs. Similarly, furan-containing compounds contributed 5% (25 compounds) of total compounds, in which 9 are aroma active. The present analysis confirmed that the aldehyde group is major aroma active volatile contributor, followed by ketone, alcohol, furans, and N-containing compounds.

Previously reported odor active value (OAV), flavor dilution factor (FD), or odor intensity (OI) of 68 OACs in various rice cultivars belonging to *aromatic* (Basmati), *indica* (scented and non-scented), *tropical japonica* (scented and non-scented), and *temperate japonica* (scented and non-scented) are depicted in Tables 8.3a and 8.3b. These OAVs, FDs, or OIs of these OACs were categorized into different scales for analysis using PCA. The OAC categories are as follows: 0 for absent of OAC in a particular cultivar; 1 for OAV/FD/OI <1 to 1; 2 for OAV/FD/OI between 1.1 to 5; 3 for OAV/FD/OI between 5.1 to 10; 4 for OAV/FD/OI between 10.1 to 25; 5 for OAV/FD/OI between 25.1 to 50; 6 for OAV/FD/OI between 50.1 to 100; 7 for OAV/FD/OI between 100.1 to 200; 8 for OAV/FD/OI between 200 to 500; and 9 for OAV/FD/OI >500.

- 1: Buttery et al. (1988), 2: Jezussek et al. (2002), 3: Ajarayasiri and Chaiseri (2008), 4: Yang et al. (2008), 5: Maraval et al. (2008), 6: Mathure et al. (2014), 7: Mahattanatawee and Rouseff (2014), 8: Hinge et al. (2015).
- B: Basmati-370, RB: Royal Basmati, A: Ambemohar-157, PB: Pusa Basmati, PS: Pusa Sughandha, I: Indica variety, J: Jasmine, JM: Jasmati, ID: Indrayani, BS: Badshabhog, T: Thai, IR: IR-64, C: California long grain rice (variety L-202), K: Khashkani, S: Improved Malagkit Sungsong, H1: Hyangminbyeo 1, H2: Hyangmibyeo 2, BP: Black pigmented rice, BG: Black glutinous rice, WG: White glutinous rice, TM: Traditional non-aromatic rice, Y: Aychade, F: Fidji, R: Ruille.
- 2-AP: 2-Acetyl-1-pyrroline; 2M35D: 2-Methoxy-3,5-dimethylpyrazine; 2AA: 2-Amino acetophenone; IN: Indole; 3M: 3-Methylindole; 2I3M: 2-Isobutyl-3-methoxypyrazine, 2A2T: 2-Acetyl-2-thiazoline; GA: Geranyl acetate; 1T: 1-Tetradecene; PE: Pentanal;

HE: Hexanal; HP: Heptanal; OC: Octanal; E2O: (E)-2-octenal; NA: Nonanal; 3Z3H:(3Z)-3-Hexenal; DE: Decanal; 2E4ED: (2E,4E)-2,4-Decadienal; EZD: (E,Z)-deca-2,4-dienal; 2Z2N: (2Z)-2-Nonenal; 2E2N: (2E)-2-Nonenal; 2E6ZN: (2E,6Z)-2,6-Nonadienal; 2Z2D: (2Z)-2-Decenal; 2E2D: (2E)-2-Decenal; 2E4EN: (2E,4E)-2,4-Nonadienal; 33P: 3-(3-Pentyl-2-oxiranyl)acrylaldehyde; DO: Dodecanal; PH: Phenylacetaldehyde; VA: Vanillin; 23B: 2,3-Butanediol; 1OCO: 1-octen-3-ol; 1O: 1-Octanol; HX: Hexanol, 2M4V: 2-Methoxy-4-vinylphenol; 2M: 2-Methoxyphenol; 4V: 4-Vinylphenol; 2PE: 2-Phenylethanol; PC: p-Cresol; MC: m-Cresol; 1O3O: 1-Octen-3-one; BI: β-Ionone; 1N3O: 1-Nonen-3-one; B23D: Butan-2,3-dione; PM3O: p-menthan-3-one; 6M5H: 6-Methyl-5-hepten-2-one; E3O2O: (E)-3-Octen-2-one; BD: b-Damascone; BDS: b-Damascenone; AL: a-Ionone; 4H25D: 4-Hydroxy-2,5-dimethyl-3(2H)-furanone; B2M3F; bis-(2-Methyl-3-furyl)-disulfide; 3H45D: 3-Hydroxy-4,5-dimethyl-2(5H)-furanone; 3457D: 3a,4,5,7a-Tetrahydro-3,6-dimethyl-2(3H)-benzofuranone; 5E3HM: 5-Ethyl-3-hydroxy-4-methyl-2(5H)-furanone; NO: γ-Nonalactone; OL: γ-Octalactone; 2PF: 2-Pentylfuran; 2M3F: 2-Methyl-3-furanthiol; AA: Acetic acid; BA: Butanoic acid; 23MA: 2- and 3-Methylbutanoic acid; PA: Pentanoic acid; PAA: Phenylacetic acid; HA: Hexanoic acid; MO: Methional; DS: Dimethyl sulfide; 3M2B1T: 3-Methyl-2-butene-1-thiol; DT: Dimethyl trisulfide.

PCA analysis was performed for 27 rice cultivars using XL-STAT (Version, 2014, Addinsoft™). Among these 27 rice cultivars, seven are *aromatic* type (three cultivars of Basmati, Royal Basmati, Pusa Basmati, Pusa Sughandha, and Ambemohar 157), seven are *indica* scented (two cultivars of Jasmine, Indica variety, Jasmati, Indrayani, Badshabhog, and Thai) and one *indica* non-scented (IR-64), eight are *tropical japonica* scented type (California long grain rice (variety L-202), Khashkani, Improved Malagkit Sungsong, Hyangminbyeo 1, Hyangmibyeo 2, Black pigmented rice, Black glutinous rice and White glutinous rice) and one non-scented *tropical japonica* type (Traditional non-aromatic rice), two are *temperate japonica* scented type (Aychade and Fidji) and one is *temperate japonica* non-scented type (Ruille). In the PCA, 27 cultivars were separated into

TABLE 8.3A Odor Active Value, Flavor Dilution Factor, or Aroma Intensity of Reported AACs in *Aromatic* and *Indica* Rice Cultivars

S. No.	Name of Comp.	Aromatic								Indica						
											Scented					NS
		B[2]	RB[4]	B[6]	A[6]	PB[6]	PS[6]	B[7]	I[2]	ID[8]	BS[6]	J[4]	J[7]	JM[7]	T[5]	IR[8]
1	2-AP	512	17	4335	7282	1612.84	1413	9.88	2	4406	865.43	191	9.93	9.5	1860	0
2	2M35D	<1							<1							
3	2AA	1024							512							
4	IN	8							1							
5	3M	4							1							
6	2I3M	8							<1							
7	2A2T												5.48			
8	GA							7.04					7.55			
9	IT															
10	PE			7	6.6	6.3	9.4			4.4						3.2
11	HE	2	232	111.56	175.48	132.71	173.86	9.96	4	119.97	36.83	117	8	9.85	24.39	47
12	HP		12	85	53	91	77			59	14	3.2				20
13	OC	2	19	237	108	349	368	9.38	2	178	49	6.2	7.97	9.98		157.65
14	E2O		1.7	12	3	13	12	7.31	2	8	0.9	0.9	3.63	3.07		1.69
15	NA		5.1	310	143	488	432	3.53		216	102	1.4	4.95	3.57		32
16	3Z3H	1							<1							
17	DE	<1	21		10	28	20	3.44	<1	17	8		4.01	4.47	53	
18	2E4ED	16	0.8					5.45	16				4.46	3.69		

TABLE 8.3A (Continued)

S. No.	Name of Comp.	Aromatic								Indica						NS
										Scented						
		B[2]	RB[4]	B[6]	A[6]	PB[6]	PS[6]	B[7]	I[2]	ID[8]	BS[6]	J[4]	J[7]	JM[7]	T[5]	IR[8]
19	EZD	<1							<1						357	
20	2Z2N	4							1							
21	2E2N	8	25	141	112	253	230	9.96	8	113	142	7.8	10	10		56.7
22	2E6ZN	2						5.1	<1				5.11			
23	2Z2D	2							1							
24	2E2D							7.72					8.28	4.54		
25	2E4EN	4	1.3	0	59	20	176	4.18	2	184	87	0.9	4.84	4.3	19	
26	33P	64							128							
27	DO							4.34					4.16	4.52		
28	PH	1	5.4		6.7				<1							1.5
29	VA	128	0.9		1.7	1.48	0		256	1.2	0				19.95	
30	23B															
31	1OCO		1.8	24	45	27	39			26	14	0.8	3.31	3.75		52
32	1O		0.05	0	0				1	0	0					
33	HX							5.87					6.28	5.53		
34	2M4V	32	11	11	7	11	14		32	11	6				187	
35	2M	2	0	0	0	8	1.4		1	0	0					
36	4V	<1	3.5	3.5	11	10.7	13		1	5	0					

S. No.	Name of Comp.	Aromatic									Indica Scented					NS
		B[2]	RB[4]	B[6]	A[6]	PB[6]	PS[6]	B[7]	I[2]	ID[8]	BS[6]	J[4]	J[7]	JM[7]	T[5]	IR[8]
37	2PE	2							1				3.23			
38	PC	2							1							
39	MC	1							<1							
40	1O3O	4						7.01	2				7.73	7.57		
41	BI	1											8.75	3.06		
42	1N3O	1						6.23	<1							
43	B23D	4							2							
44	PM3O		1.1													
45	6M5H			1.25	0.8											
46	E3O2O			157	73											31
47	BD												4.06			
48	BDS							4.64					6.5	5.44		
49	AL							4.6						4.65		
50	4H25D	8							2							
51	B2M3F	256							<1							
52	3H45D	1024							32							
53	3457D	4							<1							
54	5E3HM															
55	NO	<1							1							

TABLE 8.3A (Continued)

| S. No. | Name of Comp. | Aromatic | | | | | | | | | Indica | | | | |
| | | | | | | | | | | Scented | | | | | NS |
		B^2	RB^4	B^6	A^6	PB^6	PS^6	B^7	I^2	ID^8	BS^6	J^4	J^7	JM^7	T^5	IR^8
56	OL	<1							2							
57	2PF		5.8		7.9											2.1
58	2M3F							5.57					5.45	4.4		
59	AA	1							1							
60	BA	1							<1						1.15	
61	23MA	2							4							
62	PA	1							1							
63	PAA	32							32							
64	HA	1							<1							
65	MO	8						9.98	2				8.95	8.82		
66	DS													2.15		
67	3M2B1T												3.75			
68	DT							10					8.68	9.82		

TABLE 8.3B Odor Active Value, Flavor Dilution Factor, or Aroma Intensity of Reported AACs in *Tropical And Temperate Japonica Rice*

| S. No. | Name of Comp. | Tropical japonica | | | | | | | | | Temperate japonica | | |
| | | Scented | | | | | | | | NS | Scented | | NS |
		C[1]	K[2]	S[2]	H1[4]	H2[4]	BP[4]	BG[3]	WG[3]	TM[4]	Y[5]	F[5]	R[5]
1	2-AP	6	512	1024	153	8	246	3		5.8	2150	2640	
2	2M35D		1	4									
3	2AA		1024	2048					5				
4	IN		2	4							1.8	0.77	2.61
5	3M		4	2									
6	2I3M		4	16									
7	2A2T												
8	GA												
9	IT							2					
10	PE												
11	HE	2	2		31	44	16			167	13.94	12.27	8.47
12	HP				4	6.1	1.2			5.4			
13	OC	1	2	2	3.6	4.9	1.8			9.4			
14	E2O					0.6				1.3	16		
15	NA	3		1	1.7	2.3	1.4		1	2			
16	3Z3H		<1										
17	DE	0.7	2								46	30	
18	2E4ED	5.7	32	16									46

TABLE 8.3B (Continued)

S. No.	Name of Comp.	Tropical japonica									Temperate japonica		
		Scented								NS	Scented		NS
		C[1]	K[2]	S[2]	H1[4]	H2[4]	BP[4]	BG[3]	WG[3]	TM[4]	Y[5]	F[5]	R[5]
19	EZD		4								2328	1671	257
20	2Z2N		2	8									
21	2E2N	1	16		5	5.7	2.7			16			
22	2E6ZN		<1	1									
23	2Z2D		2	2									
24	2E2D												
25	2E4EN		8	4				3	3	1.1	74	33	38
26	33P		128	128									
27	DO		1	1									
28	PH		1	1									
29	VA		64	128							13	6.95	11.1
30	23B							1					
31	1OCO							1		1.3			
32	1O												
33	HX												
34	2M4V	0.6	16	32				133			124	84	105
35	2M		2	4									
36	4V	0.6	1	16							112	107	96

S. No.	Name of Comp.	Tropical japonica									Temperate japonica		
		Scented								NS	Scented		NS
		C^1	K^2	S^2	$H1^4$	$H2^4$	BP^4	BG^3	WG^3	TM^4	Y^5	F^5	R^5
37	2PE		4	8									
38	PC		2										
39	MC		<1										
40	1O3O		8	4									
41	BI		1										
42	1N3O		<1										
43	B23D		2	2									
44	PM3O												
45	6M5H												
46	E3O2O												
47	BD												
48	BDS												
49	AL												
50	4H25D		8	8									
51	B2M3F		256	2048									
52	3H45D		16	32									
53	3457D		1	8									
54	5E3HM			1									
55	NO		<1										

TABLE 8.3B (Continued)

S. No.	Name of Comp.	C¹	K²	S²	Tropical japonica Scented H1⁴	H2⁴	BP⁴	BG³	WG³	NS TM⁴	Temperate japonica Scented Y⁵	F⁵	NS R⁵
56	OL		<1										
57	2PF												
58	2M3F												
59	AA		<1										
60	BA		<1	1					1036		1.38	0.91	
61	23MA		2	4									
62	PA		1	1									
63	PAA		4	16									
64	HA		<1										
65	MO		8	4									
66	DS												
67	3M2B1T												
68	DT												

two components with total variance of 64.66% accounting each for PC1 (16.48%) and for PC2 (48.18%) (Figure 8.3). Twelve cultivars belonging to *indica, aromatic, tropical,* and *temperate japonica* type were grouped into positive side of PC1 and PC2. In that, the maximum cultivars from *tropical japonica* scented (6) were grouped together except (Hyangmin-byeo 1 and Hyangminbyeo 2) and one nonscented – Traditional nonaro-matic. However, all the scented and non-scented *temperate japonica* were grouped together in the positive side of PC1 and PC2. Basmati-370 from *aromatic* type and two *indica* type (Thai from Taureau Aile, Lyon, France and Indica rice variety from German super market) are placed in this group (Jezussek et al., 2002; Maraval et al., 2008). These *aromatic* and *indica* rice cultivars were grown in tropical and temperate climatic conditions; hence, probably due to similar climatic conditions of *japonica* rice type, these cultivars were grouped with *temperate* and *tropical japonica* type rice cultivars. Another 15 rice cultivars belonging to group *aromatic* (6), *indica* (5 scented and 1 nonscented) and *tropical japonica* (2 scented and 1 nonscented) were grouped into the negative side of PC1 and positive side of PC2. In this cluster, the maximum cultivars from *aromatic* and *indica* were placed together, indicating closed relationship of flavor type of them. In this group, all the cultivars belonging to different rice type were grown in tropical climatic conditions; therefore, they resemble similar rice aroma type. Bett-Garber et al. (2001) reported that environmental conditions influenced flavor and texture; hence, some cultivars were clustered into unexpected categories.

Further, to observe the contribution of OACs for the separation of rice type, OAV/FD/OI from different cultivars from similar rice type (*aromatic, indica* scented, *indica* nonscented, *tropical japonica* scented, *tropical japonica* nonscented, *temperate japonica* scented, and *temperate japon-ica* nonscented) were pulled together. Further, the PCA analysis was per-formed to check the separation of different rice types based on OACs. The results of PCA analysis are depicted in Figure 8.4. All these seven rice type groups were separated into two components with total variance of 56.84% each for PC1 (22.28%) and PC2 (34.56%). The groups of *aromatic, indica* (scented and nonscented), and *tropical japonica* (scented and nonscented) rice types were grouped together in the positive side of PC1 and PC2. However, only *temperate japonica* (scented and nonscented) rice type was separated in the negative side of PC1 and positive side of PC2. This indicated that OACs clearly separate *temperate japonica* rice type from other rice types. Based on this analysis, *aromatic, indica,* and *tropical japonica* have similar type of OACs and showed closed relationship between them. Thus, these OACs can be used for the separation of rice type and can be treated as biomarkers for distinguishing rice type.

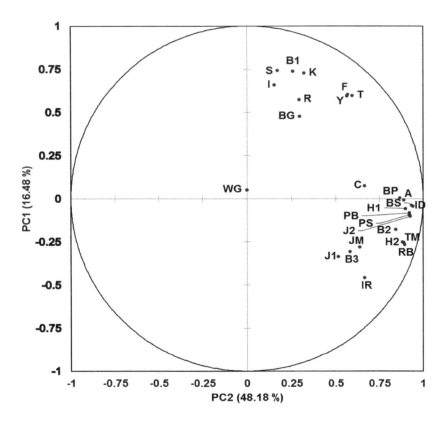

FIGURE 8.3 PCA plot of 27 rice cultivars based on 27 major OACs. B1: Basmati 370 (Jezussek et al., 2002) B2: Basmati 370 (Mathure et al., 2014), B3: Basmati (Mahattanatawee and Rouseff, 2014), RB: Royal Basmati, A: Ambemohar 157, PB: Pusa Basmati, PS: Pusa Sughandha, I: Indica variety, J1: Jasmine (Yang et al., 2008), J2: Jasmine (Mahattanatawee and Rouseff, 2014), JM: Jasmati, I. D., Indrayani, B. S., Badshabhog, T., Thai, I. R., IR-64, C: California long grain rice (variety L-202), K: Khashkani, S., Improved Malagkit Sungsong, H1: Hyangminbyeo 1, H2: Hyangmibyeo 2, BP: Black Pigmented, BG: Black glutinous, WG: White glutinous, TM: Traditional nonaromatic, Y: Aychade, F., Fidji, R., Ruille.

Environmental factors have an important role in determining the aroma in rice. The desired climatic conditions like high humidity (70–80%) with temperature range of 25 to 35°C during vegetative growth period is favorable. Bright and clear sunshine with 25 to 32°C temperature, relatively cold temperature during nights (20 to 25°C), moderate humidity, and mild wind velocity at the time of flowering and grain filling are considered necessary for proper aroma development (Shobha Rani et al., 2006). It is proved that the principal aroma compound 2-AP is mainly controlled by the *Badh2* gene, but it is largely influenced by environmental conditions such as soil type, cultural practices, and temperature during the grain filling stage, storage conditions, and storage time (Singh et al., 1997; Tang et al., 2007). Basmati rice cultivated on the foot hills of the Himalayan region in the northwestern parts of the Indian sub-continent emits special aroma. In these agro-climatic conditions, the grown Basmati rice emits specific aroma, but when the same rice cultivar is grown outside of this region, it does not produce aroma (Bhattacharjee et al., 2002). Similarly, Pusa basmati rice grown in rainy weather showed 0.030 ppm 2-AP, but it does not achieve perceivable 2-AP when grown during the summer (Nadaf et al., 2006; Yi et al., 2009). Itani et al. (2004) reported that 2-AP concentration varies with environmental conditions and found it to be maximum in brown ripened rice at low temperature (day 25°C; night 20°C) than that of high temperature (day 35°C; night 30°C) in short grain cultivar (Hieri) and long grain cultivar (Sari). In the present analysis, the separation of different rice varieties based on the OACs confirms that the geographical and agro climatic conditions have significant contributions in deciding the flavor of rice. We also confirmed that tropical region-cultivated rice types (*aromatic, indica*, and *tropical japonica*) are distinct from temperate region-cultivated rice types (*temperate japonica*). Garris et al. (2005) and Caicedo et al. (2007) demonstrated close relationship between the *aromatic, temperate japonica*, and *tropical japonica* subpopulations, whereas *indica* and *aus* subpopulations are distinct from them. The present analysis based on OACs showed that *aromatic, indica*, and *tropical japonica* have close relationship than *temperate japonica*. The analysis indicated that flavor from different rice type is due to environmental conditions than due to the genetic organization.

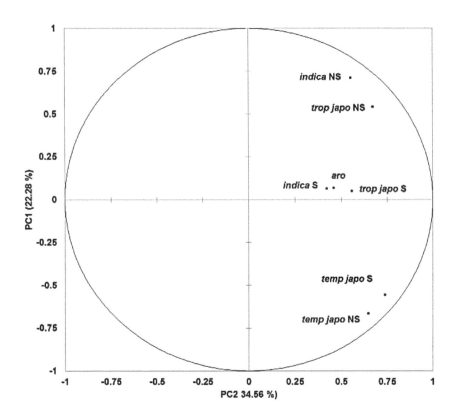

FIGURE 8.4   PCA plot showing separation of rice types based on OACs.

8.4   BIOSYNTHESIS OF ODOR ACTIVE COMPOUNDS IN RICE

A total of 484 volatile compounds are detected in rice, in which only 14% (68) compounds were found to contribute to the rice aroma. The majority of these volatile compounds are naturally synthesized in plants. Biosynthesis of plant volatiles depends on the source of carbon, nitrogen, and sulfur as well as energy provided by primary metabolism. Based on the biosynthetic origin, volatile compounds are classified into terpenoids, phenylpropanoids/benzenoids, fatty acid derivatives, and amino acid derivatives. In addition to a few species or genus specific, compounds not represented in those major classes have different biosynthetic origin. For biosynthesis of odor active or aroma contributing compounds in rice carbohydrates, fatty acids and amino acids represent the natural carbon pools. The OACs in rice are mainly biosynthesized from carbohydrate (terpenoids, furanones, and pyrones), fatty acid (aldehydes and alcohols through lipoxygenase (LOX) or α- and β-oxidation) and amino acid (acids, alcohols, aldehydes, esters, lactones, and N- and S-containing flavor molecules, benzenoid, and phenylpropanoid volatile compounds) (Figure 8.5).

8.4.1 CARBOHYDRATE-DERIVED COMPOUNDS

Carbohydrate-derived compounds are from the groups of terpenoid, furanones, and pyrones.

8.4.1.1 Biosynthesis of Terpenoid Compounds

Terpenoids compounds are enzymatically synthesized in the plastids and cytoplasm by using acetyl CoA and pyruvate. Though one of the major source for the formation of acetyl CoA is fatty acid oxidation, it probably is not involved in the formation of terpenoids as it is located in peroxisomes. Terpenoids are derived from two common five-carbon precursors, isopentenyl diphosphate (IPP) and its allylic isomer, dimethylallyl diphosphate (DMAPP) (McGarvey and Croteau, 1995). Two compartmentally separated, independent pathways (Mevalonic acid; MVA and methylerythritol phosphate; MEP) are responsible for the formation of terpenoids (C_5-isoprene building units).

The terpenoids [hemiterpenoids (C_5), monoterpenoids (C_{10}), sesquiterpenoids (C_{15}) and some diterpenoids (C_{20})] are involved in plant-insect interactions and defense or stress responses. Terpenoids from C_{10} and C_{15} members of this family were found to affect the flavor profiles of most fruits and the scent of flowers at varying levels (Pichersky and Gershenzon, 2002; Dudareva et al., 2004; Pichersky et al., 2006). Sesquiterpenes are synthesized through the MVA pathway in the cytosol, endoplasmic reticulum, and peroxisomes (Simkin et al., 2011; Pulido et al., 2012) whereas, many precursors to volatile hemiterpenes, monoterpenes, and diterpenes are synthesized through the MEP pathway exclusively in the plastid (Hsieh et al., 2008).

In addition to this, irregular volatile terpenoids (C_8 to C_{18}) are synthesized from carotenoids. Carotenoids are tetraterpenoid pigments that accumulate in the plastids of leaves, flowers, and fruits. OACs, α and β-ionone, b-damascone, b-damascenone, geranyl acetate, and 6-methyl-5-heptene-2-on, were biosynthesized from an array of carotenoid pigments *via* carotenoid cleavage dioxygenases (CCDs) (initial dioxygenase cleavage followed by enzymatic transformation and acid-catalyzed conversion) (Winterhalter and Rouseff, 2001). Sometimes, only dioxygenase cleavage step can yield a volatile product, such as α- and β-ionone in *Arabidopsis*, tomato, petunia, and melon (Simkin et al., 2004a, 2004b; Ibdah et al., 2006).

8.4.1.2 Biosynthesis of Furanones and Pyrones

Primary photosynthetic products like hexoses and pentoses serve as excellent flavor precursors for furans and pyrones through the Maillard reaction. But very few volatile compounds originate directly from carbohydrates without prior degradation of the carbon skeleton. Such compounds include the furanones and pyrones (Bood and Zabetakis, 2002).

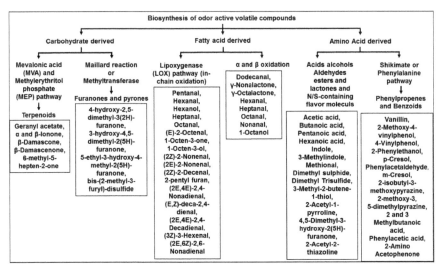

FIGURE 8.5 Biosynthetic mechanism of OACs in rice.

D-fructose-1,6-diphosphate is observed as an efficient biogenetic precursor of 4-hydroxy-2,5-dimethyl-3(2H)-furanone (furaneol). Also, hexose diphosphate is converted to 4-hydroxy-5-methyl-2-methylene-3(2H)-furanone by as yet unknown enzyme in strawberry and tomato (Klein et al., 2007). Furaneol is further metabolized by an O-methyltransferase (FaOMT) to methoxyfuraneol in strawberry (Wein et al., 2002). Carbohydrate-derived flavor molecules include 4-hydroxy-5-methyl-3(2H)-furanone (norfuraneol), 2-ethyl-4-hydroxy-5-methyl-3(2H)-furanone (homofuraneol), 4-hydroxy-2-methylene-5-methyl-3(2H)-furanone (HMMF), and 3-hydroxy-2-methyl-4H-pyran-4-on (maltol) (Schwab et al., 2008).

Among the odor active aroma components of cooked rice, the compounds derived from the Maillard reactions and some induced thermally also play an important role. In the Maillard reaction, chemical reactions between the carbonyl group of the open-chain form of a reducing sugar and the primary amino group of an amino acid, peptide, or a similar compound takes place. Though this reaction occurs during storage, the rate is greatly facilitated by heat. Some furanones are induced thermally mainly 3-hydroxy-4,5-dimethyl-2(5H)-furanone and *bis*-(2-methyl-3-furyl)-disulfide in cooked rice, having seasoning-like and meaty-like aromas, respectively (Jezussek et al., 2002).

8.4.2 FATTY ACID-DERIVED VOLATILE COMPOUNDS

The majority of OACs including aldehydes, alcohols, ketones, acids, esters, and lactones are synthesized through LOX, or α- and β-oxidation from saturated and unsaturated fatty acids (Schwab and Schreier, 2002). In plants, fatty acids are stored as triacylglycerides, and by the action of acyl hydrolase, oxidative degradation of lipids liberates the free fatty acids from acylglycerols.

8.4.2.1 Biosynthesis of Fatty Acid-Derived Volatiles Via Lipoxygenase (LOX) Pathway

Acetyl CoA from Pyruvate (final product of glycolysis) is utilized for the synthesis of fatty acids in the plastids. The unsaturated fatty acids enter the "lipoxygenase (LOX) pathway," and undergo stereospecific oxygenation to form 9-hydroperoxy and 13-hydroperoxy intermediates (Feussner and Wasternack, 2002}. These 9-hydroperoxy and 13-hydroperoxy intermediates that are further converted into C_6 and C_9 aldehydes by hydroperoxide lyases are often reduced to alcohols by alcohol dehydrogenases (Gigot et al., 2010). Further conversion leads to the formation of esters (D'Auria et al., 2007).

Aliphatic aldehydes, namely pentanal, hexanal, heptanal, octanal, (E)-2-octenal, (2Z)-2-nonenal, (2E)-2-nonenal, (2Z)-2-decenal, (2E,4E)-2,4-nonadienal, (E,Z)-deca-2,4-dienal, (2E,4E)-2,4-decadienal; alcohol 1-octen-3-ol; and ketone 1-octen-3-one are formed from their linoleic acid precursors. 2-Pentyl furan is also obtained from cyclization of 4-keto-nonanal derived from linoleic acid more likely via 9-hydroperoxide lyases (Monsoor and Proctor, 2004). The aldehyde (3Z)-3-hexenal and (2E,6Z)-2,6-nonadienal and the alcohol Hexanol are formed from their linolenic acid precursors. In cooked rice, odor-active compounds are formed during the degradation of oleic, linoleic, and linolenic acid (Zhou et al., 2002). Octanal, heptanal, nonanal, (E)-2-nonenal, decanal, and 2-heptanone formation in cooked rice was also reported from oleic acid (Monsoor and Proctor, 2004). During storage, (E)-2-nonenal (rancid), octanal (fatty), and hexanal (green) significantly increase with their odors contributing to off-flavor formation with age (Lam and Proctor, 2003). Hexanal contributes to consumer rejection due to its rancid odor (Bergman et al., 2000). Hexanal formation is greater in partially milled than in fully milled rice (Lam and Proctor, 2003).

8.4.2.2 Biosynthesis of Fatty Acid-Derived Volatiles via the a- and b-Oxidation Pathway

The free fatty acids (C_{12}–C_{18}) are enzymatically degraded via one or two intermediates to long-chain fatty aldehydes and CO_2 through α-oxidation. α-Oxidation of fatty acids in plants is catalyzed by α-dioxygenase/peroxidase and NAD^+ oxidoreductase (Saffert et al., 2000). Successive removal of C_2 units (acetyl CoA) from the parent fatty acid is carried out in β-oxidation (Goepfert and Poirier, 2007). Furan-containing OACs γ-Nonalactone and γ-octalactone belong from another major group of fatty acid-derived flavor molecules, alkanolides. Lactones usually possess 8–12 carbon atoms and are potent flavor components for a variety of fruits. Plant-derived lactones originate from their corresponding 4- or 5-hydroxy carboxylic acids. These acids are formed by either reduction of oxo acids by NAD-linked reductase, hydration of unsaturated fatty acids, epoxidation and hydrolysis of unsaturated fatty acids, or reduction of hydroperoxides (Schottler and Boland, 1996). Biosynthesis of lactones such γ-decalactone and γ-dodecalactone (aroma of peach and nectarine), δ-octalactone (pineapple), γ-octalatcone (coconut) is associated with the β-oxidation pathway (Tressl and Albrecht, 1986).

Hexanal, heptanal, octanal, and nonanal are also biosynthesized through peroxidation of (n-6) fatty acids. Dodecanal formation is reported through lipid peroxidation. The OA alkene 1-tetradecene formation is reported to be anthropogenic origin, and it is not known to be produced in nature. Similarly, the formation of other carbohydrates including tetradecene, pentadecane, hexadecadiene, and heptadecene was reported to be induced by irradiation (Schreiber et al., 1994).

8.4.3 AMINO ACID-DERIVED ODOR ACTIVE COMPOUNDS

The OACs are derived from amino acids through degradation and from phenylalanine through the shikimate/phenylalanine biosynthetic pathways.

8.4.3.1 Biosynthesis of Acids, Alcohols, Aldehydes, Esters, Lactones, and N/S-Containing Flavor Molecules

Maximum OACs in rice are derived from amino acids such as proline, alanine, valine, leucine, isoleucine, methionine, phenylalanine, and tryptophan from intermediates in their biosynthesis and contain nitrogen and sulfur (Knudsen et al., 2006). The amino acid degradation pathway leading to the formation of aroma volatiles is extensively studied in microorganisms. Biosynthetic mechanisms similar to those in bacteria or yeast might be involved in plants (Dickinson et al., 2000; Beck et al., 2002; Tavaria et al., 2002).

Three biosynthetic routes exist in plants and microorganisms for amino acid degradation to form volatiles (Figure 8.6). In the first route, decarboxylation of amino acids via amino acid decarboxylases is followed by deamination catalyzed by deaminases (Tieman et al., 2006). In the second route, aldehydes are directly synthesized in a single enzymatic step with the action of aldehyde synthase (Kaminaga et al., 2006). In the third route, initial deamination or transamination by aminotransferases leads to the formation of the corresponding α-keto acid as an intermediate. Further, these α-keto acids are subjected to decarboxylation, followed by reductions, oxidations, and/or esterifications, forming aldehydes, acids, alcohols, and esters (Reineccius, 2006). Amino acids also serve as precursors for the formation of acyl-CoAs, which are used in alcohol esterification reactions catalyzed by alcohol acyltransferases (AATs) (Beekwilder et al., 2004; Gonzalez et al., 2009).

Odor active acid compounds originate from deamination of amino acids. Acids such as 2- and 3-methyl butanoic acids are derived from isoleucine and leucine. Leucine undergoes oxidative deamination and gives 2-keto-4-methyl valeric acid, which upon action of α-ketoacid decarboxylase releases CO_2 and produces 3-methyl butanal. Finally, 3-methyl butanal dehydrogenase acts on 3-methyl butanal to give 3-methyl butanoic acids (isovaleric acid). The N-containing OAC indole is synthesized by the cleavage of indole-3-glycerol phosphate, an intermediate in tryptophan biosynthesis (Frey et al., 2000). The formation of 3-methylindole (skatole) and 4-methylphenol (p-cresol) is reported from the degradation of tryptophan and tyrosine, respectively. 3-Methylindole can also be formed nonenzymatically from tryptophan by the Strecker degradation, oxidation to indolylacetic acid, and decarboxylation. P-Cresol is also formed in citrus oil and juice by the degradation of citral. 2-Aminoacetophenone is synthesized from tryptophan, wherein it is first converted to formylkynurenine and then to N-formylacetophenone, which loses the formyl group to give 2-aminoacetophenone (Mann, 1967). 2-Aminoacetophenone is also a metabolic oxidation product derived from 3-methylindole (skatole) (Frydman et al., 1971). 2-aminoacetophenone is also reported as a Strecker degradation product of tryptophan (Christoph et al., 1998). Strecker degradation is considered a part of the overall Maillard reaction (BeMiller and Whistler, 1996). 2-Aminoacetophenone is thought to be an off-odor in brown rice in that it has a naphthalene or floor polish odor (Rapp et al., 1993).

The major AAC 2-AP is synthesized nonenzymatically through Δ1-pyrroline/γ-aminobutyraldehyde (GABald) (an immediate precursor of 2-AP) and methylglyoxal. The GABald is accumulated when there is loss of function of enzyme betaine aldehyde dehydrogenase (*BADH2*) or from proline through P5CS (pyrroline-5 carboxylate synthetase) or ornithine. Methylglyoxal is formed through triose phosphate intermediates dihydroxyacetone phosphate and glyceraldehyde-3-phosphate through elimination of the phosphate group at the active site of triose phosphate isomerase (TPI) (Figure 8.7) (Richard, 1991; Phillips and Thornalley, 1993).

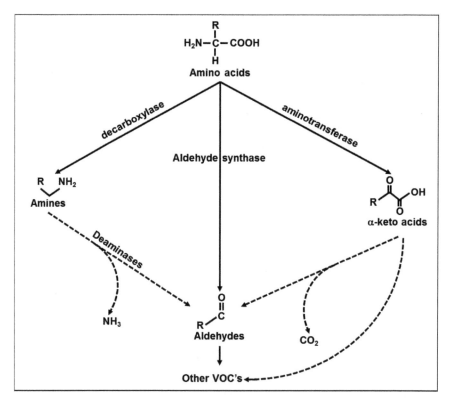

FIGURE 8.6   Biosynthesis of amino acid-derived volatile compounds.

Sulfur-containing compound methional is a thermally induced volatile originating from methionine in boiled potatoes (Jones et al., 2004) through the Maillard reaction. The methional was formed by the Strecker degradation reaction from the interaction of α-dicarbonyl compounds (intermediate products in the Maillard reaction) with methionine (Met). Furan like 4,5-dimethyl-3-hydroxy-2(5H)-furanone (sotolon) is also derived from amino acid (4-hydroxy-L-isoleucine) (Slaughter, 1999). The supposed biogenetic origin is supported by stereochemical considerations. 2-Iso-butyl-3-methoxypyrazine is synthesized from leucine in Paprika pepper (*Capsicum annum*) and chilies (*Capsicum frutescens*) (Belitz et al., 2004).

For the formation OAC like 2,3-butanediol, the observed precursors are acetaldehyde, pyruvate, and α-acetolactate in *Saccharomyces cerevisiae*. Pyruvate decarboxylases catalyze the synthesis of acetoin from the reaction between pyruvate and acetaldehyde. Acetoin is then converted into 2,3-butanediol by butanediol dehydrogenase (Sergienko and Jordan, 2001). Alkaloid diacetyl (butanedione or butane-2,3-dione) is naturally synthesized in some fermentative bacteria; it is formed via the thiamine pyrophosphate (TPP)-mediated condensation of pyruvate and acetyl CoA (Speckman et al., 1968). Acetic acid is synthesized from the β-alanine under natural condition (Mottram, 2007).

8.4.3.2   Phenylpropenes and Benzenoid Compounds

Phenylpropanoids and Benzenoid volatile compounds are derived from phenylalanine, which is synthesized in the plastid through the shikimate or phenylalanine biosynthetic pathways (Dudareva and Pichersky, 2006; Knudsen and Gershenzon, 2006; Pichersky et al., 2006). Biosynthesis of benzenoids from phenylalanine requires shortening of the carbon skeleton side chain by a C_2 unit, which can potentially occur via either the β-oxidative pathway or nonoxidatively (Boatright et al., 2004). Phenylacetaldehyde is formed from the branched chain amino acid phenylalanine by decarboxylation and oxidative removal of the amino group (Hayashi et al., 2004).

2-Phenylethanol formation from phenylacetaldehyde by the action of phenylacetaldehyde reductases (PAR) is reported in tomato (Tieman et al.,

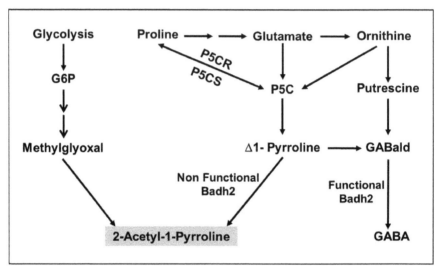

FIGURE 8.7 2-AP biosynthetic mechanism. G6P: Glucose 6 Phosphate; P5CS: Pyrroline-5 carboxylate synthetase; P5CR: Pyrroline-5 carboxylate reductase; GABald: γ- Amino Butyraldehyde; GABA: γ- Amino Butyric acid.

2007). 2-Phenylethanol and phenylacetic acid are Strecker degradation products of the amino acid L-phenylalanine (Hofmann and Schieberle, 2000; Etschmann et al., 2005). 2-Phenylethanol and phenylacetic acid in scented (Jasmine, Basmati, Goolarah, and YRF9) and non-scented rice (Pelde) contribute to a rose-like odor (Widjaja et al., 1996; Jezussek et al., 2002).

Vanillin (4-hydroxy-3-methoxybenzaldehyde) is the most widely used flavor compound in the world. Vanillin is synthesized from phenylpropanoid precursors, 4-coumaric acid. The enzyme 4-hydroxybenzaldehyde synthase converts 4-coumaric acid into 4-hydroxybenzaldehyde through a chain-shortening step (Podstolski et al., 2002). Then, 4-hydroxybenzaldehyde synthase performs hydroxylation at position 3 on the ring and converts it to p-hydroxybenzyl alcohol to 3,4-dihydroxybenzyl alcohol or aldehyde. Multifunctional O-methyltransferase catalyzes the final enzymatic step in the formation of vanillin. The enzyme O-methyltransferase from *Vanilla planifolia* has a broad substrate range, including 3,4-dihydroxybenzaldehyde (Pak et al., 2004).

Ferulic acid and trans-ferulic acid are reported as an immediate precursor for the formation of 4-vinylphenol and 2-methoxy-4-vinylphenol.

Ferulic acid is derived from L–phenylalanine with p-coumaric acid and caffeic acid as an intermediate (Huang et al., 1993; Dudareva et al., 2013). Thermal and enzymatic processes produce 2-methoxy-4-vinylphenol, 4-vinylguaiacol, and 4-vinylphenol in cooked rice through the decarboxylation of ferulic acid (Coghe et al., 2004).

8.5 METABOLIC ENGINEERING FOR ENHANCING RICE AROMA

Rice aroma improvement has been attempted through genetic and biochemical engineering for the enhancement of major aroma volatile compound 2-AP. 2-AP content was enhanced through genetic manipulation of the *BADH2* and P5CS genes. The nonfunctional nature of the *BADH2* gene is responsible for 2-AP accumulation in fragrant rice varieties (Berner et al., 1986; Bradbury et al., 2005). Therefore, many researchers have attempted downregulation of the *BADH2* gene in nonaromatic rice varieties using hairpin RNA (Vanavichit et al., 2005; Niu et al., 2008), artificial microRNA (Chen et al., 2012), and transcription activator-like effector nucleases (TALEN) (Shan et al., 2015) for synthesis of 2-AP. The P5CS gene expression was positively correlated with 2-AP content in aromatic rice (Huang et al., 2008). Overexpression of the P5CS gene in aromatic *indica* rice cultivars Ambemohar 157 and Indrayani resulted in more than 2-fold increase in 2-AP. In addition to 2-AP, increased expression of other positively associated OACs (nonanal, 2-nonenal, nonanal and 2-nonenal 1-hexanol, 2-octenal and guaiacol) were observed in transgenic rice, whereas hexanal content was found to be reduced upon overexpression of P5CS (Kaikavoosi et al., 2015). Hexanal is an important off flavor compound related to aroma in rice (Buttery et al., 1988; Widjaja et al., 1996).

Specific rice aroma or flavor is a result of interaction of different OACs and their quantity. 2-AP is the principle OAC that attributes characteristic pandan, cooked rice, sweet, pleasant, and popcorn-like aroma in scented rice cultivars (Buttery et al., 1983; Jezussek et al., 2002; Ajarayasiri and Chaiseri, 2008; Maraval et al., 2008; Yang et al., 2008). In addition to 2-AP, other positively contributing aroma components are 2-phenylethanol and phenylacetic acid having rose-like odor (Widjaja et al., 1996;

Jezussek et al., 2002), heptanal (floral tone) (Yang et al., 2008), (E,E)-nona-2,4-dienal (Ajarayasiri and Chaiseri, 2008; Yang et al., 2008), decanal (Maraval et al., 2008; Mathure et al., 2011), pentanal (nutty and sweet) (Yang et al., 2008; Mathure et al., 2014), [E]-2-octenal (Mathure et al., 2014), nonanal (floral odor) (Lam and Proctor, 2003; Mathure et al., 2014; Kong and Zhao, 2014), 2-heptanone (Zhou et al., 2002), vanillin (Schwab, 2000), and 3-hydroxy-4,5-dimethyl-2(5H)-furanone and bis-(2-methyl-3-furyl)-disulfide, seasoning-like odor (Sanz et al., 1997; Jezussek et al., 2002). Guaiacol (key odorant in black rice) (Yang et al., 2008), 1-octen-3-ol having mushroom type odor, and 1-hexanol (vegetal, green odor) contributed positively to rice aroma (Mathure et al., 2014).

Maximum OACs are aldehydes that are synthesized through the LOX pathway (Dudareva et al., 2013). The expression of *OsLOX3* (rice lipoxygenase 3) and *OsHPL1* (rice hydroperoxide lyase 1) is positively associated with fruity or floral odor of nonanal (Kong and Zhao, 2014). However, hexanal and 3-hexanal are off-flavor content compounds that were highly reduced with antisense-mediated HPL depletion in transgenic potato (Vancanneyt et al., 2001). Therefore, careful manipulation of *OsLOX3* and *OsHPL1* should be done for the enhancement of selected aliphatic aldehydes and thus enhance rice aroma. Similarly, fatty acid-derived odor active volatile (1-hydroxy-2-butanone, 1-penten-3-ol, heptanal, 3-hexen-1-ol, 2-octanol, cis-3-hexenal, hexanal, and 2-nonenal) synthesis was accelerated more than 3-fold by expression of the yeast $\Delta 9$ desaturase gene in tomato (*Lycopersicon esculentum*). Also, some volatiles such as 2-ethyl-furan, 5-ethyl-2-[5H]-furanone, eugenol, and 2-ethyl-thiophene that are not known to be biochemically derived from fatty acids showed sharp increase in transgenic leaves (Wang et al., 2001). Thus, for the enhancement of fatty acid-derived aroma volatiles in rice, the yeast $\Delta 9$ desaturase gene transformation could be attempted.

Overexpression of aromatic L-amino acid decarboxylases (LeAADC1A or LeAADC2) in transgenic tomato plants resulted in up to 10-fold increased emissions of 2-phenylacetaldehyde, 2-phenylethanol, and 1-nitro-2-phenylethane. Antisense reduction of LeAADC2 significantly reduced emissions of these volatiles. These results showed that it is possible to change phenylalanine-based flavor and aroma volatiles in plants by manipulating the expression of a single gene (Tieman et al., 2006). The

PAR catalyzed the last step in the synthesis of the aroma volatile 2-phenyl-ethanol and increased expression of LePAR1 or LePAR2 that resulted in obtaining significantly higher levels of 2-phenylethanol in tomato (Tieman et al., 2007). For enhancing floral-like aroma emitted by phenylacetalde-hyde, the overexpression of aromatic l-amino acid decarboxylases can be done in rice. In order to increase 2-phenylethanol content having rose, honey-like odor, the manipulation of rice PAR could be attempted.

The carotenoid cleavage dioxygenase gene could be a possible target gene for enhancing floral and violet-like odor of β-ionone and α- ionone in rice. It was reported previously that the expression of β-ionone, pseu-doionone, and geranyl acetone was controlled by the carotenoid cleavage dioxygenase 1 genes in tomato (Simkin et al., 2004a) and *Petunia* flowers (Simkin et al., 2004b). Hexanol and Z-3-hexenol (having green, grassy odor) levels were increased in fruit with increased alcohol dehydrogenase (ADH) activity and reduced with low ADH activity in fruit (Speirs et al., 1998). Some compounds are associated with off-flavor of cooked rice. During the storage, the content of off-flavor rice components hexanal, (E)-2-nonenal, and octanal increases significantly (Lam and Proctor, 2003; Wongpornchai et al., 2004; Mathure et al., 2011). 2-Aminoacetophenone is thought to be an off-odor in brown rice in that it has a naphthalene or floor polish odor (Rapp et al., 1993). Thus, for improvement of aroma, 2-AP and other selected positive OACs could targeted via the increased expres-sion of related genes, whereas off-flavor compounds could be reduced by downregulation of specific genes involved in their biosynthesis.

8.6 CONCLUSION

More than 480 volatile compounds from different chemical classes (alde-hydes, ketones, hydrocarbons, alcohols, phenols, N-containing aromatic, ester, furans, and carboxylic acid) are present in various rice types. Among these, only 14% (68) volatile compounds share as OACs and contribute to aroma in various scented rice. The 2-AP serves as a principle OAC in all rice types. Based on aroma active values, different rice types, namely *aromatic, indica*, and *tropical japonica* can be grouped together, thus sep-arating *temperate japonicas*. Thus, aroma volatiles can very well act as biomarkers for the separation of rice types. Although ice aroma is genetically regulated, environmental and geographic climatic conditions have a major role in deciding the particular flavor type. The type of scented rice grown in similar climatic conditions is likely to exhibit similarity in their odor or flavor. The OACs are biosynthesized from carbohydrate, fatty acids, and amino acids. Rice aroma improvement through metabolic engineering can be achieved for enhancing 2-AP content. Besides 2-AP, off-flavor compounds could also be downregulated and positive OACs could be increased through metabolic engineering of specific genes for the development of aromatic rice favored by consumers.

8.7   SUMMARY

More than 480 volatile compounds were detected in various types of fragrant rice cultivars. Among these volatile compounds, only 14% (68 compounds) were identified as OACs contributing to rice aroma. The principle aroma compound 2-acetyl-1-pyrroline (2-AP) has the highest odor active value/flavor dilution factor/odor intensity and is reported as a major key volatile present in all types of rice flavors. The content of 2-AP and other OACs varies among the different rice types. The contribution of these volatiles into overall aroma and their interaction generates characteristic flavor blend for each rice type. Thus, these volatile act as biomarkers and distinguishing different flavor types (Basmati, Jasmine, Thai, Jasmati, Black rice, etc.) and rice types (*aromatic, indica, tropical japonica,* and *temperate japonica*). Further, understanding biosynthetic pathway of these OACs in rice will open door toward the metabolic engineering of rice aroma.

ACKNOWLEDGMENTS

Altafhusain Nadaf acknowledges the financial assistance from the Department of Atomic Energy, Board of Research in Nuclear Sciences, Mumbai, India (Sanction No. 2006/37/45/BRNS), Rahul Zanan acknowledges the Science and Engineering Research Board (SERB), Department of Science and Technology, India (Sanction No. SR/FT/LS-350/2012), and Vidya Hinge acknowledges the WOS-A scheme of the Department of Science and Technology, India (Sanction No. SR/WOS-A/LS433/2011(G)) for financial assistance. Deo Rashmi acknowledges the UGC-BSR research fellowship (Sanction No. F7-144/2007(BSR).

KEYWORDS

- aroma biosynthesis
- aroma volatiles
- biomarkers
- flavor
- isozyme markers
- odor active compounds (OAC)
- odor active value (OAV)
- rice aroma
- α- and β-oxidation pathway

REFERENCES

Acree, T. E. (1997). GC/Olfactometry: GC with a sense of smell. *Anal. Chem. News Features., 69*(3), 170–175.

Agrama, H. A., Yan, W., Jia, M., Fjellstrom, R., & McClung, A. M. (2010). Genetic structure associated with diversity and geographic distribution in the USDA rice world collection. *Nat. Sci., 2*(04), 247.

Ajarayasiri, J., & Chaiseri, S. (2008). Comparative study on aroma-active compounds in thai, black and white glutinous rice varieties. *Kasetsart J. (Nat. Sci.), 42*, 715–722.

Ayano, Y., & Furuhashi, T. (1970). Volatiles from cooked Kaorimai rice'. Chiba Diagaku. *Engeigakubu Gakujutsu Hokoku, 18*, 53.

Beck, H. C., Hansen, A. M., & Lauritsen, F. R. (2002). Metabolite production and kinetics of branched-chain aldehyde oxidation in Staphylococcus xylosus. *Enzyme and Microbial Technology, 31*, 94–101.

Beekwilder, J., Alvarez-Huerta, M., Neef, E., Verstappen, F. W. A., Bouwmeester, H. J., & Aharoni, A. (2004). Functional characterization of enzymes forming volatile esters from strawberry and banana. *Plant Physio., 135*, 1865–1878.

Belitz, H. D., Grosch, W., & Schieberle, P. (2004). *Food Chemistry,* 4th revised and extended revision. pp. 1–989.

BeMiller, J. N., & Whistler, R. L. (1996). Carbohydrates. In: *Food Chemistry*. Fennema, O. R. (ed.). Marcel Dekker, Inc. New York.

Bergman, C. J., Delgado, J. T., Bryant, R., Grimm, C., Cadwallader, K. R., & Webb, B. D. (2000). Rapid gas chromatographic technique for quantifying 2-acetyl-1-pyrroline and hexanal in rice (*Oryza sativa* L.). *Cereal Chem., 77*, 454–458.

Berner, D. K., & Hoff, B. J. (1986). Inheritance of scent in American long grain rice. *Crop Science, 26*, pp. 876–878.

Bett-Garber, K. L., Champagne, E. T., McClung, A. M., Moldenhauer, K. A., Linscombe, S. D., & McKenzie, K. S. (2001). Categorizing rice cultivars based on cluster analysis of amylose content, protein content, and sensory attributes. *Cereal Chem., 78*, 551–558.

Bhattacharjee, P., Singhal, R. S., & Kulkarni, P. R. (2002). Basmati rice: A review. *Intl. J. Food Sci. Tech., 37*, 1–12.

Blank, I. (1997). Gas chromatography-olfactometry in food aroma analysis. In: Marsily, R., (ed.). *Techniques for Analyzing Food Aroma*. Dekker, New York, pp. 293–330.

Boatright, J., Negre, F., Chen, X., Kish, C. M., Wood, B., Peel, G., Orlova, I., Gang, D., Rhodes, D., & Dudareva, N. (2004). Understanding in vivo benzenoid metabolism in petunia petal tissue. *Plant Physiol., 135*, 1993–2011.

Bood, K. G., & Zabetakis, I. (2002). The biosynthesis of strawberry flavor (II): biosynthetic and molecular biology studies. *J. Food Sci., 67*, 2–8.

Bradbury, L. M. T., Fitzgerald, T. L., Henry, R. J., Jin, Q., & Waters, D. L. E. (2005). The gene for fragrance in rice. *Plant Biotech. J., 3*, 363–370.

Bryant, R. J., & McClung, A. M. (2011). Volatile profiles of aromatic and non-aromatic rice cultivars using SPME/GC–MS. *Food Chem., 124*, 501–513.

Bullard, R. W., & Holguin, G. (1977). Volatile components of unprocessed rice (*Oryza sativa* L.). *J. Agric. Food Chem., 25*(1), 99–103.

Buttery, R. G., Ling, L. C., & Juliano, B. O. (1982). 2-Acetyl-1-pyrroline – an important aroma component of cooked rice. *Chem. Ind. (London), 23*, 958–959.

Buttery, R. G., Ling, L. C., & Mon, T. R. (1986). Quantitative analysis of 2-acetyl-1-pyrroline in rice. *J Agric Food Chem., 34*(1), 112–114.

Buttery, R. G., Ling, L. C., Juliano, B. O., & Turnbaugh, J. G. (1983). Cooked rice aroma and 2-acetyl- 1-pyrroline. *J. Agric. Food Chem., 31*, 823–826.

Buttery, R. G., Orts, W. J., Takeoka, G. R., & Nam, Y. (1999). Volatile flavor components of rice cakes. *Agric. Food Chem., 47* (10), 4353–4356.

Buttery, R. G., Turnbaugh, J. G., & Ling, L. C. (1988). Contribution of volatiles to rice aroma. *J. Agric. Food Chem., 36*, 1006–1009.

Caicedo, A. L., Williamson, S. H., Hernandez, R. D., Boyko, A., Alon, A. F., York, T. L., Polato, N. R., Olsen, K. M., Nielsen, R., McCouch, S. R., Bustamante, C. D., & Purugganan, M. D. (2007). Genome-wide patterns of nucleotide polymorphism in domesticated rice. *PLoS Genet., 3*, 163.

Calingacion, M., Fang, L., Quiatchon-Baeza, L., Mumm, R., Riedel, A., Hall, R. D., & Fitzgerald, M. (2015). Delving deeper into technological innovations to understand differences in rice quality. *Rice, 8*, 6 DOI 10.1186/s12284–015–0043–8.

Champagne, E. T. (2008). Rice aroma and flavor: A literature review. *Cereal Chem., 85*(4), 445–454.

Chen, M., Wei, X. I., Shao, J., Tang, S., Ju Luo, S., & Hu, P. (2012). Fragrance of the rice grain achieved via artificial microRNA-induced down-regulation of OsBADH2. *Plant Breeding, 131*(5), 584–590.

Cheng, A., Xiang, C., Li, X., Yang, C. Q., Hu, W. L., Wang, L. J., Lou, Y. G., & Chen, X. Y. (2007). The rice (E)-b-caryophyllene synthase (OsTPS3) accounts for the major inducible volatile sesquiterpenes. *Phytochemistry, 68*, 1632–1641.

Chikubu, S. (1970). Stale flavor of stored rice. *Jap. Agril. Res. Quar., 5*, 63–8.

Cho, S. (2012). *Aroma Chemistry of Wild Rice (Zizania palustris) and African Species (Oryza sativa, Oryza glaberrima*, and interspecific hybrids). A dissertation submitted to the graduate faculty of the University of Georgia in partial fulfillment of the requirements for the degree Doctor of Philosophy Athens, Georgia.

Chou, S. L. (1948). China is the place of origin of rice (in Chinese). *J. Rice Soc. China, 7*, 53–54.

Christoph, N., Bauer-Christoph, C., Gessner, M., Koehler, H. J., Simat, T. J., & Hoenicke, K. (1998). Formation of 2-aminoacetophenone and formylaminoacetophenone in wine by reaction of sulfurous acid with indole-3-acetic acid. *Wein-Wissenschaft., 53*, 79–86.

Coghe, S., Benoot, K., Delvaux, F., Vanderhaegen, B., & Delvaux, F. R. (2004). Ferulic acid release and 4-vinylguaiacol formation during brewing and fermentation: indications for feruloyl esterase activity in Saccharomyces cerevisiae. *J. Agric. Food Chem., 52*, 602–608.

D'Auria, J. C., Pichersky, E., Schaub, A., Hansel, A., & Gershenzon, J. (2007). Characterization of a BAHD acyltransferase responsible for producing the green leaf volatile (Z)-3-hexen-1-yl acetate in Arabidopsis thaliana. *Plant Journal, 49*, 194–207.

Dickinson, J. R., Harrison, S. J., Dickinson, J. A., & Hewlins, M. J. (2000). An investigation of the metabolism of isoleucine to active amyl alcohol in Saccharomyces cerevisiae. *J. of Bio. Chem., 275*, 10937–10942.

Dudareva, N., & Pichersky, E. (2006). Floral scent metabolic pathways: their regulation and evolution. In: *Biology of Floral Scent* (Dudareva, N., & Pichersky, E., eds.). Boca Raton, FL, CRC Press, pp. 55–78.

Dudareva, N., Klempien, A., Muhlemann, J. F., & Kaplan, A. (2013). Biosynthesis, function and metabolic engineering of plant volatile organic compounds. *New Phytologist, 198*, 16–32.

Dudareva, N., Pichersky, E., & Gershenzon, J. (2004). Biochemistry of plant volatiles. *Plant Physiology, 135*, 1893–1902.

Endo, K., Aoki, T., Yoda, Y., Kimura, K. I., & Hama, C. (2007). Notch signal organizes the Drosophila olfactory circuitry by diversifying the sensory neuronal lineages. *Nat. Neurosci., 10*(2), 153–160.

Etschmann, M. M. W., Sell, D., & Schrader, J. (2005). Production of 2-phenylethanol and 2-phenylethylacetate from L-phenylalanine by coupling whole-cell biocatalysis with organophilic pervaporation. *Biotech. Bioengineering, 92*, 624–634.

FAO, (2015). www.fao.org/economic/est/publications/rice./the-fao-rice-price./en/.

Feussner, I., & Wasternack, C. (2002). The lipoxygenase pathway. *Annual Review of Plant Biology, 53*, 275–297.

Fitzgerald, M. A., McCouch, S. R., & Hall, R. D. (2009). Not just a grain of rice: The quest for quality. *Trends Plants Sci., 14*, 133–139.

Frey, M., Stettner, C., Pare, P. W., Schmelz, E. A., Tumlinson, J. H., & Gierl, A. (2000). An herbivore elicitor activates the gene for indole emission in maize. *PNAS, 97,* 14801–14806.

Frydman, R. B., Tomaro, M. L., & Frydman, B. (1971). Pyrrolooxygenases: The biosynthesis of 2-aminoacetophenone. *FEBS Letters, 17,* 273–276.

Garris, A., Tai, T., Coburn, J., Kresovich, S., & McCouch, S. R. (2005). Genetic structure and diversity in *Oryza Sativa* L. *Genetics., 169,* 1631–1638.

Gigot, C., Ongena, M., Fauconnier, M. L., Wathelet, J. P., Jardin, P. D., & Thonart, P. (2010). The lipoxygenase metabolic pathway in plants: potential for industrial production of natural green leaf volatiles. *Biotechnologie, Agronomie, Soc. Environ., 14,* 451–460.

Givianrad, M. H. (2012). Characterization and assessment of flavor compounds and some allergens in three Iranian rice cultivars during gelatinization process by HS-SPME/GC-MS. *E-Journal of Chemistry, 9*(2), 716–728.

Glaszmann, J. C. (1987). Isozymes and classification of Asian rice varieties. *Theor. Appl. Genet., 74,* 21–30.

Goepfert, S., & Poirier, Y. (2007). B-Oxidation in fatty acid degradation and beyond. *Curr. Opin. Plant Biol., 10,* 245–251.

Gonzalez, M., Gaete-Eastman, C., Valdenegro, M., Figueroa, C. R., Fuentes, L., Herrera, R., & Moya-Le, M. A. (2009). Aroma development during ripening of *fragaria chiloensis* fruit and participation of an *Alcohol Acyltransferase* (FcAAT1) gene. *J. Agril. and Food Chem., 57,* 9123–9132.

Grimm, C. C., Champagne, E. T., Lioyd, S. W., Easson, M., Condon, B., & McClung, A. (2011). Analysis of 2-acetyl-l-pyrroline in rice by HSSE/GCMS. *Cereal Chemistry, 88*(3), 271–277.

Grosch, W. (2001). Evaluation of the key odorants of foods by dilution experiments, aroma models and omission. Chem. Senses 26, 533–545.

Hayashi, S., Yagi, K., Ishikawa, T., Kawasaki, M., Asai, T., Picone, J., Turnbull, C., Hiaratake, J., Sakata, K., Takada, M., Ogawa, K., & Watanabe, N. (2004). Emission of 2-phenylethanol from its b-D-glucopyranoside and the biogenesis of these compounds from [2H8]L-phenylalanine in rose flowers. *Tetrahedron, 60,* 7005–7013.

Hernandez, P., Thomas, C., Hsieh, Y., Smith, C., & Fischer, N. (1989). Foliage volatiles of two rice cultivars. *Photochemistry, 28,* 2959–2962.

Hien, N. L., Yoshihashi, T., Sarhadi, W. A., Thanh, V. C., Oikawa, Y., & Hirata, Y. (2006). Evaluation of aroma in rice (*Oryza Sativa* L.) using KOH method, molecular marker and measurement of 2-acetyl-1-pyrroline concentration. *Jpn. J. Trop. Agric., 50,* 190–198.

Hinge, V., Patil, H., & Nadaf, A. (2015). Comparative characterization of aroma volatiles and related gene expression analysis at vegetative and mature stages in basmati and non-basmati rice (*Oryza sativa* L.) cultivars. *Appl. Biochem. Biotechnol.,* DOI 10.1007/s12010–015–1898–2.

Hofmann, T., & Schieberle, P. (1998). Flavor contribution and formation of the intense roastsmelling odorants 2-propionyl-1-pyrroline and 2-propionyltetrahydropyridine in Maillard-type reactions. *J. Agric. Food Chem., 46,* 2721–2726.

Hsieh, M. H., Chang, C. Y., Hsu, S. J., & Chen, J. J. (2008). Chloroplast localization of methylerythritol 4-phosphate pathway enzymes and regulation of mitochondrial genes in IspD., & IspE albino mutants in Arabidopsis. *Plant Molecular Biology, 66,* 663–673.

Huang, T. C., Teng, C. S., Chang, J. L., Chuang, H. S., Ho, C. T., & Wu, M. L. (2008). Biosynthetic mechanism of 2-acetyl-1-pyrroline and its relationship with Δ1-pyrroline-5-carboxylic acid and methylglyoxal in aromatic rice (*Oryza sativa* L.) callus. *Journal of Agricultural and Food Chemistry, 56*(16), 7399–7404.

Huang, Z., Dostal, L., & Rosazza, J. P. (1993). "Microbial transformations of ferulic acid by Saccharomyces cerevisiae and Pseudomonas fluorescens". *Applied and Environmental Microbiology, 59* (7), 2244–2250.

Hussain, A., Naqvi, S. H. M., & Hammerschimdt, F. J. (1987). Isolation and identification of volatile components from Basmati rice (*Oryza Sativa* L). In: Martens, M., Dalen, G. A., Russwurm, H. (eds.), *Flavor Sci. Tech. Wiley*, pp. 95–100.

Ibdah, M., Azulay, Y., Portnoy, V., Wasserman, B., Bar, E., Meir, A., Burger, Y., Hirschberg, J., Schaffer, A. A., & Katzir, N. (2006). Functional characterization of CmCCD1, a carotenoid cleavage dioxygenase from melon. *Phytochemistry, 67,* 1579–1589.

Ishitani, K., & Fushimi, C. (1994). Influence of pre- and post-harvest conditions on 2-acetyl-1-pyrroline concentration in aromatic rice. *Koryo., 183,* 73–80.

Itani, T., Tamaki, M., Hayata, Y., Fushimi, T., & Hashizume, K. (2004). Variation of 2-acetyl-1-pyrroline concentration in aromatic rice grains collected in the same region in Japan and factors affecting its concentration. *Plant Prod. Sci., 7,* 178–183.

Jezussek, M., Juliano, B. O., & Schieberle, P. (2002). Comparison of key aroma compounds in cooked brown rice varieties based on aroma extract dilution analyses. *J. Agric. Food Chem., 50,* 1101–1105.

Jones, M. G., Hughes, J., Tregova, A., Milne, J., Tomsett, A. B., & Collin, H. A. (2004). Biosynthesis of the flavor precursors of onion and garlic. *J. Exp. Bot., 55,* 1903–1918.

Kaikavoosi, K., Kad, T. D., Zanan, R. L., & Nadaf, A. B. (2015). 2-Acetyl-1-pyrroline augmentation in scented indica rice (*Oryza sativa* L.) varieties through Δ(1)-pyrroline-5-carboxylate synthetase (P5CS) gene transformation. *Appl. Biochem. Biotechnol., 177*(7), 1466–1479.

Kaminaga, Y., Schnepp, J., Peel, G., Kish, C. M., Ben-Nissan, G., Weiss, D., Orlova, I., Lavie, O., Rhodes, D., & Wood, K. (2006). Plant phenylacetaldehyde synthase is a bifunctional homotetrameric enzyme that catalyzes phenylalanine decarboxylation and oxidation. *J. of Bio. Chem., 281,* 23357–23366.

Khorheh, N. A., Givianrad, M. H., Ardebili, M. S., & Larijani, K. (2011). Assessment of flavor volatiles of iranian rice cultivars during gelatinization process. *J. of Food Biosci. and Tech., 1,* 41–54.

Khush, G. S. (2000). Taxonomy and origin of rice. In: *Aromatic Rices* (Singh, R. K., Singh, U. S., & Khush, G. S., eds.). Oxford & IBH Publishing Co. Pvt. Ltd., New Delhi, India, pp. 5–13.

Klein, D., Fink, B., Arold, B., Eisenreich, W., & Schwab, W. (2007). Functional characterization of enone oxidoreductases from strawberry and tomato fruit. *J. Agric. Food Chem., 55,* 6705–6711.

Knudsen, J. T., & Gershenzon, J. (2006). The chemical diversity of floral scent. In: *Biology of Floral Scent* (Dudareva, N., & Pichersky, E., eds.). Boca Raton, FL., CRC Press, pp. 27–52.

Knudsen, J. T., Eriksson, R., Gershenzon, J., & Stahl, B. (2006). Diversity and distribution of floral scent. *Botanical Review, 72*, 1–120.

Kong, Z., & Zhao, D. (2014). The inhibiting effect of abscisic acid on fragrance of kam sweet rice. *Journal of Food and Nutrition Research, 2*(4), 148–154.

Kovach, M. J., Calingacion, M. N., Fitzgerald, M. A., & McCoucha, S. R. (2009). The origin and evolution of fragrance in rice (*Oryza sativa* L.) *PNAS., 106*(34), 14444–14449.

Laguerre, M. L., Mestres, C., Davrieux, F., Ringuet, J., & Boulange, R. (2007). Rapid discrimination of scented rice by solid-phase microextraction, mass spectrometry, and multivariate analysis used as a mass sensor. *J. Agric. Food Chem., 55*, 1077–1083.

Lam, H. S., & Proctor, A. (2003). Milled rice oxidation volatiles and odor development. *J. Food Sci., 68*, 2676–2681.

Legendre, M. G., Dupuy, H. D., Ory, R. L., & McIlrath, W. L. (1978). Instrumental analysis of volatiles from rice and corn products. *J. Agric. Food Chem., 26*(5).

Liyanaarachchi, G. D., Kottearachchi, N. S., & Samarasekera, R. (2014). Volatile profiles of traditional aromatic rice varieties in Sri Lanka. *J. Natn. Sci. Foundation Sri. Lanka, 42*(1), 87–93.

Lorieux, M., Petrov, M., Huang, N., Guiderdoni, A., & Ghesquière, E. (1996). Aroma in rice: genetic analysis of a quantitative trait. *Theor. Appl. Genet., 93*, 1145–1151.

Maga, J. A. (1984). Rice product volatiles: A review. *J. Agric. Food Chem., 32*(5), 964–970.

Mahatheeranont, S., Keawsaard, S., & Dumri, K. (2001). Quantification of the rice aroma compound, 2-acetyl-1-pyrroline, in uncooked Khao Dawk Mali 105 brown rice. *J. Agric. Food Chem., 49*(2), 773–779.

Mahatheeranont, S., Promdang, S., & Chiampiriyakul, A. (1995). Voaltile aroma compounds of Khao Dawk Mali 105 rice. *Kasetsart J. (Nat. Sci.), 29*, 508–514.

Mahattanatawee, K., & Rouseff, L. R. (2014). Comparison of aroma active and sulfur volatiles in three fragrant rice cultivars using GC–Olfactometry and GC–PFPD. *Food Chem., 154*, 1–6.

Mann, S. (1967). Quinazoline derivatives in pseudomonads. *Arch. Mikrobiol., 56*(4), 324–329.

Maraval, I., Mestres, C., Pernin, K., Ribeyre, F., Boulanger, R., Guichard, E., & Gunata, Z. (2008). Odor-active compounds in cooked rice cultivars from camargue (France) analyzed by Gc-O and Gc-Ms. *J. Agric. Food Chem., 56*, 5291–5298.

Marsili, R. (2007). Application of sensory-directed flavor-analysis techniques. In: *Sensory-Directed Flavor Analysis*, Marsili, R., editor. Boca Raton, FL., CRC/Taylor & Francis, pp. 55–80.

Mathure, S. V., Jawali, N., Thengane, R. J., & Nadaf, A. B. (2014). Comparative quantitative analysis of headspace volatiles and their association with BADH2 marker in non-basmati scented, basmati and non-scented rice (*Oryza sativa* L.) cultivars of India. *Food Chem., 1*(142), 383–391.

Mathure, S. V., Wakte, K. V., Jawali, N., & Nadaf, A. B. (2011). Quantification of 2-acetyl-1-pyrroline and other rice aroma volatiles among Indian scented rice cultivars by HS-SPME/GC-FID. *Food Anal. Methods, 4*(3), 326–333.

McGarvey, D. J., & Croteau, R. (1995). Terpenoid metabolism. *Plant Cell, 7*, 1015–1026.

Meullenet, J. F., Griffin, V. K., Carson, K., Davis, G., Davis, S., Gross, J., Hankin, J. A., Sailor, E., Sitakalin, C., Suwansri, S., & Caicedo, V. (2001). Rice external preference mapping for Asian consumers living in the United States. *J. Sensory Study, 16,* 73–94.

Mitsuda, H., Yasumoto, K., & Iwami, K. (1968). Analysis of volatile components in rice bran. *Agric. Biol. Chem., 32*(4), 453–458.

Monsoor, M. A., & Proctor, A. (2004). Volatile component analysis of commercially milled head and broken rice. *J. Food Sci., 69,* 632–636.

Mottram, D. S. (2007). Chemistry, bioprocessing and sustainability, In: *Flavors and Fragrances–*(Berger, R. G., ed.), Spinger, Berlin, pp. 269–283.

Nadaf, A. B., Krishnan, S., & Wakte, K. V. (2006). Histochemical and biochemical analysis of major aroma compound (2-acetyl-1-pyrroline) in basmati and other scented rice (*Oryza sativa* L.). *Curr. Sci., 91*(11), 1533–1537.

Nagaraju, M., Mohanty, K. K., Chowdhury, D., & Gangadharan, C. (1991). A simple technique to detect scent in rice. *Oryza, 28,* 109–110.

Niu, X., Tang, W., Huang, W., Ren, G., Wang, Q., et al. (2008) RNAi-directed downregulation of OsBADH2 results in aroma (2-acetyl-1-pyrroline) production in rice (*Oryza sativa* L.). *BMC Plant Biol. 8,* 100.

Pak, F. E., Gropper, S., Dai, W. D., Havkin-Frenkel, D., & Belanger, F. C. (2004). Characterization of a multifunctional methyltransferase from the orchid Vanilla planifolia. *Plant Cell Rep., 22,* 959–966.

Paule, C. M., & Powers, J. J. (1989). Sensory and chemical examination of aromatic and nonaromatic rices. *J. Food Sci., 54,* 343–346.

Petrov, M., Danzart, M., Giampaoli, P., Faure, J., & Richard, H. (1996). Rice aroma analysis: Discrimination between a scented and a nonscented rice. *Sci. Alim., 16,* 347–360.

Phillips, S. A., & Thornalley, P. J. (1993). The formation of methylglyoxal from triose phosphates: Investigation using a specific assay for methylglyoxal. *Euro. J. of Biochem., 212*(1), 101–105.

Pichersky, E, Noel, J. P., & Dudareva, N. (2006). Biosynthesis of plant volatiles: nature's diversity and ingenuity. *Science, 311,* 808–811.

Pichersky, E., & Gershenzon, J. (2002). The formation and function of plant volatiles: perfumes for pollinator attraction and defense. *Curr. Opin. Plant Biol., 5,* 237–243.

Piyachaiseth, T., Jirapakkul, W., & Chaiseri, S. (2011). Aroma compounds of flash-fried rice. *Kasetsart J. (Nat. Sci.), 45,* 717–729.

Podstolski, A., Havin-Frenkel, D., Malinowski, J., Blount, J. W., Kourteva, G., & Dixon, R. A. (2002). Unusual 4–hydroxybenzaldehyde synthase activity from tissue cultures of vanilla orchid Vanilla planifolia. *Phytochemistry, 61,* 611–620.

Pulido, P., Perello, C., & Rodriguez-Concepcion, M. (2012). New insights into plant isoprenoid metabolism. *Molecular Plant, 5,* 964–967.

Rapp, H., Versini, G., & Ullemeyer, H. (1993). 2-Aminoacetophenone: Causing components of the untypical aging note (Naphtalinton, "hybrid tone") in wine. *Vitis., 32,* 61–62.

Reineccius, G. A. (2006). *Flavor Chemistry and Technology.* 2nd edition, Taylor & Francis group, pp. 139–220.

Richard, J. (1991). Kinetic parameters for the elimination reaction catalyzed by triosephosphate isomerase and an estimation of the reaction's physiological significance. *Biochemistry, 30,* 4581–4585.

Rothe, M. (1976). Aroma values-a useful concept? *Die Nahrung., 20*(3), 259–266.

Rychlik, M., Schieberle, P., & Grosch, W. (1998). Compilation of odor thresholds, odor qualities and retention indices of key food odorants; German Research Center for Food Chemistry and Institute of Food Chemistry of the Technical University Munchen: Garching, Germany.

Saffert, A., Hartmann-Schreier, J., Schon, A., & Schreier, P. (2000). A dual function a-dioxygenase-peroxidase and NAD+ oxidoreductase active enzyme from germinating pea rationalizing a-oxidation of fatty acids in plants. *Plant Physiol., 123,* 1545–1551.

Sanz, C., Olias, J. M., & Perez, A. G. (1997). Aroma biochemistry of fruits and vegetables.. In: *Phytochemistry of Fruit and Vegetables*, Tomas-Barberan, F. A., & Robins, R. J. (eds.). Clarendon Press, Oxford, pp. 125–155.

Sarkarung, S., Somrith, B. S. Chitrakorn, Das, T., Baqui, M. A. Buu, B. C., Nematzadeh, G. A., Karbalaie, M. T., & Farrokhzad, F. (2000). Aromatic rice's of other country, In: *Aromatic Rices,* Singh, R. K., Singh, U. S., & Khush, G. S. (eds.). Science Publishers Inc., Enfield, NH, USA, pp. 179–200.

Schottler, M., & Boland, W. (1996). Biosynthesis of dodecano-4- lactone in ripening fruits: crucial role of an epoxide-hydrolase in enantioselective generation of aroma components of the nectarine (Prunus persica var nucipersica) and the strawberry (Fragaria ananassa). *Helv. Chim. Acta., 79,* 1488–1496.

Schreiber, G. A., Schulzki, G., Spiegelberg, A. M., Helle, N., & Boegl, K. W. (1994). Evaluation of a gas chromatographic method to identify irradiated chicken, pork, and beef by detection of volatile hydrocarbons. *Journal of AOAC International, 77,* 1202–1217.

Schwab, W., & Schreier, P. (2002) Enzymic formation of flavor volatiles from lipids. In: *Lipid Biotechnology* (Kuo, T. M., & Gardner, H. W., eds.). New York, Marcel Dekker, pp. 293–318.

Schwab, W. (2000). Biosynthesis of plants flavors: analysis and biotechnological approach. In: *Flavor Chemistry: Industrial and Academic Research,* Risch, S. J., & Ho, C. T. (eds.). Oxford Univ. Press, Washington DC, pp. 72–86.

Schwab, W., Rikanati, D. R., & Lewinsohn, E. (2008). Biosynthesis of plant-derived flavor compounds. *The Plant J., 54,* 712–732.

Sergienko, E. A., & Jordan, F. (2001). Catalytic acid–base groups in yeast pyruvate decarboxylase. 2. Insights into the specific roles of D28 and E477 from the rates and stereospecificity of formation of carboligase side products. *Biochem., 40,* 7369–7381.

Shan, Q., Zhang, Y., Chen, K., Zhang, K., & Gao, C. (2015). Creation of fragrant rice by targeted knockout of the OsBADH2 gene using, *TALEN Technology, 13*(6), 791–800.

Shobha, R. N., Pandey, M. K., Prasad, G. S. V., & Sudharshan, S. (2006). Historical significance, grain quality features and precision breeding for improvement of export quality basmati varieties in India. *Indian J. Crop Science, 1*(1–2), 29–41.

Simkin, A. J., Guirimand, G., Papon, N., Courdavault, V., Thabet, I., Ginis, O., Bouzid, S., Giglioli-Guivarc'h, N., & Clastre, M. (2011). Peroxisomal localisation of the final steps of the mevalonic acid pathway in planta. *Planta, 234,* 903–914.

Simkin, A. J., Schwartz, S. H., Auldridge, M., Taylor, M. G., Klee, H. J. (2004a). The tomato carotenoid cleavage dioxygenase 1 genes contribute to the formation of the flavor volatiles beta-ionone, pseudoionone, and geranylacetone. *Plant J., 40,* 882–892.

Simkin, A. J., Underwood, B. A., Auldridge, M., Loucas, H. M., Shibuya, K., Schmelz, E., Clark, D. G., & Klee, H. J. (2004b). Circadian regulation of the PhCCD1 carotenoid cleavage dioxygenase controls emission of beta-ionone, a fragrance volatile of petunia flowers. *Plant Physiol., 136*, 3504–3514.

Singh, R. K., Singh, U. S., & Khush, G. S. (2000). *Aromatic Rices.* Oxford & IBH Publ., New Delhi, India. pp. 1–281.

Singh, R., & Srivastava, V. K. (1997). *Report on Basmati and Scented Fine Grained-Types,* at Varanasi Centre.' BHU. Varanasi, UP, pp. 30.

Slaughter, J. C. (1999). The naturally occurring furanones: formation and function from pheromone to food. *Biol. Rev., 74*, 259–276.

Speckman, R. A., & Collins, E. B. (1968). "Diacetyl biosynthesis in Streptococcus diacetilactis and Leuconostoc citrovorum". *J. Bacteriol., 95*(1), 174–80. PMC 251989. PMID 5636815.

Speirs, J., Lee, E., Holt, K., Yong-Duk, K., Scott, N. S., Loveys, B., Schuch, W. (1998). Genetic manipulation of alcohol dehydrogenase levels in ripening tomato fruit affects the balance of some flavor aldehydes and alcohols. *Plant Physiol., 117*, 1047–1058.

Sukhonthara, S., Theerakulkait, C., & Miyazawa, M. (2009). Characterization of volatile aroma compounds from red and black rice bran. *J. Oleo. Sci., 58*(3), 155–161.

Suvarnalatha, G., Narayan, M. S., Ravishankar, G. A., & Venkataraman, L. V. (1994). 'Flavor production in plant cell cultures of Basmati rice'. *J. Sci. Food Agric., 66,* 439–442.

Tanaka, Y., Resurreccion, A. P., Juliano, B. O., & Bechtel, D. B. (1978). Properties of whole and undigested fraction of protein bodies of milled rice. *Agric. Biol. Chem., 42*, 2015–2023.

Tang, T., Lu, J., Huang, J., He, J., McCouch, S. R., Shen, Y., Kai, Z., Purrugganan, M. D., Shi, S., & Wu, C. (2007). Genomic variation in rice: genesis of highly polymorphic linkage blocks during domestication. *PLoS Genet., 2*, e199.

Tava, A., & Bocchi, S. (1999). Aroma of cooked rice (*Oryza sativa*): Comparison between commercial Basmati and Italian line B5–3. *Cereal Chem., 76*, 526–529.

Tavaria, F. K., Dahl, S, Carballo, F. J., & Malcata, F. X. (2002). Amino acid catabolism and generation of volatiles by lactic acid bacteria. *Journal of Dairy Science, 85,* 2462–2470.

Tieman, D. M., Loucas, H. M., Kim, J. Y., Clark, D. G., & Klee, H. J. (2007). Tomato phenylacetaldehyde reductases catalyze the last step in the synthesis of the aroma volatile 2-phenylethanol. *Phytochemistry, 68*, 2660–2669.

Tieman, D. M., Taylor, M. G., Schauer, N, Fernie, A. R., Hanson, A. D., & Klee, H. J. (2006). Tomato aromatic amino acid decarboxylases participate in synthesis of the flavor volatiles 2-phenylethanol and 2-phenylacetaldehyde. *Proceedings of the National Academy of Sciences, USA, 103*, 8287–8292.

Ting, Y. (1957). The origin and evolution of cultivated rice in China (in Chinese with English abstract). *Acta. Agron Sinica., 8*, 243–260.

Tressl, R., & Albrecht, W. (1986). Biogenesis of aroma compounds through acyl pathways. In: *Biogeneration of Aromas*, Parliament, T. H., & Croteau, R., eds., ACS. Washington, DC, USA, pp. 114–133.

Tsugita, T. (1985). Aroma of cooked rice. *Food Reviews International., 1*(3), 497–520.

Tsuzuki, E., Morinaga, K., & Shida, S. (1977). 'Studies on the characteristics of scented rice III. Varietal differences of some carbonyl compounds evolved in cooking scented rice varieties'. *Bull. Fac. Agric. Miyazaki Univ., 24,* 35.

Vachhani, M. V., Butany, W. T., & Nair, C. P. K. (1962). A tentative commercial classification of rice. *Rice Newsl., 10,* 15.

Van Ruth, S. M. (2001). Methods for gas chromatography-olfactometry: a review. *Biomol. Eng., 17*(4–5), 121–128.

Vanavichit, A., Tragoonrung, S., Theerayut, T., Wanchana, S., & Kamolsukyunyong, W. (2005). Transgenic rice plants with reduced expression of Os2AP and elevated levels of 2-acetyl-1-pyrroline. United States Patent, *Patent No. US 7,319,181 B2.*

Vancanneyt, G., Sanz, C., Farmak, T., Paneque, M., Ortego, F., Castañera, P., & Sánchez-Serrano, J. J. (2001). Hydroperoxide lyase depletion in transgenic potato plants leads to an increase in aphid performance. *Proc. Natl. Acad. Sci. USA, 98,* 8139–8144.

Verma, D. K., & Srivastav, P. P. (2017). Proximate composition, mineral content and fatty acids analyses of aromatic and non-aromatic Indian rice. *Rice Science, 24*(1), 21–31.

Verma, D. K., Mohan, M., & Asthir, B. (2013). Physicochemical and cooking characteristics of some promising basmati genotypes. *Asian Journal of Food Agro-Industry,* 6(2), 94–99.

Verma, D. K., Mohan, M., Prabhakar, P. K., & Srivastav, P. P. (2015). Physico-chemical and cooking characteristics of Azad basmati. *International Food Research Journal,* 22(4), 1380–1389.

Verma, D. K., Mohan, M., Yadav, V. K., Asthir, B., & Soni, S. K. (2012). Inquisition of some physico-chemical characteristics of newly evolved basmati rice. *Environment and Ecology,* 30(1), 114–117.

Wang, C., Xing, J., Chin, C. K., Ho, C. T., & Martin, C. E. (2001). Modification of fatty acids changes the flavor volatiles in tomato leaves. *Phytochemistry, 58(2),* 227–32.

Weber, D. J., Rohilla, R., & Singh, U. S. (2000). Chemistry and biochemistry of aroma in scented rice. In: *Aromatic Rices*, Singh, R. K., Singh, U. S., & Khush, G. S. (eds.). Science Publishers Inc., Enfield, NH, USA, pp. 29–46.

Wein, M., Lavid, N., Lunkenbein, S., Lewinsohn, E., Schwab, W., & Kaldenhoff, R. (2002). Isolation, cloning and expression of a multifunctional O-methyltransferase capable of forming 2,5-dimethyl-4-methoxy-*3*(2H)-furanone, one of the key aroma compounds in strawberry fruits. *Plant J., 31,* 755–765.

Widjaja, R., Craske, J. D., & Wootton, M. (1996). Comparative studies on volatile components of non-fragrant and fragrant rices. *J. of Sci. Food and Agril., 70*(2), 151–161.

Wilkie, K., & Wootton, M. (2004). *Flavor Qualities of New Australian Fragrant Rice Cultivars*, Publication No. 04/160 Project No. UNS-12A.

Winterhalter, P., & Rouseff, R. (2001). Carotenoid-derived aroma compounds: an introduction. In: Winterhalter, P., & Rouseff, R. (eds.). Carotenoid-derived aroma compounds. Washington, DC, USA: *American Chemical Society*, 1–17.

Withycombe, D. A., Lindsay, R. C., & Stuiber, D. A. (1978). Isolation and identification of volatile components from wild rice grain (*Zizania aquatica*). *J. Agric. Food Chem.,* 26(4), 816–822.

Wongpornchai, S., Dumri, K., Jongkaewwattana, S., & Siri, B. (2004). Effects of drying methods and storage time on the aroma and milling quality of rice (*Oryza sativa* L.) cv. Khao Dawk Mali 105. *Food Chem., 87,* 407–414.

Yajima, I., Yanai, T., Nakamura, M., Sakakibara, H., & Habu, T. (1978). Volatile flavor components of cooked rice, *Agril and Biol. Chem., 42*(6), 1229–1233.

Yajima, I., Yanai, T., Nakamura, M., Sakakibara, S., & Hayashi, K. (1979). Volatile flavor components of cooked kaorimai (Scented Rice, *O. Sativa japonica*). *Agril. Bio. Chem., 43*(12), 2425–2429.

Yang, D. S., Shewfelt, R. L., Lee, K., & Kays, S. J. (2008). Comparison of odor-active compounds from six distinctly different rice flavor types. *J. Agric. Food Chem., 56*(8), 2780–2787.

Yi, M., New, K. T., Vanavichit, A., Chai-arree, W., & Toojinda, T. (2009). Marker assisted backcross breeding to improve cooking quality traits in Myanmar rice cultivar Manawthukha. *Field Crops Res., 113*, 178–186.

Yoshihashi, T., Huong, N. T. T., & Inatomi, H. (2002). Precursors of 2-acetyl-1-pyrroline, a potent flavor compound of an aromatic rice variety. *J. Agric. Food Chem., 50*, 2001–2004.

Zeng, Z., Zhang, H., Chen, J. Y., Zhang, T., & Matsunag, M. (2007). Direct extraction of volatiles of rice during cooking using solid-phase microextraction. *Cereal Chem., 84*(5), 423–427.

Zeng, Z., Zhang, H., Zhang, T., Tamogami, S., & Chen, J. Y. (2009). Analysis of flavor volatiles of glutinous rice during cooking by combined gas chromatography–mass spectrometry with modified headspace solid-phase microextraction method. *J. of Food Com. Anal., 22*(4), 347–353.

Zhou, Z., Robards, K., Helliwell, S., & Blanchard, C. (2002). Composition and functional properties of rice. *International Journal of Food Science & Technology, 37*, 849–868.

CHAPTER 9

BIOCHEMICAL, GENETIC AND GENOMIC PERSPECTIVES OF FLAVOR AND FRAGRANCE IN RICE

ARPIT GAUR,[1] SHABIR H. WANI,[2] SAROJ KUMAR SAH,[3] and ASIF B. SHIKARI[4]

[1]Research Scholar, Division of Genetics and Plant Breeding, Sher-e-Kashmir University of Agricultural Sciences and Technology of Kashmir, Shalimar, Srinagar–190025, Jammu and Kashmir, India, E-mail: arpitgaur@skuastkashmir.ac.in

[2]Assistant Professor-cum-Scientist, Division of Genetics and Plant Breeding, Sher-e-Kashmir University of Agricultural Sciences and Technology of Kashmir, Shalimar, Srinagar–190025, Jammu and Kashmir, India, E-mail: shabirhussainwani@gmail.com

[3]Research Scholar, Department of Biochemistry, Molecular Biology, Entomology and Plant Pathology, Mississippi State University, MS 39762, USA, E-mail: saroj1021@gmail.com

[4]Associate Professor-cum-Scientist, Division of Genetics and Plant Breeding, Sher-e-Kashmir University of Agricultural Sciences and Technology of Kashmir, Shalimar, Srinagar–190025, Jammu and Kashmir, India, E-mail: asifshikari@gmail.com

CONTENTS

9.1   INTRODUCTION

Rice is a part of routine diet for more than 50% of the world's population (Brar and Khush, 2002; Sabouri et al., 2012). Two critical parameters, i.e., aroma and cooked kernel elongation determine the preference of the rice variety to cook and eat and its market value (Verma et al., 2012, 2013, 2015; Kioko et al., 2015). Fragrant or scented or aromatic rice is a general term used for rice cultivars that have a perfumed and nutty flavor (Goufo et al., 2010a; Verma and Srivastav, 2016). Aromatic rice varieties possess a characteristic aroma called nutty or popcorn-like flavor (Bryant and McClung, 2010; Verma et al., 2012, 2013, 2015; Mo et al., 2015). Owing to its favorable flavor, the global demand for fragrant rice is increasing day by day (Myint et al., 2012; Hashemi et al., 2013). In the rice-producing countries, premium price and high consumer acceptance of fragrant rice have increased more topic of discussion for the rice producers (Shao et al., 2013; Shi et al., 2014). Fragrance and flavor of food affect the desirability and value of that food since time immortal. This convenience of human race enforced them to lay down the best example of selecting a homozygous recessive trait, i.e., aroma of rice, during rice domestication of several thousand years. The fragrance and flavor of rice remain one of the most desirable quality traits among all, which directly affect the choice of end consumers and marketability of rice, internationally. Unceasingly, an increasing number of fragrant rice lovers are shifting the aromatic rice toward the clutch of essential commodities of international food markets. As aromatic rice is gaining much more popularity with higher trade value, especially in those corners of the world where rice is not a part of traditional food habit, the trait becomes an important one for the rice breeders to aim with. Mainly in the Middle East and western communities, aromatic rice is gaining popularity among all the consumers around the world (Dong et al., 2001; Cordeiro et al., 2002; Bourgis et al., 2008; Chen et al., 2008). According to Kovach et al. (2009), aroma characteristics have been determined in three different types of subpopulations of rice, including, Group V (Sadri and Basmati), Indonesia (Jasmine), and tropical japonica. There are two groups of aromatic rice, the long-grained basmati type and the small and medium-grained indigenous aromatic varieties (Rai et al., 2015).

The biochemical foundations of aroma indicate the existence of more than 200 volatiles in aromatic rice (Champagene, 2008; Sakthivel et al., 2009; Mahattanatawee and Rouseff, 2014). Among these, the main characteristic fragrance of rice is associated with the level of synthesis of a novel compound, namely 2-acetyl-1-pyrroline (2-AP). 2-AP is considered as the primary aroma compound in aromatic rice (Mahalingam and Nadarajan, 2005; Fitzgerald et al., 2009; Jewel et al., 2011; Hashemi et al., 2013). The 2-AP compound is found in leaf and seed tissues (Kadaru, 2007), but few scientists have also reported a considered level of 2-AP production in roots of the rice plant. The accumulation of 2-AP is a result of the loss of function in the *BADH2* gene, which is located in the metacentric region of chromosome 8. Beside this major locus *BADH2*, few more Quantitative trait locus (QTLs) on chromosomes 3, 4, and 12 have also been reported with a significant role in rice aroma. Further, *BADH2*-independent biosynthetic pathways have also been reported (Huang et al., 2008), which makes aroma an intricate trait. The advancements in rice research, genomics, and genetics have opened new avenues in the array of understanding the secrets of aroma with higher precisions.

9.2 BIOCHEMISTRY OF FLAVOR AND FRAGRANCE

Fragrance and flavor in plants are the results of numerous volatile and semi-volatile compounds. Plant chemists have recognized over 250 volatile and semi-volatile compounds in both fragrant and nonfragrant types of rice varieties; however, none of these compounds have been found to be significantly associated with the presence or absence of fragrance and flavor of rice, except 2-AP. So far, it has been reported that all these volatiles and semi-volatiles are obtainable from both types of rice, but a comparison between nonfragrant and fragrant rice types revealed a significant difference in their respective concentrations. The magical compound 2-AP, which imparts a characteristic popcorn-like-aroma, has been identified as the chief compound that imparts the natural fragrance of rice by Buttery et al. (1983) and hundreds of other plant scientists (Paule and Powers, 1989; Lin et al., 1990; Tanchotikul and Hsieh,1991; Ahmed et al., 1996). Besides the major aromatic compound 2-AP, Widjaja et al. (1996) identified 70 more volatiles through a competitive study between aromatic and nonaromatic rice; among those, alkanals, alk-2-enals, alka(E)-2,4-dienals, 2-pentylfuran, 2-AP, and 2-phenylethanol are the major volatile components, whereas the remaining were found to contribute a small role in the total aroma profile. Compared to aromatic rice, volatile compounds, viz., n-hexanal, (E)-2-heptanal, (E)-2-octenal, (E)-2-(E)-4-decadienal, n-nonanal, 1-octen-3-ol, 2-pentylfuran, 4-vinylphenol, and 4-vinylguaiacol, are dominantly available in nonaromatic rice varieties. Later, more than 300 volatiles were recognized from different cultivars of scented and nonscented rice (Yang et al., 2008). In perfumed rice varieties, Kim (1999) identified 4 acids, 15 alcohols, 16 hydrocarbons, 16 aldehydes and ketones, and 10 other miscellaneous compounds; among those, paraffin was obtained as the most common hydrocarbon, whereas n-heptanol, n-nonanal, and n-pentanol were the most common aromatic (alcohol) compounds. n-butanol and n-hexanol were detected in aromatic rice only after cooking. In 1999, no significant difference in term of the presence of hydrocarbons between fragrant and nonfragrant rice was reported by Kim (Kim, 1999). He found that aromatic rice had higher levels of alcohol (mainly n-pentanol, 1-octen-3-ol, menthol, and estragol), aldehydes and ketones (e.g., n-pentanal, n-heptanal, and n-nonanal), acids, and few other compounds. Kim (1999) also suggested that 2-AP is available in both aromatic and nonaromatic rice varieties, and the only difference is in the concentration of this magical molecule, which is almost 10–15 times higher in aromatic rice variety than in nonaromatic ones.

Taste of some foods like cheese, popcorn, corn tortillas, green tea, and wine is attributed to the availability of 2-AP (Bradbury, 2009). More recently, in addition to 2-AP, 2-amino acetophenone and

3-hydroxy-4,5-dimethyl-2(5H)-furanone in Basmati 370 (Jezussek et al., 2002), and guaiacol, indole, and p-xylene in Black rice (Yang et al., 2008) have been found to be responsible for their unique flavor. It is also suggested that 2-AP content of rough rice decreased during storage (Champagne et al., 2010). Consequently, it can be concluded that every variety has a unique scent resulting from some volatile compounds, which might differ from well-characterized popcorn-like 2-AP-associated aroma although little is known about their relationships with aroma/flavor (Champagne, 2008).

9.2.1 2-ACETYL-1-PYRROLINE(2-AP):APOTENTFRAGRANCE COMPOUND

2-Acetyl-1-pyrroline [2-AP; IUPAC: 5-acetyl-3,4-dihydro-2H-pyrrole and 1-(3,4-dihydro-2H-pyrrol-5-yl)-ethanone] was first described as a major contributor for muskiness of aromatic rice by Buttery et al. (1983) (shown in Figure 9.1), and their reports were strongly supported by a large number of cereal chemists. Odor quality evaluation of 2-AP was described by Tsugita (1985–86) as popcorn-like, and the assessment order of the amount of popcorn-like odor in 10 different rice varieties ranked them in general order of the concentration of this compound. However, the odor threshold level of rice is 0.1 ppb. Its concentration in aromatic rice verities varies from 6 ppb to 90 ppb in the milled rice and 100 ppb to 200 ppb in brown rice, depending upon the variety such as in Basmati rice (0.34 ppm), Jasmine rice (0.81 ppm), and Texmati (0.53 ppm).

Apart from playing a major role in imparting the characteristic fragrance to rice, this magical molecule has been reported widely to play a similar role in many other plants species such as in leaves of Paddan

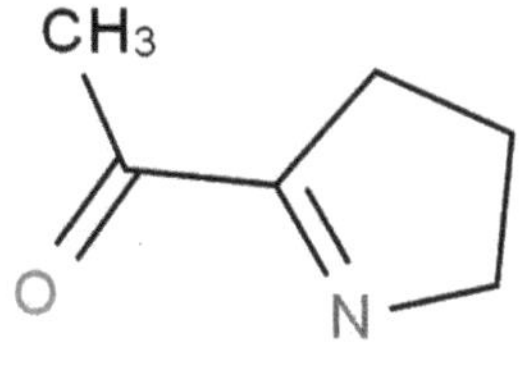

FIGURE 9.1 2-acetyl-1-pyrroline.

(*Pandanus amaryllifolius* Roxb.), *Bassia latifolia*, spinach (*Spinacia oleracea*), bread flowers (*Vallaris glabra* Ktze.), soybean, mung bean (*Vigna radiate*), sorghum (*Sorghum bicolor*), cucumber, *Bacillus cereus* (Romanczyk et al., 1995), *Lactobacillus hilgardii* (Costello and Henschke, 2002), and fungi (Nagsuk et al., 2003). 2-AP has also been reported in many cereal- and vegetable-derived products such as rye bread, boiled potato, popcorn, and sweet corn-based confectionaries and animal-derived products such as cooked American lobsters and cray fish, heat-treated fat-free milk, etc. interestingly, this compound is not only reported in living organisms or their derived products but also in nonliving products such as processed food. The highest amount of 2-AP is found in *Pandanus amaryllifolius*, which is near about 10-12 ppm and almost 50% higher than that of the highest aroma-possessing rice varieties. The leaves of Pandan is widely used as a condiment in most of southeast Asia while cooking non-aromatic rice and in flavoring in many food products such as sweets, confectioneries, and drinks.

9.2.2   BIOSYNTHESIS OF 2-AP IN AROMATIC RICE VARIETIES

Although it is more than a decay since 2-AP is known as the principle compound for aroma in scented rice and L-proline as a precursor of aroma in rice, researchers have still not got an appreciable success in elucidating the biosynthetic pathway of 2-AP in rice. In 1998, an attempt was made by Hofmann and Schieberle (1998) to develop a method for synthesizing 2-AP and 2-acetyltetrahydropyridine in vitro.

An appraisable attempt in elucidating the biosynthetic pathway of 2-AP in rice was made by Vanavichit et al. (2005); they proposed that 2-AP is synthesized via the polyamine pathway. The immediate precursor of 2-AP was found to be 1-pyrroline (1P), which is formed from 4-amino-butyraldehyde (AB-ald), an immediate precursor of γ-aminobutyric acid (GABA).

The major breakthrough in the direction of understanding the biosynthetic pathway of 2-AP came with the identification of the candidate gene controlling the synthesis of 2-AP in rice by Bradbury et al. (2005). After the identification of the candidate gene, Chen and his team (2008) proposed that AB-ald is known to be maintained in an equimolar ratio with an immediate 2-AP precursor Δ^1-pyrroline, and the AB-ald levels plays an important role in the regulation of the rate of 2-AP biosynthesis. The functional *BADH2* enzyme (encoded by the dominant *BAD2* gene) hinders 2-AP biosynthesis in non-fragrant rice by converting AB-ald, which is a presumed precursor of 2-AP to GABA, but the nonfunctional *BADH2* enzyme (encoded by the recessive *BAD2* gene) leads to the increment in the concentration of 2-AP in scented rice by enhancing the AB-ald accumulation. Bradbury et al. (2008) also suggested that γ-amino butyraldehyde (GABald) is an effective substrate for *BADH2* and that accumulation and spontaneous cyclization of to form Δ^1-pyrroline due to a nonfunctional *BADH2* enzyme as the likely cause of 2-AP accumulation in rice. However, in another study, increased expression of Δ^1-pyrroline-5-carboxylate synthetase in fragrant varieties compared with non-fragrant varieties, as well as associated elevated concentrations of its product, led to the conclusion that Δ^1-pyrroline-5-carboxylate, usually the immediate precursor of proline synthesized from glutamate, reacts directly with methylglyoxal to form 2-AP (Huang et al., 2008), with no direct role proposed for *BADH2*. The biosynthesis of 2-AP is shown in Figure 9.2.

9.3 GENETICS OF FRAGRANCE AND FLAVOR: CLASSICAL VIEW

The modern aromatic rice is a result of the desirability of human, its strong preference, and selection for this unique trait during evolution and domestication. After the identification of the principle compound for the aroma in aromatic rice, plant breeders have deliberately engaged themselves in understanding the trend of inheritance of fragrance in scented rice. In the early era of breeding for improved aroma in rice, plant breeders developed many techniques to evaluate and detect the fragrance such as chewing the seed, cooking seeds, or dipping leaves in KOH. However, with the report presented by Buttery et al., in 1983 and 1986 that revealed the presence of 2-AP in both fragrant and traditional rice with significantly higher proportion of 2-AP in scented rice than in non-scented rice, breeders reached a conclusion that for transformation of a nonaromatic rice to an aromatic rice, one would require an alteration in the existing metabolic pathway

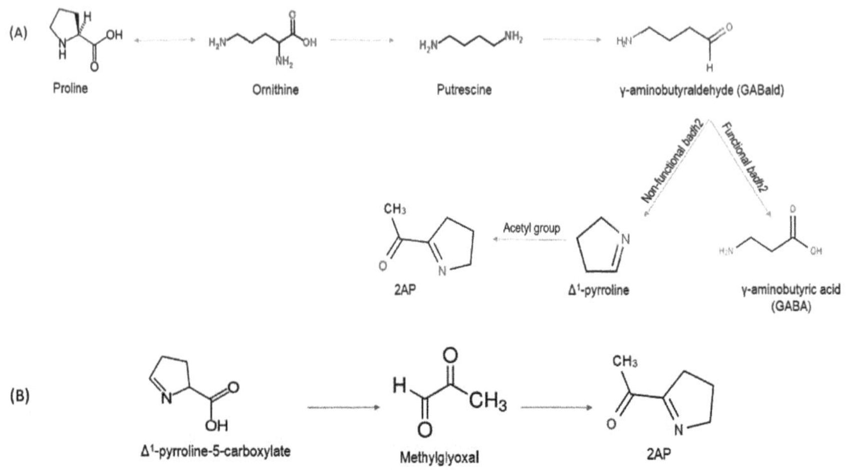

FIGURE 9.2 Biosynthesis of 2-AP in aromatic rice varieties: A) BADH2-dependent pathway (Bradbury et al., 2008) and B) BADH2-independent pathway (Huang et al., 2008).

instead of developing a new one. It was hypothesized that an alternate enzyme in aromatic rice boosts the synthesis of 2-AP, which is directed by a sole dominant gene with one or more alternative alleles.

At the edge of the 20th century, plant breeders crossed T142 (aromatic) × IR20 (non-aromatic) and found high yielding lines possessing strong aroma. These newly developed lines with strong aroma undoubtedly clarify the scenario of the inheritance of scent as an extremely inheritable trait in front of plant breeders. Further, the studies on the inheritance of scent revealed that though the inheritance is highly vigorous, it also depends on the type of crosses, genotypes, and nature of gene action, significantly. Until now, several ambiguous suggestions by a number of researchers were reported. Sood and Siddiq (1980) indicated the engrossment of a single recessive gene in the inheritance of aroma in rice and were supported by Shekhar and Reddy (1981); Bemer and Hoff (1986); Bollich et al. (1992); Ali et al. (1993); Li et al. (1996); Li and Gu (1997); and Sadhukhan et al. (1997). Sadhukhan et al. (1997) reported that aroma in scented varieties Govindobhog and Kataribhog were allelic and controlled by a recessive nuclear gene.

Tsuzuki and Shimokava (1990) observed 13:3 (non-aromatic:aromatic) ratio in F_2 that indicates the single gene for aroma and one inhibitor gene to be responsible for aroma development; they confirmed their results by seg-regation in F_3 generation. On the other hand, several other studies supported the involvement of more than one recessive gene in controlling the presence scent in rice. Dhulappanavar (1976) advocated four complementary reces-sive aroma genes, whereas Reddy and Sathyanarayanaiah (1980) reported three complementary recessive aroma genes. Siddiq et al. (1986) reported three recessive genes, of which 2 were found essential for the presence of aroma. Geetha (1994) reported two recessive genes controlling the aroma trait, and in the same year, Vivekandan and Giridharan (1994) suggested that both monogenic and digenic recessive gene control the trait of scent in rice depending upon the material used for the study. In contrast, Kadam and Patankar (1983) advocated a single dominant gene. Dhulappanavar and Mensinkai (1969) interpreted their result to indicate two dominant aroma genes that intermingled in a duplicate or complimentary manner.

In recent years, it has been found that aroma is initially governed by a single recessive gene, which is a major QTL located on chromosome 8 and three minor QTLs on chromosome 3, 4, and 12 (Lorieux et al., 1996; Amarawathi et al., 2008). Many studies reported that *fgr* as the recessive gene located on chromosome 8 (Lorieux et al., 1996; Garland et al., 2000; Jin et al., 2003). The *fgr* gene is a fragrant-related gene, with an 8-bp dele-tion in exon 7; it was also reported that *fgr* encodes *BADH2* (Bradbury et al., 2005). Chen and his colleague cloned the gene and found *fgr* within 69 kb on chromosome 8 and reported that a variation of the *fgr* gene leads to functional loss of the *BADH2* protein that is associated with fragrance in rice (Bradbury et al., 2005; Chen et al., 2006, 2008). The *fgr* gene encodes the *BADH2* enzyme and is usually referred to as *BADH2*. Numerous mark-ers have been developed for fragrant rice breeding. Many of them reported *BADH2* alleles (Shao et al., 2013; Shi et al., 2014). Recently, He and Park (2015) discovered a novel fragrant allele (*BADH2*-E12) and developed a functional marker for fragrance in rice. The novel allele was a 3-bp dele-tion in exon 12. These readings clearly indicate the complicit genetic basis of aroma in rice. In recent years, Sakthivel et al. (2009) tried to short the possible reasons that are significant in all those contradictory reports for the inheritance of scent in rice. They stated that, "much of this conflicting information on the inheritance of fragrance might have arisen due to the following reasons:

1. Unreliable and cumbersome phenotyping methods used for fragrance determination (Sood and Siddiq, 1978),
2. Failure to consider the endosperm fragrance in rice seeds (Bemer and Hoff, 1986), and
3. Segregation distortion (Lorieux et al., 1996).

Recently, Kaikavoosi et al. (2015) developed transgenic rice by introducing the P5CS gene, which leads to a significant increase in proline, P5CS enzyme activity, and 2-AP levels, indicating the role of proline as a precursor amino acid in the biosynthesis of 2-AP in aromatic rice. Similarly, Kredphol et al. (2015) developed a new rice variety through MAS breeding that contains non-glutinous and semi-dwarf traits from IR64 and the aromatic and black pericarp characteristics of Leum Pua and high yield potential than Leum Pua rice variety.

9.4 EFFECT OF ENVIRONMENTAL FACTORS ON AROMA

The quantity of aroma of rice is based on genetic as well as environmental elements (Jewel et al., 2011; Hashemi et al., 2013). The evidence is supported by Tava and Bocchi (1999), who reported that due to differences in genotype, environment, and interaction between genotype and environment, 2-AP highly exist in Italian and Basmati rice. The fragrance may be controlled by a major gene, but the environmental conditions and cultivation practices could easily influence the intensity of fragrance or the concentration of 2-AP (Yi et al., 2009). There are many environmental factors that influence the aroma quality, for example, storage time and temperature, the effect of planting density, and harvesting time (Goufo et al., 2010a). Goufo et al. (2010b) reported that planting at low density and early harvesting could improve aroma content and other seed qualities. Other studies showed that Basmati variety seems to contain stronger aroma if the day temperature remains cold between 25 to 32°C, whereas overnight temperature should be between 20 to 25°C. The humidity should be 70–80 % in grain-filling and primordial phases (Yi et al., 2009; Golam et al., 2010). Bradbury et al. (2005) reported that the level of 2-AP is higher in plants exposed to water stress due to the contribution of *BADH* with stress tolerance.

There are many other environmental factors that affect the quantity of aroma, like soil type (Yi et al., 2009); abiotic stress (Itani et al., 2004); and conditions of storage, time of harvest, and flowering time (Itani et al., 2004; Yi et al., 2009). Recently, Mo et al. (2015) reported that shading during the grain filling period increases 2-AP content in fragrant rice. Another factor affecting the quality of aroma is the milling process of rice. The extent to which bran is removed from rice kernels during the milling process is referred to as the degree of milling (DOM). Recently, Rodriguez-Arzuaga et al. (2015) reported that DOM affects the raw rice appearance and aroma-related attributes. Likewise, the study conducted by Griglione et al. (2015) reported that the storage temperature (5°C vs. 25°C) does not significantly influence aroma preservation. They also reported that heptanal/1-octen-3-ol and heptanal/octanal acted as indices of aroma quality for the Italian cultivars investigated and more in general, of aroma rice quality. In this way, many environmental attributes affect the aroma of rice.

9.5   MAPPING THE AROMATIC GENES

While considering 2-AP as a potent aromatic compound in rice, some attempts were made to the array of mapping the gene governing 2-AP synthesis in different varieties of aromatic rice, including Della (Ahn et al., 1992), Azucena (Lorieux et al., 1996; Bourgis et al., 2008), Suyunuo (Chen et al., 2006; Shi et al., 2008), Wuxianjing (Chen et al., 2006), KDML105 (Lanceras et al., 2000; Tragoonrung et al., 1996), and Kyeema (Bradbury et al., 2005). However, the varying level of 2-AP among different aromatic rice verities and its expensive assay limited the mapping experiments.

Ahn et al. (1992) took the very first systematic initiative for mapping the gene for grain aroma. He mapped a single recessive gene to chromosome 8 using the RFLP technique based on its closely linked clone RG28 near 4.5 cM map distance and produced a consequence genetic map within the region between two SSR markers SRJ02 and RM515. Some selected list of inheritance of gene in rice is shown in Table 9.1. The gene responsible for 2-AP synthesis was identified with the first mapped based cloning between the flanking regions RG1 and RG28 (Vanavichit et al., 2004) in a segregating population developed from a Jasmine variety KDML105

TABLE 9.1 Inheritance and Gene Action of Aroma Genes in Aromatic Rice

Gene Action	Number of Genes	Phenotyping	References
Complementary, Dominant	3	Leaf analysis in boiling water	Nagaraju et al. (1975)
	4	Chewing milled rice	Dhulappanvar (1976)
	2	Leaf analysis in KOH	Tripathi and Rao (1979)
	3	Leaf analysis in KOH	Reddy and Sathyanaryanaiah (1980)
	1	Leaf analysis in KOH	Kuo et al. (2005)
	2	-	Chaut et al. (2010)
Duplicate, Dominant	1 or 2	Leaf analysis in KOH	Sarawagi and Bisene (2006)
Recessive	1	Leaf analysis in KOH	Sood and Siddiq (1978)
	1	Chewing	Bemer and Hoff (1986)
	1	Leaf analysis in KOH	Ahn et al. (1992)
	1	Leaf analysis in KOH	Ali et al. (1993)
	2	Chewing	Pinson (1994)
	2	Leaf analysis in KOH	Vivekanadan and Giridharan (1994)
	1	Chewing	Tragoonrug et al. (1996)
	1	Chewing	Corderio et al. (2002)
	1	Leaf in KOH	Dartey et al. (2006)
	2	Leaf in KOH	Hien et al. (2006)
	1	Leaf analysis in KOH	Sun et al. (2008)
	1	Leaf in KOH	Sarhadi et al. (2009)
	1	Leaf in KOH	Asante et al. (2010)
	1	Leaf in KOH	Vazirzanjani et al. (2011)
Polygenic	3	Chewing	Loreiux et al. (1996)
	3	Seed in KOH	Amarawathi et al. (2008)

and nonaromatic rice as parents. In the segregating generation, the original region of 1.13 Mb flanking between RG1 and RG28 was narrowed down to 82.2 Kb. Within this narrowed region of 82.2 kb, three KDML BACs were cloned, including the identification of three new candidate genes (Vanavichit et al., 2005). Among these three newly identified genes, a single recessive gene, namely Os2-AP, was identified as the major gene determining 2-AP synthesis in the rice plant at different growth stages. The comparative analysis of Os2-AP gene sequences between KDML105 and Nipponbare revealed two important mutational events within exon 7 of Os2-AP of aromatic KDML105, at positions 730 (A to T) and 732 (T to A), followed by the 8-bp deletion "GATTAGGC" starting at position 734.

Another attempt to map the gene was done in an aromatic rice variety Kyeema (Bradburry et al., 2005), where a gene responsible for 2-AP synthesis was identified within the flanking region of RM515 and SSRJ07 on the longer arm of chromosome 8. The *in silico* physical map was developed using four BAC of Nipponbare spanning within a region of 386 bp from RM515 to SSRJ07. The data produced by physical map advised for a BAC clone (clone AP004463) as most likely to be having the gene. The resequencing of all 17 genes laying within the BACs revealed the significant level of polymorphism only in one gene, which was having three newly recognized SNPs along with the 8-bp deletion in exon 7, while the remaining were having little or no polymorphism. By recombining ability, the gene was identified *BAD2* as it was showing a significant homologue with a *BAD1* gene located on chromosome 4. Later, the competitive studies of sequences and amino acids revealed that both Os2-AP and *BAD2* are the same gene governing the 2-AP synthesis and may be considered as synonyms.

Recently, many studies have been done on molecular mapping of the *fgr* gene with different markers (Wanchana et al., 2005; Chen et al., 2006; Hien et al., 2006; Amarawathi et al., 2008; Lang and Buu, 2008; Sun et al., 2008; Bradbury, 2009; Yi et al., 2009; Chaut et al., 2010). So far, many of them presented markers that are highly associated with the *fgr* locus (Shi et al., 2008; Sun et al., 2008; Madhav et al., 2009; Sakthivel et al., 2009). Mapping and fine mapping of *fgr* were done by Chen et al. (2006); they found the *fgr* gene to a 69-Kb interval. Sequence analysis, QTL mapping, and fine mapping revealed that *BADH2* locus of rice constituting the *fgr* gene has been recognized as an important genetic determinant of fragrance (Yi et al., 2009; Fitzgerald et al., 2010; Siddiq et al., 2012). Selected list of molecular mapping of genes in rice is shown in Table 9.2.

Beside these two significant mutations (Figure 9.3), a discontinuous 8-bp deletion and an SNP of the fragrant allele, a 7-bp insertion in exon 8 (Amarawathi et al., 2008), 7-bp deletion in exon 2 (Shi et al., 2008), a MITE

TABLE 9.2 QTL(s)/Genes for Aroma in Rice

Gene(s)	Markers	Chromosome number	References
1	RFLP	8	Ahn et al. (1992); Yano et al. (1992); Lorieux et al. (1996)
1	EST	8	Wanchan et al. (2005)
1	SSR	8	Garland et al. (2000); Cordeiro et al. (2002); Jin et al. (2003); Bradburry et al. (2005, 2009); Wanchana et al. (2005); Chen et al. (2006); Li et al. (1996); Amarawathi et al. (2008); Sun et al. (2008); Yi et al. (2009)
2	SSR	-	Hien et al. (2006)
QTL(s)	STS	4 and 12	Lorieux et al. (1996)
1	SSR, RFLP, STS	8	Lang and Buu (2008)
3 QTLs	SSR	QTLs on 3, 4 and 8	Amarawathi et al. (2008)

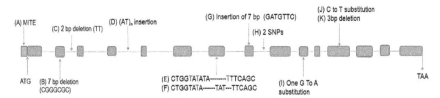

FIGURE 9.3 Different allelic variations (mutations) in the BADH2 gene: A) a MITE insertion in the promoter (green) (Bourgis et al., 2008); B) deletion of 7 bp (CGGGCGC) in exon (blue) 2 (Shi et al., 2008); C) & D) a 2 bp (TT) deletion in the 2nd intron and (AT)$_n$ insertion in the 4th intron; E) a continuous 8-bp (GATTAGGC) deletion with three SNPs (red) in exon 7 (Bradbury et al., 2005), F) & G) discontinuous mutation and one SNP in exon 7 and 7-bp insertion (GATGTTC) in exon 8 (Amarawathi et al., 2008), H) two SNPs in intron 8 (Sun et al., 2008), I) & J) SNPs in exons 10 and 12; and K) 3-bp deletion in exon 12 (He and Park, 2015).

insertion in promoter (Bourgis et al., 2008), a 2-bp (TT) deletion in intron 2, a (AT)$_n$ insertion in intron 4 (Chen et al., 2008), and two SNPs in the central section of intron 8 (Sun et al., 2008) were also reported. For the larger aromatic gene Os2-AP, it was found that three minor QTLs located on chromosome 3, 4, and 12 significantly play a role in imparting the aroma; however, the direct function of these three minor QTLs are completely unknown.

9.6 REGULATION OF FRAGRANCE IN RICE

Fine mapping confirmed the aroma as a recessive trait to nonaromatic rice that eventually occurs due to a mutation leading to an 8-bp deletion in exon 7 of the Os2-AP gene. This mutational event causes a premature stop codon that disrupts the function of the gene and ultimately leads to the increased level synthesis of a truncated protein 2-AP, the major volatile compound imparting the fragrance and aroma in rice.

The transcriptomic analysis of the Os2-AP gene reveals a differential level of its expression in all plant parts of the rice plant. The gene starts expressing itself from the very young age of seedling to the grain maturing stage and the expression lasts in the harvested grains. The level of expression differs from the plant part to age of the plant. Exceptionally, the expression of Os2-AP has not been reported in roots so far by most of the scientists except by Vanavichit et al. (2005) who observed a very low level of a transcript of the Os2-AP gene in the root of few aromatic varieties. A common hypothesis postulated about the expression of the Os2-AP gene is that this gene shows a suppressive expression that resulte from a premature stop codon at position 753, which shortened the full-length peptide to 252 amino acids in aromatic rice (Bradbury et al., 2005; Vanavichit et al., 2005). This short, incomplete peptide was reported to trigger nonsense-mediated decay (NMD) in several cases (Chang et al., 2007).

The confirmation of the suppressive expression of Os2-AP in 2-AP synthesis was later done by using RNAi technology. The first-ever experiment reported in the literature in this context was done by Vanavichit et al., in 2008. The RNAi was constructed from the genomic sequence spanning exons 6 to 8 in the opposite direction from the corresponding cDNA. This allowed the transcript to create double-stranded RNA, resulting in NMD and aromatic Nipponbare that could accumulate 2-AP in a range of 0.05–0.20 ppm (Vanavichit et al., 2005). In this experiment, the strongest RNAi expression gave the strongest suppression and the highest accumulation of 2-AP, comparable to the 2-AP content in Jasmine rice (Vanavichit et al., 2005). In an independent study, transgenic rice containing RNAi by an inverted repeat of cDNA encoding Os2-AP accumulated 2-AP in considerable amounts (Niu et al., 2008).

9.7   FUTURE PERSPECTIVES

World population is increasing day by day, and thus, the demand for rice. People always choose the best, and thus, fragrant rice is highly appealing to human beings. The global demand for aromatic rice is also increasing. Moreover, most of the rice breeders are focusing on to developing a better quality of aromatic rice to fulfill the demands. For aroma, there are many volatile compounds that are available and among them, 2-AP is the major one. Some studies also suggested that every variety has its own unique aroma, but little information is available on this area. Thus, scientists should focus to find the more understanding on this interaction of volatile compounds.

9.7.1   RESEARCH NEED IN RICE

There are several methods available to develop an aromatic rice, and among them, the genetic transformation method could be a method of choice. Recently, Kaikavoosi et al. (2015) developed transgenic rice by introducing the P5CS gene, which leads to a substantial increase in proline, P5CS enzyme activity, and 2-AP levels, thus indicating the role of proline as a precursor amino acid in the biosynthesis of 2-AP in aromatic rice. Another approach is by using molecular marker analysis. Among the many molecular markers, SSR is useful for this area of research. It is very beneficial for rice breeders to expand new aromatic rice varieties. This technique is a cost-effective method. Presently, many tools and novel markers are already available for accurate genetic mapping. Recently, Kredphol et al. (2015) developed a new rice variety through molecular-assisted selection breeding that contains non-glutinous and semi-dwarf traits from IR64 and the aromatic and black pericarp characteristics of Leum Pua and showed high yield potential than Leum Pua rice variety. Recently, He and Park discovered a new allele and developed functional markers for the fragrance in rice, which adds more flavor in this area of research. This functional allele can be used as selection tools to developed improved fragrant rice varieties. Thus, understanding of these techniques has great potential for the optimization of production of aromatic rice in the near future.

9.8 CONCLUSION AND SUMMARY

Aromatic rice is highly demanding for human beings and cooked kernel elongation and aroma are crucial quality traits that determine the global market value and cooking and eating qualities of rice. Therefore, aroma is considered a prominent characteristic of numerous breeding programs and demand of this rice is increasing day by day. The flavor or fragrance of rice is associated with the presence of 2-AP. Among many characteristics, shading during grain filling has more effects on quality and yield of rice, which leads to the accumulation of GABA and 2-AP. These results further emphasize the important relationship between stress and 2-AP production in fragrant rice. Although 2-AP is the major aromatic compound in aromatic rice, a range of volatile components is available in addition to 2-AP. These compounds are responsible for aroma; thus, every rice variety has a unique aroma. The SSR marker analysis is a method of choice for QTL analysis for aroma, which is crucial for rice breeders to develop new aromatic rice varieties. Consequently, identifying the most relevant genes underlying the fragrance trait seems troublesome. Therefore, there is a need to map new dense marker to identify major and minor genes that control rice aroma. Of course, for a deeper understanding, biochemical, genetic, and genomic perspectives that may affect this important trait should also be considered.

KEYWORDS

- abiotic stress
- aroma
- *BADH2*
- biosynthesis
- *fgr* gene
- fine mapping
- fragrance
- GABA
- genetics

- **genotypes**
- **marker**
- **QTL**
- **recessive gene**
- **rice**
- **SSR**

REFERENCES

Ahmed, S. A., Borua, I., Sarkar, C. R., & Thakur, A. C. (1996). Volatile component (2-ace-tyl-1-pyrroline) in scented rice. *Proceedings of the Seminar on Problems and Prospects of Agricultural Research and Development in North-east India*, Assam Agricultural University, Jorhat, India, *27–28*, pp. 55–57.

Ahn, S. N., Bollich, C. N., & Tanksley, S. D. (1992). RFLP tagging of a gene for aroma in rice. *Theor. Appl. Genet., 84*, 825–828.

Ali, S. S., Jafri, S. J. H., Khan, M. J., & Butt, M. A. (1993). Inheritance studies on aroma in two aromatic varieties of Pakistan. *Intl. Rice Res. Newsl., 18*(2), 6.

Amarawathi, Y., Singh, R., Singh, A. K., Singh, V. P., Mohapatra, T., Sharma, T. R., & Singh, N. K. (2008). Mapping of quantitative trait loci for basmati quality traits in rice (*Oryza sativa* L.). *Mol. Breed., 21*(1), 49–65.

Asante, M. D., Kovach, M. J., Huang, L., Harrington, S., Dartey, P. K., Akromah, R., Semon, M., & McCouch, S. (2010). The genetic origin of fragrance in NERICA1. *Mol. Breed., 26*, 419–424.

Bemer, D. K., & Hoff, B. J. (1986). Inheritance of scent in American long grain rice. *Crop Sci., 26*, 876–878.

Bollich, C. N., Rutger, J. N., & Webb, B. D. (1992). Development in rice research in the United States. *Int. Rice Comm. Newslt., 41*, 32–34.

Bourgis, F., Guyot, R., Gherbi, H., Tailliez, A. I., Salse, J., Lorieux, M., & Delseny, G. A. (2008). Characterization of the major fragrance gene from an aromatic japonica rice and analysis of its diversity in Asian cultivated rice. *Theor. Appl. Genet., 117*, 353–368.

Bradbury, L. M. T. (2009). Identification of the gene responsible for fragrance in rice and characterization of the enzyme transcribed from this gene and its homologs. *PhD thesis*, Southern Cross University, Lismore, Australia.

Bradbury, L. M. T., Gillies, S. A., Brushett, D. J., Waters, D. L. E., & Henry, R. J. (2008). Inactivation of an amino aldehyde dehydrogenase is responsible for fragrance in rice. *Plant Mol. Biol., 68*, 439–449.

Bradbury, L. M., Fitzgerald, T. L., Henry, R. J., Jin, Q., & Waters, D. L. (2005) The gene for fragrance in rice. *Plant Biotechnol. J., 3*(3), 363–370.

Brar, D. S., & Khush, G. S. (2002). Transferring genes from wild species into rice. In: *Quantitative Genetics Genomics and Plant Breeding* (Kang, M. S., ed.). CAB International, UK. pp. 197–217.

Bryant, R. J., & McClung, A. M. (2010). Volatile profiles of aromatic and non-aromatic rice cultivars using SPME/GC–MS. *Food Chem., 124*(2), 501–513.

Buttery, R. G., Ling, L. C., & Mon, T. R. (1986). Quantitative analysis of 2-acetyl-1-pyrroline in rice. *J. Agric. Food Chem., 34*, 112–114.

Buttery, R. G., Ling, L. C., Juliano, B. O., & Turnbaugh, J. G. (1983). Cooked rice aroma and 2-acetyl-l- pyrroline. *J. Agric. Food Chem., 31*, 823–826.

Champagne, E. T. (2008). Rice aroma and flavor: a literature review. *Cereal Chemistry, 85*(4), 445–454.

Champagne, E. T., Bett-Garber, K. L., Fitzgerald, M. A., Grimm, C. C., Lea, J., Ohtsubo, K., Jongdee, S., Xie, L., Bassinello, P. Z., Resurreccion, A., Ahmad, R., Habibi, F., & Reinke, R. (2010). Important sensory properties differentiating premium rice varieties. *Rice J., 3*, 270–281.

Chang, Y. F., Saadi, I. J., & Wilkinson, M. F. (2007). The nonsense-mediated decay RNA surveillance pathway. *Ann. Rev. Biochem., 76*, 51–74.

Chaut, A. T., Yutaka, H., & Vo, C. T. (2010). Genetic analysis for the Fragrance of aromatic rice variety. In: *3rd International Rice Congress*, Hanoi, Vietnam, pp. 3898.

Chen, S. H., Wu, J., Yang, Y., Shi, W. W., & Xu, M. L. (2006). The FGR gene responsible for rice fragrance was restricted within 69kb. *Plant Sci., 171*, 505–514.

Chen, S., Yang, Y., Shi, W., Ji, Q., He, F., Zhang, Z., Cheng, Z., Liu, X., & Xu, M. (2008). *BADH2*, encoding betaine aldehyde dehydrogenase, inhibits the biosynthesis of 2-acetyl-1-pyrroline, a major component in rice fragrance. *Plant Cell, 20*, 1850–1861.

Cordeiro, G. M., Christopher, M. J., Henry, R. J., & Reinke, R. F. (2002). Identification of microsatellite markers for fragrance in rice by analysis of the rice genome sequence. *Mol. Breed., 9*, 245–50.

Costello, P. J., & Henschke, P. A. (2002). Mousy off-flavor of wine: Precursors and biosynthesis of the causative N-heterocycles 2-ethyltetrahydropyridine, 2-acetyltetrahydropyridine, and 2-acetyl-1- pyrroline by Lactobacillus hilgardii DSM 20176. *J. Agril. Food Chem., 50*, 7079–7087.

Dartey, P., Asante, M., & Akromah, R. (2006). Inheritance of aroma in two rice cultivars. *Agric. Food Sci. J. Ghana, 5*, 375–379.

Dhulappanavar, C. V., & Mensinkai, S. W. (1969). Inheritance of scent in rice. *Karnataka Univ. J.* (cited by Tripathi and Rao, 1979), *14*, 125–129.

Dhulappanavar, C. V. (1976). Inheritance of scent in rice. *Euphytica, 25*, 659–662.

Dong, Y., Tsuzuki, E., & Terao, H. (2001). Genetic analysis of aroma in three rice cultivars (*Oryza sativa* L.). *J. Genet. Breed., 55*, 39–44.

Fitzgerald, M. A., McCouch, S. R., & Hall, R. D. (2009). Not just a grain of rice: the quest for quality. *Trends Plant Sci., 14*, 133–139.

Fitzgerald, T. L., Waters, D. L. E., Brooks, L. O., & Henry, R. J. (2010). Fragrance in rice (*Oryza sativa*) is associated with reduced yield under salt treatment. *Environ. Exp. Bot., 68*, 292–300.

Garland, S., Lewin, L., Blakeney, A., Reinke, R., & Henry, R. (2000). PCR-based molecular markers for the fragrance gene in rice (Oryza sativa. L.). *Theor. Appl. Genet., 101*, 364–371.

Geetha, S. (1994). Inheritance of aroma in two rice crosses. *Intl. Rice Res. Notes, 19*(2), 5.

Golam, F., Norzulaani, K., Jennifer, A. H., Subha, B., Zulqarnain, M., Osman, M., & Mohammad, O. (2010). Evaluation of kernel elongation ratio and aroma association in global popular aromatic rice cultivars in tropical environment. *Afric. J. Agrl. Res., 5*(12), 1515–1522.

Goufo, P., Duan, M., Wongpornchai, S., & Tang, X. (2010a). Some factors affecting the concentration of the aroma compound 2-acetyl-1-pyrroline in two fragrant rice cultivars grown in South China. *Front. Agrl. China, 4*(1), 1–9.

Goufo, P., Wongpornchai, S., & Tang, X. (2010b). Decrease in rice aroma after application of growth regulators. *Agron. Sustain. Dev., 31*(2), 349–359.

Griglione, A., Liberto, E., Cordero, C., Bressanello, D., Cagliero, C., Rubiolo, P., Bicchi, C., & Sgorbini, B. (2015). High-quality Italian rice cultivars: chemical indices of ageing and aroma quality. *Food Chem., 172*, 305–313.

Hashemi, F. S. G., Rafi, M. Y., Ismail, M. R., Mahmud, T. M. M., Rahim, H. A., Asfaliza, R., Malek, M. A., & Latif, M. A. (2013). Biochemical, genetic and molecular advances of fragrance characteristics in rice. *Crit. Rev.Plant Sci., 32*(6), 445–457.

He, Q., & Park, Y. J. (2015). Discovery of a novel fragrant allele and development of functional markers for fragrance in rice. *Mol. Breed., 35*(11), 1–10.

Hien, N., Yoshihashi, T., Sarhadi, W. A., Thanh, V., Oikawa, Y., & Hirata, Y. (2006). Evaluation of aroma in rice (*Oryza sativa* L.) using KOH method, molecular markers and measurement of 2-acetyl-1-pyrroline concentration. *Nettai Nogyo Jpn. J. Tropical Agric., 50*, 190–198.

Hofmann, T., & Schieberle, P. (1998). Flavor contribution and formation of the intense roastsmelling odorants 2-propionyl-1-pyrroline and 2-propionyltetrahydropyridine in Maillard-type reactions. *J. Agric. Food Chem., 46*, 2721–2726.

Huang, T. C., Teng, C. S., Chang, J. L., Chuang, H. S., Ho, C. T., & Wu, M. L. (2008). Biosynthetic mechanism of 2- acetyl-1-pyrroline and its relationship with Δ1-pyrroline-5-carboxylic acid and methylglyoxal in aromatic rice (*Oryza sativa* L.) callus. *J. Agric. Food. Chem., 56*, 7399–7404.

Itani, T., Tamaki, M., Hayata, Y., Fushimi, T., & Hashizume, K. (2004). Variation of 2-acetyl-1-pyrroline concentration in aromatic rice grains collected in the same region in Japan and factors affecting its concentration. *Plant Prod. Sci., 7*, 178–183.

Jewel, Z. A., Patwary, A. K., Maniruzzaman, S., Barua, R., & Begum, S. N. (2011). Physico-chemical and genetic analysis of aromatic rice (*Oryza sativa* L.) Germplasm. *The Agriculturist, 9*, 82–88.

Jezussek, M., Juliano, B. O., & Schieberle, P. (2002). Comparison of key aroma compounds in cooked brown rice varieties based on aroma extract dilution analyses. *J. Agric. Food. Chem., 50*(5), 1101–1105.

Jin, Q., Waters, D., Corderio, G. M., Henry, R. J., & Reinke, R. F. (2003). A single nucleotide polymorphism (SNP) marker linked to the fragrance gene in rice (*Oryza sativa* L.). *Plant Sci., 165*, 359–364.

Kadam, B. S., & Patankar, V. K. (1938). Inheritance of aroma in rice. *Chron. Bot., 4*, 32.

Kadaru, S. B. (2007). Identification of molecular markers and association mapping of selected loci associated with agronomic traits in rice. *PhD Thesis*, Louisiana State University, Baton Rouge, USA.

Kaikavoosi, K., Kad, T. D., Zanan, R. L., & Nadaf, A. B. (2015). 2-acetyl-1-pyrroline augmentation in scented indica rice (*Oryza sativa* L.) varieties through delta(1)-pyrroline-5-carboxylate synthetase (P5CS) gene transformation. *Appl. Biochem. Biotechnol., 177*(7), 1466–1479.

Kerdphol, R., Sreewongchai, T., Sripichitt, P., Uckarach, S., & Worede, F. (2015). Obtaining a black pericarp and improved aroma using genetic resources from Leum Pua rice. S*cience Asia, 41*(2), 93.

Kim, C. Y. (1999). A study on the growth characteristics and analysis of chemical compounds of grains as affected by cultivation method in colored and aromatic rice varieties'. *PhD Thesis*, submitted to Chungnam National University, Taejeon, Korea. pp. 112.

Kioko, W. F., Musyoki, M. A., Piero, N. M., Muriira, K. G., Wavinya, N. D., Rose, L., Felix, M., & Ngithi, N. (2015). Genetic diversity studies on selected rice (*Oryza sativa* L) populations based on aroma and cooked kernel elongation. *J. Phylogen Evolution. Biol., 3*, 158. doi:10.4172/2329–9002.1000158.

Kovach, M. J., Calingacion, M. N., Fitzgerald, M. A., & McCouch, S. R. (2009). The origin and evolution of fragrance in rice (*Oryza sativa* L.). *PNAS USA, 106*(34), 14444–14449.

Kuo, S. M., Chou, S. Y., Wang, A. Z., Tseng, T. H., Chueh, F. S., Yen, H. E., & Wang, C. S. (2005). The betaine aldehyde dehydrogenase (*BAD2*) gene is not responsible for the aroma trait of SA0420 rice mutant derived by sodium azide mutagenesis. In: *5th International Rice Genetics Symposium and 3rd International Rice Functional Genomics Symposium.* International Rice Research Institute Press, Manila, Philippines.

Lanceras, J. C., Huang, Z. L., Naivikul, O., Vanavichit, A., Ruanjaichon, V., & Tragoonrung, S. (2000). Mapping of genes for cooking and eating qualities in Thai jasmine rice (KDML 105). *DNA Res., 7*, 93–101.

Lang, N. T., & Buu, B. C. (2008). Development of PCR-based markers for aroma (*fgr*) gene in rice (*Oryza sativa* L.). *Omonrice, 16*, 16–23.

Li, J., & Gu, D. (1997). Analysis of inheritance of scented rice variety Shenxiangjing. *China Rice Res. Newsl., 5*(1), 4–5.

Li, J., Ku, D., & Li, L. (1996). Analysis of fragrance inheritance in scented rice variety, Shenxiangjing 4. *Acta. Agriculturae Shanghai, 12*(3), 78–81.

Lin, C. F., Hsieh, R. C. Y., & Hoff, B. J. (1990). Identification and quantification of the popcorn-like aroma in Louisiana aromatic Della rice (*Oryza sativa* L.). *J. Food Sci., 35*, 1466–1467.

Lorieux, M., Petrov, M., Huang, N., Guiderdoni, E., & Ghesquiere, A. (1996). Aroma in rice: genetic analysis of a quantitative trait. *Theor. Appl. Genet., 93*, 1145–1151.

Madhav, M. S., Pandey, M. K., Kumar, P. R., Sundaram, R. M., Prasad, G. S. V., Sudarshan, I., & Rani, N. S. (2010). Identification and mapping of tightly linked SSR marker for aroma trait for use in marker assisted selection in rice. R*ice Genetics Newsletter, 25*, 38–39.

Mahalingam, L., & Nadarajan, N. (2005). Inheritance of scentedness in two-line rice hybrids. *Intl. Rice Res. Notes, 30*, 13–14.

Mahattanatawee, K., & Rouseff, R. L. (2014). Comparison of aroma active and sulfur volatiles in three fragrant rice cultivars using GC–Olfactometry and GC–PFPD. *Food Chem., 154*, 1–6.

Mo, Z., Li, W., Pan, S., Fitzgerald, T. L., Xiao, F., Tang, Y., Wang, Y., Duan, M., Tian, H., & Tang, X. (2015). Shading during the grain filling period increases 2-acetyl-1-pyrroline content in fragrant rice. *Rice (NY), 8*, 9.

Myint, K. M., Arikit, S., Wanchana, S., Yoshihashi, T., Choowongkomon, K., & Vanavichit, A. (2012). A PCR-based marker for a locus conferring the aroma in Myanmar rice (Oryza sativa L.). *Theor. Appl. Genet., 125*(5), 887–896.

Nagaraju, M., Chaudhary, D., & Balakxishna Rao, M. J. (1975). A simple technique to identify scent in rice andinheritance pattern of scent. *Cum. Sci., 44*, 599.

Nagsuk, A., Winichphol, N., & Rungsarthong, V. (2003). Identification of 2-acetyl- 1-pyrroline, the principal aromatic rice flavor compound, in fungus cultures. In: *Proceedings of the 2nd International Conference on Medicinal Mush- rooms & International Conference on Biodiversity and Bioactive Compounds*, Pattaya Exhibition Center, Cholburi, Thailand, pp. 395–400.

Niu, X., Tang, W., Huang, W., Ren, G., Wang, Q., Luo, D., Xiao, Y., Yang, S., Wang, F., & Lu, B. R. (2008). RNAi-directed downregulation of Os*BADH2* results in aroma (2-acetyl-1-pyrroline) production in rice (*Oryza sativa* L.). *BMC Plant Biol., 8*, 1–10.

Paule, C. M., & Powers, J. J. (1989). Sensory and chemical examination aromatic and nonaromatic rices. *J. Food Sci., 54*, 343–346.

Pinson, S. R. M. (1994). Inheritance of aroma in six rice cultivars. *Crop Sci., 34*, 1151–1157.

Rai, V. P., Singh, A. K., & Jaiswal, H. K. (2015). Evaluation of molecular markers linked to fragrance and genetic diversity in Indian aromatic rice. *Turkish J. Bot., 39*, 209–217.

Reddy, P. R., & Sathyanarayanaiah, K. (1980). Inheritance of aroma in rice. *Indian J. Genet. Pl. Breed., 40*, 327–329.

Rodriguez-Arzuaga, M., Cho, S., Billiris, M. A., Siebenmorgen, T., & Seo, H. S. (2015). Impacts of degree of milling on the appearance and aroma characteristics of raw rice. *J. Sci. Food. Agric.,* doi: 10.1002/jsfa.7471.

Romanczyk, L. J., McClelland, C. A., Post, L. S., & Aitken, W. M. (1995). Formation of 2-acetyl-1-pyrroline by several Bacillus cereus strains isolated from cocoa fermentation boxes. *J. Agric. Food Chem., 43*, 469–475.

Sabouri, A., Rabiei, B., Toorchi, M., Aharizad, S., & Moumeni, A. (2012). Mapping quantitative trait loci (QTL) associated with cooking quality in rice (*Oryza sativa* L.). *Aust. J. Crop Sci., 6*, 808.

Sadhukhan, R. N., Roy, K., & Chattopadhyay, P. (1997). Inheritance of aroma in two local aromatic rice cultivars. *Environ. Ecol., 15*(2), 315–317.

Sakthivel, K., Sundaram, R., Shobha Rani, N., Balachandran, S., & Neeraja, C. (2009). Genetic and molecular basis of fragrance in rice. *Biotechnol. Adv., 27*, 468–473.

Sarawagi, A. K., & Bisne, R. (2006). Inheritance of aroma in indigenous short grain aromatic rice cultivars. *J. Rice Res., 1*, 180–1.

Sarhadi, W. A., Ookawa, T., Yoshihashi, T., Madadi, A. K., Yosofzai, W., Oikawa, Y., & Hirata, Y. (2009). Characterization of aroma and agronomic traits in Afghan native rice cultivars. *Plant Prod. Sci., 12*, 63–69.

Shao, G., Tang, S., Chen, M., Wei, X., He, J., Luo, J., Jiao, G., Hu, Y., Xie, L., & Hu, P. (2013). Haplotype variation at *BADH2*, the gene determining fragrance in rice, *Genomics, 101*(2), 157–162.

Shekhar, B. P. S., & Reddy, G. M. (1981). Genetic basis of aroma and flavor components studies in certain scented cultivars of rice'. IV International SABRO Congress, (Abstr.).

Shi, W., Yang, Y., Chen, S., & Xu, M. (2008). Discovery of a new fragrance allele and the development of functional markers for the breeding of fragrant rice varieties. *Mol. Breed.*, *22*(2), 185–192.

Shi, Y., Zhao, G., Xu, X., & Li, J. (2014). Discovery of a new fragrance allele and development of functional markers for identifying diverse fragrant genotypes in rice. *Mol. Breed.*, *33*(3), 701–708.

Siddiq, E. A., Sadananda, A. R., & Zaman, F. U. (1986). Use of primary trisomic of rice in genetic analysis. In: *Rice Genetics, Proceedings of the International Rice Genetics Symposium*, International Rice Research Institute, Manila, Philippines, pp. 185–197.

Siddiq, E. A., Vemireddy, L. R., & Nagaraju, J. (2012). Basmati rices: Genetics, breeding and trade. *Agric. Res.*, *1*(1), 25–36.

Sood, B. C., & Siddiq, E. A. (1978). A rapid technique for scent determination in rice. *Indian J. Genet. Plant Breed.*, *38*, 268–271.

Sood, B. C., & Siddiq, E. A. (1980). Studies on component quality attributes of basmati rice. *Z. Plunzenzuecht.*, *84*, 299–301.

Sun, S. H., Gao, F. Y., Lu, X. J., Wu, X. J., Wang, X. D., Ren, G. J., & Luo, H. (2008). Genetic analysis and gene fine mapping of aroma in rice (*Oryza sativa* L. Cyperales, Poaceae). *Genet. Mol. Biol.*, *31*, 532–538.

Tanchotikul, U., & Hsieh, T. C. Y. (1991). An improved method for quantification of 2-acetyl-1-pyrroline a popcorn-like aroma, in aromatic rice by high-resolution gas chromatography/mass spectrometry/selected ion monitoring. *J. Agri. Food Chem.*, *39*, 944–947.

Tava, A., & Bochi, S. (1999). Aroma of cooked rice (*Oryza sativa*): Comparison between commercial basmati and Italian line B5–3. *Cereal Chemistry*, *76*(4), 526–529.

Tragoonrung, S., Sheng, J. Q., & Vanavichit, A. (1996). Tagging an aromatic gene in lowland rice using bulk segregant analysis. *Rice Genetics III. IRRI*, pp. 613.

Tripathi, R. S., & Rao, M. J. B. K. (1979). Inheritance and linkage relationship of scent in rice. *Euphytica*, *28*, 319–323.

Tsugita, T. (1985–1986). Aroma of cooked rice. *Food Rev. Intl.*, *1*, 497–520.

Tsuzuki, E., & Shimokawa, E. (1990). Inheritance of aroma in rice. *Euphytica*, *46*, 157–159.

Vanavichit, A., Kamolsukyurnyong, W., Wanchana, S., Wongpornchai, S., Ruengphayak, S., Toojinda, T., & Tragoonrung, S. (2004). Discovering genes for rice grain aroma. In: *Proceedings of the 1st International Conference on Rice for the Future, 31 August–3 September*, Kasetsart University, Bangkok, Thailand, pp. 71–80.

Vanavichit, A., Tragoonrung, S., Toojinda, T., Wanchana, S., & Kamolsukyunyong, W. (2008). Transgenic rice plants with educed expression of Os2-AP and elevated levels of 2-acetyl-1-pyrroline. *US Patent, 7*, 319–181.

Vanavichit, A., Yoshihashi, T., Wanchana, S., Areekit, S., Saengsraku, D., Kamolsukyunyong, W., et al. (2005). Cloning of Os2-AP, the aromatic gene controlling the biosynthetic switch of 2- acetyl-1-pyrroline and gamma-aminobutyric acid (GABA) in rice. *5th International Rice Genetics Symposium. Philippines: IRRI*, November 19–23, pp. 44.

Vazirzanjani, M., Sarhadi, W. A., New, J., Amirhosseini, M. K., Siranet, R., Trung, N. Q., Kawai, S., & Hirata, Y. (2011). Characterization of aromatic rice cultivars from Iran and surrounding regions for aroma and agronomic traits. *SABRAO J. Breed. Genet., 43*, 15–26.

Verma, D. K., & Srivastav, P. P. (2016). Extraction technology for rice volatile aroma compounds. In: *Food Engineering Emerging Issues, Modeling, and Applications* (eds. Meghwal, M., & Goyal, M. R.). In book series on Innovations in Agricultural and Biological Engineering, Apple Academic Press, USA, pp. 245–291.

Verma, D. K., Mohan, M., & Asthir, B. (2013). Physicochemical and cooking characteristics of some promising basmati genotypes. *Asian Journal of Food Agro-Industry, 6*(2), 94–99.

Verma, D. K., Mohan, M., Prabhakar, P. K., & Srivastav, P. P. (2015). Physico-chemical and cooking characteristics of Azad basmati. *International Food Research Journal, 22*(4), 1380–1389.

Verma, D. K., Mohan, M., Yadav, V. K., Asthir, B., & Soni, S. K. (2012). Inquisition of some physico-chemical characteristics of newly evolved basmati rice. *Environment and Ecology, 30*(1), 114–117.

Vivekanandan, P., & Giridharan, S. (1994). Inheritance of aroma and breadthwise grain expansion in Basmati and non-Basmati rices. *Intl. Rice Res. Notes, 19*(2), 4–5.

Wanchana, S., Kamolsukyunyong, W., Ruengphayak, S., Toojinda, T., Tragoonrung, S., & Vanavichit, A. (2005). A rapid con-construction of a physical contig across a 4.5 cm region for rice grain aroma facilitates marker enrichment for positional cloning. *Sci. Asia., 31*, 299–306.

Widjaja, R. I., Craske, J. D., & Wootton, M. (1996). Comparative studies on volatile components of non-fragrant and fragrant rices. *J. Sci. Food Agric., 70*, 151–161.

Yang, D. S., Shewfelt, R. L., Lee, K. S., & Kays, S. J. (2008). Comparison of odor-active compounds from six distinctly different rice flavor types. *J. Agric. Food Chem., 56*(8), 2780–2787.

Yano, M., Shimosaka, E., Saito, A., & Nakagahra, M. (1992). Lirikage analysis of a gene for scent in indica rice variety, Surjamkhi, using restriction fragment length polymorphism markers. *Jpn. J. Breed., 41*(Suppl. 1), 338–339.

Yi, M., Nwe, K. T., Vanavichit, A., Chai-arree, W., & Toojinda, T. (2009). Marker assisted backcross breeding to improve cooking quality traits in Myanmar rice cultivar Manawthukha. *Field Crops Res., 113*(2), 178–186.

CHAPTER 10

GENETIC ENGINEERING FOR FRAGRANCE IN RICE: AN INSIGHT ON ITS STATUS

SUBRAMANIAN RADHESH KRISHNAN,[1]
PANDIYAN MUTHURAMALINGAM,[2]
BHAGAVATHI SUNDARAM SIVAMARUTHI,[3]
CHAKRAVARTHI MOHAN,[4] and MANIKANDAN RAMESH[5]

[1]*Research Scholar, Department of Biotechnology, Alagappa University, Karaikudi–630003, India, Tel.: +91-4565 225215, Mobile: +91-9566422094, Fax: +91 4565 225202, E-mail: radheshkrishnan.s@gmail.com*

[2]*Research Scholar, Department of Biotechnology, Alagappa University, Karaikudi–630 003, India, Tel.: +91-4565 225215, Mobile: +91-9597771342, Fax: +91-4565 225202, E-mail: pandianmuthuramalingam@gmail.com*

[3]*Postdoctoral Fellow, Faculty of Pharmacy, Department of Pharmaceutical sciences, Chiang Mai University, Chiang Mai, Thailand, Tel.: +66-5394-4341; Fax: +66-5389-4163, Mobile: +66-61-3216651, E-mail: sivasgene@gmail.com*

[4]*Postdoctoral Fellow, Molecular Biology Laboratory (LBM), Department of Genetics and Evolution (DGE), Federal University of Sao Carlos (UFSCar), SP, Brazil – 13565905, Tel.: +551633518378, Mobile: +5516996196465, E-mail: chakra3558@gmail.com*

[5]*Associate Professor, Department of Biotechnology, Alagappa University, Karaikudi–630003, India, Tel.: +91-4565 225215, Mobile: +91-9442318200, Fax: +91-4565 225202, E-mail: mrbiotech.alu@gmail.com*

CONTENTS

10.1   INTRODUCTION

Rice is a cereal grain that holds the majority in crop production, and hence, it is definitely the staple food crop for nearly two-third of the world population. China and India are the major contributors of its large production and distribution worldwide. Around 25% (478.18 million tons) of rice produced during 2014–2015 was contributed by India, of which scented rice alone was traded for Rs. 275,978.7 million to different countries of the world (Wakte et al., 2016). Of the 481 metric tons total rice production in the world during 2011, scented rice contributed to 18% of the total trade by dominating the total rice trade in the market (Giraud, 2013). The world's rice forecast by FAO for the year 2016–17 marketing year is 165.5 million tons (Table 10.1). Aroma or fragrance is one of the highly valued grain qualities in rice (Kovach et al., 2009; Verma et al., 2012, 2013, 2015) as the consumers prefer more odor and taste (Figure 10.1). Thus, scented rice is of high market value that determines the local and national identity (Fitzgerald et al., 2009). Currently, scented rice is the niche in the market for the major section of rice traders.

TABLE 10.1 Importance of the Main Rice Producers and Exporters

Major Producer and Exporter	2011-13 Average	2014	2015 Estimate	2016 Forecast	Annual Change 2016/15		2016	
							Previous	Revision
	MT				%		MT	%
World	735.7	744.7	738.8	746.8	8.1	1.1	745.5	13
Asia	667.6	673.4	668	675.6	7.6	1.1	675	0.6
Africa	26.8	28.7	29	29.9	0.9	3.1	29.1	0.8
Central America	2.8	2.9	2.6	2.8	0.1	5.7	2.8	0
South America	24.5	24.7	25.5	23	-2.5	-10.0	23.7	-0.7
North America	8.7	10.1	8.7	11.1	2.4	27.4	10.4	0.7
Europe	4.3	4.0	4.2	4.2	0.0	0.9	4.2	0.0

Source: FAO (2016); MT = Million tones.

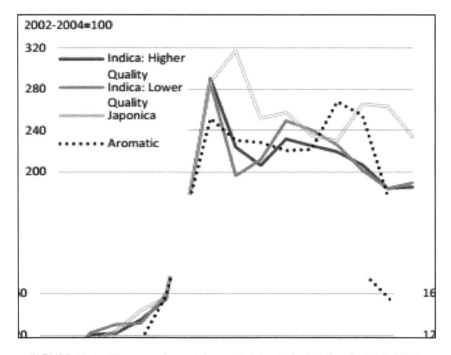

FIGURE 10.1 Consumer demand for scented rice in the last decade (FAO, 2016).

10.2   GENETIC DIVERSITY OF AROMA IN RICE AND ITS EXPRESSION ANALYSIS

Rice is one of the major crops in Asian countries with more than 400,000 rice germplasm accessions (Toriyama, 2005). For an efficient understanding of the origin and evolution of the crop, characterizing the genetic diversity and studying the relationship is inevitable. The scented rice cultivars share a small but special group among the Poaceae family with a greater importance in the rice trade (Figure 10.1). A good scented rice is selected not only based on its aroma but also on certain key characteristics like grain quality, stickiness, better amino acid profiles, and the presence of active components in it. In India, Basmati 370 stands atop owing to its higher leucine, methionine, and phenylalanine content (Shekhar and Reddy, 1982). As the fragrance gene (fgr) is recessive in aromatic rice varieties, many wild-type cultivars lack the fragrance. A proper domestication of fragrant and nonfragrant rice varieties could pave way for new traits in scented rice (Bradbury et al., 2005). The fragrance in rice is due to production of many aromatic compounds (Petrov et al., 1996). The most determined molecule is 2-acetyl-1-pyrroline (2-AP) whose accumulation alters the growth, production, grain quality, and yield of the plant (Buttery et al., 1982; Buttery et al., 1983a). Although there are numerous reports on 2-AP, their specific role in controlling the fragrance is not deciphered yet (Myint et al., 2012). Bradbury et al. (2005) and Vanavichit et al. (2015) reported the presence of the recessive gene betaine aldehyde dehydrogenase (*BADH2*) in the synthesis of 2-AP. The *BADH2* gene was subjected to Oryzabase database (NBPR, 2016) and Rice genome annotation project (RGAP) for retrieving the gene details based on previous literatures (Table 10.2).

Gene ontology analysis showed the presence of this gene in peroxisomes, and mostly, it does respond to abiotic stresses, flooding, and biological processes like glycine betaine biosynthetic process or phenylpropanoid metabolic process. The genetic diversity of this gene has been shown in numerous varieties. A frame shift was observed in most of the aromatic accessions by deletion of 8 bp on exon 7 of *BADH2*. Several studies have reported the diversity of the *BADH2* gene that causes a reading frame shift

TABLE 10.2 Aroma Gene Details. Retrieved from Oryzabase (NBPR, 2016) and RGAP Database (Kawahara et al., 2013)

Details of the Aroma Gene	Remarks
CGSNL Gene Symbol	SK2(T) (SCL, FGR)
Gene Symbol Synonym	sk2(t) (scl, fgr), fgr, scl, BAD2, BADH2, sk2, Os-BADH2, Badh2, badh2, Os2-AP, BADH-2
CGSNL Gene Name	SCENTED KERNEL 2
Gene Name Synonym	scented kernel2, fragrance, betaine-aldehyde dehydro-genase 2, scented kernel-2, betaine aldehyde dehydro-genase-2
Protein Name	BETAINE-ALDEHYDE DEHYDROGENASE 2
Allele	badh2-E2, badh2-E7, badh2.1
Chromosome Number	8
Trait class	Biochemical character Seed – Physiological traits – Taste Tolerance and resistance – Stress tolerance
RAP ID	Os08g0424500
Gene Ontology	Peroxisomes Response to stress Betaine-aldehyde de-hydrogenase activity Response to flooding Phenylpro-panoid metabolic process Glycine betaine biosynthetic process
Trait ontology	Aroma, submergence sensitivity
CDS coordinates (5'-3')	20379794 – 20386061
Nucleotide length	1512
Predicted protein length	504
Predicted molecular weight	54682.7
pI	5.1127

mutation and varies depending on the cultivar (Bourgis et al., 2008; Shi et al., 2008; Kovach et al., 2009; Sakthivel et al., 2009a). In silico developmental stage expression analysis revealed that the *BADH2* gene was highly expressed in the pregermination stage and developing seed 1, 2, 4, 5 (Figure 10.2). Tissue-specific expression analysis showed the *BADH2* gene is mainly expressed in dry seed, radicle, palea, and endosperm tissues, and other tissues showed negligible expression. This developmental stage and tissue-specific expression results strongly report that the gene is present in mature seed, developing seed, and endosperm.

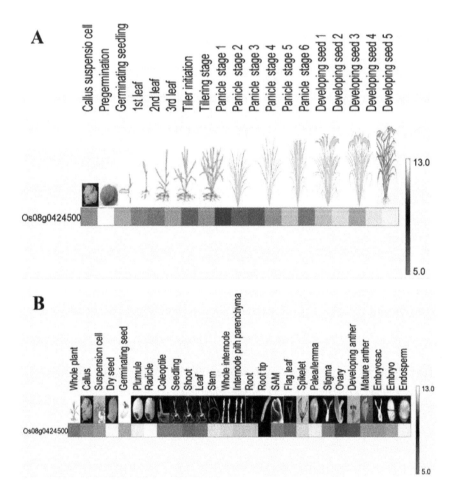

FIGURE 10.2   Expression of the *BADH2* gene. A) Developmental stage expression and B) Tissue-specific expression.

10.3   PATHWAY AND CHEMICAL CONS OF AROMATIC RICE

Various volatile compounds contribute to the fragrance in rice. Several researchers have reported the presence of acids, aldehydes, alcohols, esters, hydrocarbons, ketones, phenols, pyrazines, pyridines, and other compounds (Yajima et al., 1978; Tsugita et al., 1980; Maga, 1984; Paule and Powers, 1989). 2-Acetyl-1-pyrroline (2-AP) was isolated and characterized from the steam distillation extract of scented rice. Buttery et al. (1982, 1983) first reported that the cooked rice aroma was due to the presence of the aromatic compound 2-AP; eventually, researchers were eager to know the gene responsible for it (Figure 10.1). The first-ever mapping of the fragrance gene was reported by Ahn et al. (1992). It was a decade later that Vanavichit et al. (2004, 2005) successfully mapped the gene controlling the rice fragrance and named it *Os2-AP*. Disorder/mutation in amino aldehyde dehydrogenase (AMADH) results in the disruption of *Os2-AP*, resulting in 2-AP accumulation through 4-aminobutanal (Figure 10.3B). This disorder triggers the aroma in rice, wherein γ-amino butyric acid (GABA) starts to accumulate due to abiotic stresses (Taylor et al., 2010). The GABA transaminase, glutamate decarboxylase, and succinate semialdehyde dehydrogenase forms the base of GABA shunt (Figure 10.4). Hitherto, it was believed that GABA is predominantly accumulated by GABA shunt. Turano et al. (1997) reported putrescine-mediated GABA accumulation through bypassing of the Krebs cycle by AMADH. As such the AMADH cannot degrade betaine aldehyde as the

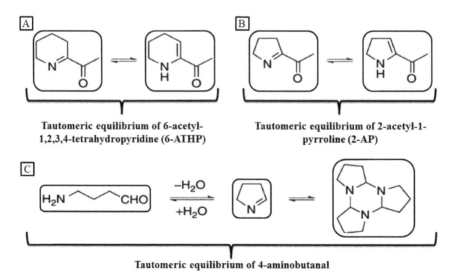

FIGURE 10.3 Tautomeric equilibrium to precede the pathway for 2-acetyl-1-pyrroline (2-AP) synthesis in aromatic rice. (Reprinted with permission from Vanavichit, A., & Yoshihashi, T. (2010). Molecular aspects of fragrance and aroma in rice. *Advances in Botanical* Research, 56, 50–73. © 2010. With permission from Elsevier.)

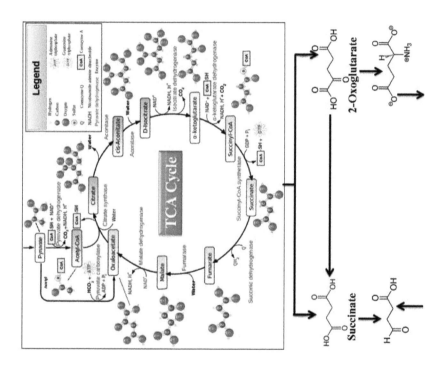

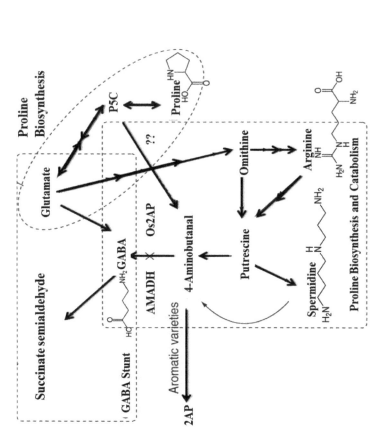

FIGURE 10.4 Formation pathway of 2-AP in aromatic rice (Vanavichit and Yoshihashi (2010). Aromatic varieties lack AMADH enzyme activity, which converts 4-aminobutanal to GABA to yield 2-AP. (Reprinted with permission from Vanavichit, A., & Yoshihashi, T. (2010). Molecular aspects of fragrance and aroma in rice. *Advances in Botanical Research, 56,* 50–73. © 2010. With permission from Elsevier.)

substrate in its native state; this triggers stress resulting in the accumulation of the polyamine compound 2-AP. A high accumulation of 2-AP was observed in various samples of rice cultivated in drought-hit and rain-fed areas (Yoshihashi et al., 2004). The whole genome sequencing of rice has led to the discovery of gene responsible for the fragrance (*fgr*) by comparing the fragrant and nonfragrant genotypes. The re-sequencing of fragrant genotypes of rice located the *fgr* at chromosome 8, which controls 2-AP. (Bradbury et al., 2005, 2008). The *fgr* gene was found to be recessive as the RNA interference against *Os*2-AP resulted in nonaromatic rice. A detailed study on this gene revealed the presence of 15 exons interrupted by 14 introns.

The pathway networks were determined by input of the *BADH2* (Osa00260) gene into KEGG (Figure 10.5). Glycine, serine, and threonine metabolism that control and (or) organize the biosynthesis of BADH2 gene were predicted. Betaine-aldehyde dehydrogenase enzyme codes for aroma in rice. KEGG BRITE database is used for the molecular mapping process, significantly large-scale datasets in Omics studies like genomics, transcriptomics, proteomics and metabolomics, to the brite functional hierarchies used for biological interpretation. KO (KEGG Orthology) and enzyme order were given (Table 10.3).

A set of 8-bp deletion was observed in the mutant form of RNAi crops in exon 7. A second mapping was reported by crossing between aromatic Kyeema and a cultivar of non-aromatic rice, which resulted in the presence of the *fgr* gene on chromosome 8 between SSR markers SSRJ07 and RM515 (Bradbury et al., 2005). Four Nipponbare BAC clones when predicted by *in* silico physical mapping depicted flanked SSR markers SSRJ07 and RM515 of 386 kb. Seventeen genes were retrieved from the re-sequenced data with a significant variation in sequence of BAD2, a homologue of betaine aldehyde dehydrogenase (BADH). Eventually a third set of BAC clones were sequenced, and *in silico* analysis of their critical region in Nipponbare (Wanchana et al., 2005) and 93–11 (Chen et al., 2008). Based on the sequence similarity of gene and protein, both *Os2A* and *BAD2* were considered to be similar. Bourgis et al. (2008) identified BAD2 as superior aroma candidate in the Azucena japonica cultivar, which was also acquired by crossing Azucena × IR64.

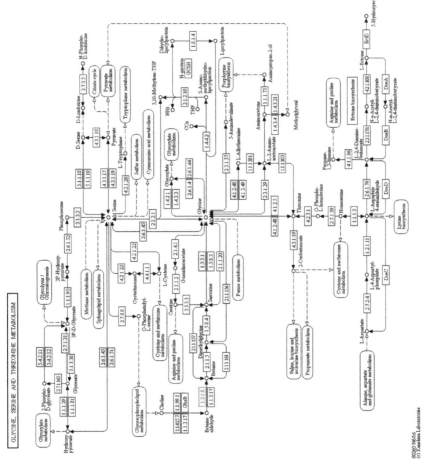

FIGURE 10.5 BADH2 (Osa00260) gene metabolic pathway predicted by KEGG.

TABLE 10.3   KEGG BRITE BADH2 Gene Details

KEGG Or-thology (KO) [BR:osa00001]	Metabolism Amino acid metabolism 00260 Glycine, serine and threonine metabolism 4336081 (BADH)
Enzymes [BR:osa01000]	1. Oxidoreductases
	1.2 Acting on the aldehyde or oxo group of donors
	1.2.1 With NAD+ or NADP+ as acceptor
	1.2.1.8 Betaine-aldehyde dehydrogenase 4336081 (BADH)

The cultivable rice, *O. sativa* L. (spp. *japonica* or *indica*) of Asia is presumed to be derived from one or both *O. rufipogon* Griff and *O. nivara*, which are prevalent throughout the monsoon (Prathepha, 2009). The most highly valued grain quality trait of rice is its pleasant fragrance. Hitherto, the Asian cultivated rice comprised several groups that are genetically distinct, but with this complex evolutionary history, there remains a question as how they came to exist across the highly divergent rice.

10.4   THE GENETICS OF RICE FRAGRANCE

The studied rice varieties, and experimental approaches like binary system, sensory panelists scaling fragrance, and gas chromatographic methods make the researchers to focus on a single gene, which determines the aroma of the rice (Garland et al., 2000; Cordeiro et al., 2002; Jin et al., 2003), or two to four genetic loci responsible for the fragrance in rice (Lorieux et al., 1996).

A recessive gene that contributes to rice aroma is located on chromosome 8, which was determined in Della, a fragrant cultivar (Berner and Hoff, 1986). Berner and Hoff (1986) revealed the inheritance of aroma by performing reciprocal crosses using Dawn, a nonfragrant cultivar, and Della. They screened the strength of the fragrance by smelling the grains and leaf of F1 and F2 progenies, which was smelled after placing them in KOH (0.3 mol/L), and the grains were chewed to rate the fragrance. The monogenic recessive inheritance responsible for the aroma was confirmed by the segregation of aroma in the F3 grains (seeds from F2 plants). Ahn

et al. (1992) used 126 cDNA clones of rice and oat genome as hybridiza-
tion probes and identified a DNA marker gene (*fgr*) that is located closely
to the linkage group of chromosome 8, at 4.5 cM distance from the restriction fragment length polymorphism marker (RFLP) "RG28."

The main component responsible for the fragrance of rice was identified as 2-AP, which occurs in leaf, grain, and stem except root (Buttery
et al., 1983b). Lorieux et al. (1996) developed a new approach by 2-AP
quantification by gas chromatography and molecular marker mapping of
chromosome 8 using several molecular markers (RFLP, STS, RAPD, and
isozyme) with a gap of 20-25 cM. Both analyses were performed on a
population of 135 double haploid lines. They found that two regions (RG
28 at 6.4 ± 2.6 and RG 1 at 5.3 ± 2.7 cM distance, respectively) of chromosome 8 are involved in the regulation of 2-AP and evidenced that this
aromatic compound is the main component for the fragrance of rice. The
study also identified two additional quantitative trait loci located on chromosome 4 and 12 that play a role in the aroma of rice. Garland et al. (2000)
identified a single nucleotide polymorphism (SNP) in the homologous
regions of RG28, an RFLP marker located on chromosome 8 and developed a PCR-based molecular marker for differentiating the nonscented
and scented rice cultivars. Cordeiro et al. (2002) reported a microsatellite
marker (single sequence repeat) that is located near the fragrance gene
"*fgr*" using the rice genome sequence. Jin et al. (2003) identified a single
molecular marker (SNP), RSP04 at 2 cM from the fragrance gene "*fgr*,"
by using pyrosequence analysis. The fragrance BADH2 protein encoded
by the *badh2* gene renders nonfragrance by inhibiting 2-AP (Kovach et al.,
2009; Sakthivel et al., 2009b; Bao, 2014). The closely related BADH2
neighborhood proteins interactions were assumed by STRING analysis.
Now, the PPIs are predicted based on the publicly available datasets.
Neighborhood protein functions are uncharacterized (Figure 10.6).

10.5 MARKERS FOR IDENTIFYING THE FRAGRANCE GENE

Apart from a few minor segments of rice varieties, the major organization
of heredity is usually based on the geographical distribution or particular
adaptation and its characteristics that have been identified with genetic

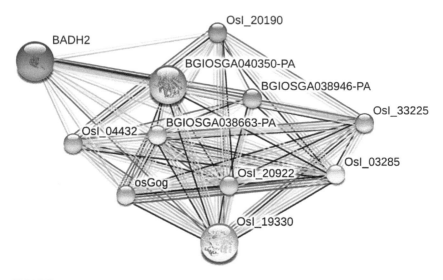

FIGURE 10.6   BADH2 protein and its neighborhood proteins predicted by STRING analysis.

markers (Second, 1982; Glaszmann, 1987), and the relationship between wild type and cultivated rice was studied by many groups (Bautista et al., 2001; Garris et al., 2005; Londo et al., 2006). Small groups of strongly linked varieties with strong similarity but genetically distinct were confirmed by specific marker studies, and the presence of rare allelic association was noted. The advancement and availability of whole genome sequence and molecular markers of rice have paved way to identify and map the quantitative trait loci (QTLs) that decipher the aroma trait. Moreover, it is a prerequisite to exploit molecular markers to identify QTLs from nonadapted germplasm or wild species that are competent of developing rice grain yields (Brar and Khush, 1997). Molecular marker analysis pointed out that the aromatic traits are driven by monogenic recessive inheritance, which is a self-governing cytoplasmic gene. The fragrance gene is associated with many genes that are directly involved in rice metabolism (Hien and Yoshihashi, 2006), and hence, the complications due to genetic inheritance of this fragrance and the key locus point could be revealed (Sakthivel et al., 2009b; Chaut and Yutaka, 2010). Systematic fine mapping, QTL mapping, sequence analysis, and complementation testing were used to find some important fragrance genes (Table 10.4) like

TABLE 10.4 List of Fragrance Genes and Their Functions

Genes	Function	Reference
Fgr	Betain aldehyde dehydrogenase (BADH) codes for aroma in rice	Bradbury et al. (2005); Chen et al. (2006); Kovach et al. (2009); Jin et al. (2010a); Jantaboon et al. (2011)
2-AP	Strong flavor constituent in rice aroma	Buttery et al. (1983b); Hashemi et al. (2013)
SSIIa	Improve the food cooking and eating quality	Win et al. (2012)
Wx	Quality improvement	Win et al. (2012); Win et al., 2012)

BADH2 (Bradbury et al., 2005; Chen et al., 2006) and its two null aromatic alleles badh2-E7 (Kovach et al., 2009) and badh2-E2 (Chen et al., 2008), and 2-AP (Buttery et al., 1983b; Hashemi et al., 2013), which are localized on chromosome 8 in rice. Even though, a single recessive gene for aroma has been detected, recent findings have revealed QTLs that possess an inherent nature to regulate the fragrance in rice (Lorieux et al., 1996; Amarawathi et al., 2008; Chaut and Yutaka, 2010). Predicting the relationship between the fragrance and its relevant gene locus using potential markers may unravel the key players.

10.6 IMPROVEMENT BY BREEDING

The grain superiority/quality could be deciphered using QTLs, which highlight naturally occurring variations in grain traits. Only a few of them have been exploited in existing rice breeding programs. Recent findings reveal improving the grain quality to enhance the cooking and eating quality. The most conventional genes are *Wx*, *SSIIa*, and *fragrance* (Table 10.5). Functional markers well known from GS3 are also known for their grain length development (Wang et al., 2011; Agarwal et al., 2016). There are two approaches to conduct marker-assisted selection (MAS) in the breeding system. The foreground selection of a potential marker that defines the grain quality and various phenotypic characteristics denotes the acquired trait. Secondly, yield potential and the aroma of rice such as Basmati or Jasmine fall under background selection. Thus, the loss of crops with good trait is evaded during introduction.

TABLE 10.5 Some Key Markers for Selecting Cooking and Eating Quality in Rice

Gene marker	Forward primer	Reverse primer	Reference
fgr	1F:GGAGCTTGCTGATGTGTGTAAA	1R:GGAAACAAACCTTAACCATAG	Chen et al. (2008); Jin et al. (2010b)
	2F:CCTCTGCTTCTGCCTCTGAT	2R:GATTGCGCGGAGGTACTTG	
SSIIa	NF1 :CGAGGCGCAGCACAACAG	NR1: GGCCGTGCAGATCTTAACCAT	Bao et al. (2006)
	F22: CAAGGAGAGCTGGAGGGGGC	R21: ACATGCCGCGCACCTGGAAA	Ayres et al. (1997)
Wx	484: CTTTGTCTATCTCAAGACAC	485: TTGCAGATGTTCTTCCTGATG	

Primers for analysis of fragrance in rice (Bradbury et al., 2005; Bradbury et al., 2008)

Primer name		5' Primer sequence 3'
External Sense Primer (ESP)		TTGTTTGGAGCTTGCTGATG
Internal Fragrant Antisense Primer (IFAP)		CATAGGAGCAGCTGAAATATATACC
Internal Non-fragrant Sense Primer (INSP)		CTGGTAAAAAGATTATGGCTTCA
External Antisense Primer (EAP)		AGTGCTTTACAAAGTCCCGC

MAS is one of the chief methods to improve the superiority of the maintainer line. Few noteworthy maintainer lines such as Zhenshan 97B, Longtefu B (Zhou et al., 2003; Wang et al., 2007), G46B (Gao et al., 2009), and II32B (Jin et al., 2010a) have been the objective of transferring the *Wx* allele, which has a low amylose content. The restorer lines possess better quality, thereby making the hybrid rice a superior trait that was derived by crossing with an improved maintainer line. Furthermore, MAS with the *Wx* gene marker for quality improvement of the conventional rice has been reported (Jairin et al., 2009; Jantaboon et al., 2011). Users generally prefer fragrant rice to nonfragrant rice. Functional markers for *fgr* have been developed and uninterruptedly used to transfer this gene from fragrant rice to the non-fragrant ones. SSIIa is responsible for the variation of gelatinization temperature; the functional markers for SSIIa have been developed and used in MAS to improve the cooking quality (Jairin et al., 2009; Jantaboon et al., 2011; Salgotra et al., 2012).

The breeding methodology is conducive for many traits in developing a novel variety. Added to this, there is a remarkable growth in the agronomic resistant traits like grain yield potential, quality or resistance to infections such as blast (Singh et al., 2012; Win et al., 2012; Pandey et al., 2013) or brown plant hopper (Jairin et al., 2009). Thus, MAS is dealt by combining two or more quality traits like submergence tolerance (Jantaboon et al., 2011) and plant stature (Pandey et al., 2013) in the Jasmine or Basmati rice types.

10.7 MOLECULAR IMPROVEMENT FOR AROMA IN RICE

MAS is predominantly accompanied with QTLs for plant molecular breeding to produce new and desired traits by genetic transformation. Furthermore, gene linking and silencing methods have paved a remarkable role in increasing rice aroma. Niu et al. (2008) proposed an increase in the aroma production of fragrant rice by RNAi-mediated *OsBADH2* downregulation with the hp-RNA approach. The *OsBADH2* expression was observed to be constitutive with lesser expression on mature roots of rice. Chen et al. (2012) developed fragrance in nonfragrant rice by

RNAi-mediated downregulation of *OsBADH2* expression using artificial miRNA (amiRNA) technology. Researchers have developed a mutant plant that contains a stable expression of transgene over generations. They have identified it as transgenic after screening a large number of transgenic plants. Zhang et al. (2013) constructed transcription activator-like effector nucleases (TALENs)-coding plasmid to achieve a more efficient and rapid modification in the gene of monocot systems like rice (targeted genes: *OsBADH2*, *OsDEP1*, *OsCKX2*, and *OsSD1*) and *Brachypodium* (eight targeted genes). A recent study reported the development of fragrant rice from a rice variety (non-fragrant) by targeted knockout of the *OsBADH2* using TALENs approach. They also demonstrated the efficiency of targeted multiple gene knockouts in rice via TALENs method (Shan et al., 2015).

With the advent of gene editing and sequencing technologies like Clustered regularly interspaced short palindromic repeats (CRISPR/Cas) and next generation sequencing (NGS) systems, the trait improvement has become easier in monocot systems (Matsumoto et al., 2016). Both genetic engineering (GE) and MAS have greatly enriched the grain quality and nutrition in rice. MAS, being predominant, has an advantage for rice, and various traits have been marked with molecular markers. On the other hand, crop domestication after trait prediction has been historically criticized and could be accomplished using NGS technologies wherein a greater hypothesis could be built in the trait conservation for subsequent generation. In this case, GE is useful to trigger or introduce a new trait into rice by transferring the gene of target from other species and to express the new gene in the host. Kovach et al. (2009) suggested *BADH2* as one of the domesticated gene that upregulates aroma in rice. Through NGS technologies, domestication of *BADH1* and *BADH2* was proposed by extended haplotype homozygocity (EHH).

Geneticists have unraveled many unexplored QTLs and genes for grain quality improvement. Functions of majority of these genes have been predicted to be associated with specific pathways related to flavonoid, lipid, protein/amino acid, and starch biosynthesis. Some of these have been successfully cloned through map-based technologies to improve the breeding and selection processes through MAS (Bao, 2014).

10.8   POTENTIAL FUTURE PERSPECTIVES

We discussed various features and characteristics of scented rice, its development, use and genetic improvisation for new traits to feed the need of the society. Estimating the nature of the aromatic substance in the diverse rice species could lead to understand the new algorithms. Many researchers have already focused on rice seed, rice bran, and bran oil as high nutritional value products (Schwab et al., 2008; Pengkumsri et al., 2015a, 2015b). The estimation and validation of the health improving properties like antihypercholestoremic, antiglycemic, anti-inflammatory, and antioxidant have carved a new era of phytochemical genomics (Fernie and Schauer, 2009). In silico molecular breeding is a new era of plant history, wherein alleles of different markers are deliberated using bioinformatic tools; the morphology of new rice can also be predicted and displayed. Although there are many reports on molecular breeding and genetically modified/produced aromatic/scented rice, field experiments are a prerequisite to determine the efficacy of the transgene (Fukushima and Kusano, 2014).

The coming years may foresee a change in the acceptance of genetically modified rice, which may have a high impact and can fill the wide gap that is present between the scanty production of fragrant rice and its ever-growing demand. Another challenging area is the use of genome engineering techniques that can be exploited in aromatic rice to generate elite varieties, which may lead to new horizons in aromatic rice biotechnology. Recent research has demonstrated great feasibilities in plant biotechnology by exploiting a combinatorial approach that couples advanced molecular tools, NGS approaches, in silico bioinformatic analysis, and proteomic tools, which aid in a better understanding of the molecular mechanisms that have not yet been deciphered. With the rice genome available, its much easy to use such a combinatorial technique to elucidate and understand several unknown mechanisms that would enhance our knowledge on aromatic rice and benefit the mankind as well.

10.9   SUMMARY AND CONCLUSION

To sum up, the core aim of this chapter entitled "Genetic engineering for fragrance in rice: An insight on its status" was to provide a short and brief understanding about the scented rice and the science behind its fragrance. The role of BADH2 and Os2-AP in the production of 2-AP and its general mechanism were explained. Furthermore, the molecular development with respect to high yield and aroma was discussed. Functional genomics has simplified in delineating the molecular mechanism underlying the fragrance of rice. With the availability of rice genome coupled with gene editing technologies, it is possible to elucidate the intricacies behind rice aroma and its development.

KEYWORDS

- 2-acetyl-1-pyrroline
- abiotic stresses
- aroma
- background selection
- chromosome
- complementation testing
- cytoplasmic genes
- foreground selection
- fragrance
- genetic markers
- mutation
- physical mapping
- RNA interference
- scented rice
- submergence tolerance
- whole genome sequencing

REFERENCES

Agarwal, P., Parida, S. K., Raghuvanshi, S., Kapoor, S., Khurana, P., Khurana, J. P., & Tyagi, A. K. (2016). Rice improvement through genome-based functional analysis and molecular breeding in India. *Rice* (New York), *9*, 1.

Ahn, S. N., Bollich, C. N., & Tanksley, S. D. (1992). RFLP tagging of a gene for aroma in rice. *Theoretical and Applied Genetics, 84*, 825–828.

Amarawathi, Y., Singh, R., Singh, A. K., Singh, V. P., Mohapatra, T., Sharma, T. R., & Singh, N. K. (2008). Mapping of quantitative trait loci for basmati quality traits in rice (Oryza sativa L.). *Molecular Breeding, 21*, 49–65.

Ayres, N. M., McClung, A. M., Larkin, P. D., Bligh, H. F. J., Jones, C. A., & Park, W. D. (1997). Microsatellites and a single nucleotide polymorphism differentiate apparent amylase classes in an extended pedigree of US rice germplasm. *Theor Appl Genet, 94*, 773–781.

Bao, J. S., Corke, H., & Sun, M. (2006). Nucleotide diversity in starch synthase IIa and validation of single nucleotide polymorphisms in relation to starch gelatinization temperature and other physicochemical properties in rice (*Oryza sativa* L.). *Theoretical and Applied Genetics, 113*, 1171–1183.

Bao, J. (2014). Genes and QTLs for rice grain quality improvement. In: *Rice – Germplasm, Genetics and Improvement* (Yan, W. and Bao, J. Eds.), InTech, DOI: 10.5772/56621. Available from: https://www.intechopen.com/books/rice-germplasm-genetics-and-improvement/genes-and-qtls-for-rice-grain-quality-improvement.

Bautista, N. S., Solis, R., Kamijima, O., & Ishii, T. (2001). RAPD, RFLP and SSLP analyses of phylogenetic relationships between cultivated and wild species of rice. *Genes & Genetic Systems, 76*, 71–79.

Berner, D. K., & Hoff, B. J. (1986). Inheritance of scent in American long grain rice. *Crop Sci., 26*, 876–878.

Bourgis, F., Guyot, R., Gherbi, H., Tailliez, E., Amabile, I., Salse, J., Lorieux, M., Delseny, M., & Ghesquière, A. (2008). Characterization of the major fragrance gene from an aromatic japonica rice and analysis of its diversity in Asian cultivated rice. *Theoretical and Applied Genetics, 117*, 353–368.

Bradbury, L. M. T., Fitzgerald, T. L., Henry, R. J., Jin, Q., & Waters, D. L. E. (2005). The gene for fragrance in rice. *Plant Biotechnology Journal, 3*, 363–370.

Bradbury, L. M. T., Gillies, S. A., Brushett, D. J., Waters, D. L. E., & Henry, R. J. (2008). Inactivation of an amino aldehyde dehydrogenase is responsible for fragrance in rice. *Plant Molecular Biology, 68*, 439–449.

Brar, D. S., & Khush, G. S. (1997). Alien introgression in rice. *Plant Molecular Biology, 35*, 35–47.

Buttery, R. G., Ling, L. C., & Juliano, O. B. (1982). 2-Acetyl-1-pyrroline: An important aroma component of cooked rice. *Chemical Industry, (London), 12*, 958–959.

Buttery, R. G., Ling, L. C., Juliano, B. O., & Turnbauhg, J. G. (1983). Cooked rice aroma and 2-acetyl-1-pyrroline. *Journal of Agricultural and Food Chemistry, 31*, 823–826.

Buttery, R. G., Ling, L. C., Juliano, B., & Turnbaugh, J. G. (1983). Cooked rice aroma and 2-acetyl-1-pyrroline. *Journal of Agricultural and Food Chemistry, 31*, 823–826.

Chaut, A. T., Yutaka, H., & Vo, C. T. (2010). *Genetic Analysis for the Fragrance of Aromatic Rice Varieties*. 3rd International Rice Congress.

Chen, M., Wei, X., Shao, G., Tang, S., Luo, J., & Hu, P. (2012). Fragrance of the rice grain achieved via artificial microRNA-induced down-regulation of OsBADH2. *Plant Breeding, 131*, 584–590.

Chen, S., Wu, J., Yang, Y., Shi, W., & Xu, M. (2006). The FGR gene responsible for rice fragrance was restricted within 69 kb. *Plant Science, 171*, 505–514.

Chen, S., Yang, Y., Shi, W., Ji, Q., He, F., Zhang, Z., Cheng, Z., Liu, X., & Xu, M. (2008). Badh2, encoding betaine aldehyde dehydrogenase, inhibits the biosynthesis of 2-acetyl-1-pyrroline, a major component in rice fragrance. *The Plant Cell Online, 20*, 1850–1861.

Cordeiro, G. M., Christopher, M. J., Henry, R. J., & Reinke, R. F. (2002). Identification of microsatellite markers for fragrance in rice by analysis of the rice genome sequence. *Molecular Breeding, 9*, 245–250.

FAO (2016). FAO Rice Market Monitor (RMM). volume XIX, Issue No. 4 December 2016, Food and Agriculture Organization of the United Nations, Rome, Italy. URL: http://www.fao.org/economic/est/publications/rice-publications/rice-market-monitor-rmm/en/. Accessed on 15 July 2016.

Fernie, A. R., & Schauer, N. (2009). Metabolomics-assisted breeding: a viable option for crop improvement? *Trends in Genetics, 25*, 39–48.

Fitzgerald, M. A., McCouch, S. R., & Hall, R. D. (2009). Not just a grain of rice: the quest for quality. *Trends in Plant Science, 14*, 133–139.

Fukushima, A., & Kusano, M. (2014). A network perspective on nitrogen metabolism from model to crop plants using integrated "omics" approaches. *Journal of Experimental Botany, 65*, 5619–5630.

Gao, F. Y., Lu, X. J., Wang, W. M., Sun, S. S., Li, Z. H., Li, H. J., & Ren, G. J. (2009). Trait-specific improvement of a cytoplasmic male-sterile line using molecular marker-assisted selection in rice. *Crop Sci.,* 99–106.

Garland, S., Lewin, L., Blakeney, A., Reinke, R., & Henry, R. (2000). PCR-based molecular markers for the fragrance gene in rice (*Oryza sativa. L.*). *TAG Theoretical and Applied Genetics, 101*, 364–371.

Garris, A. J., Tai, T. H., Coburn, J., Kresovich, S., & McCouch, S. (2005). Genetic structure and diversity in Oryza sativa L. *Genetics, 169*, 1631–1638.

Giraud, G. (2013). The world market of fragrant rice, main issues and perspectives. *International Food and Agribusiness Management Review, 16*(2), 1–20.

Glaszmann, J. (1987). Isozymes and classifications of Asian rice varieties. *Theoretical and Applied Genetics, 74*, 21–30.

Grimm, C. C., Bergman, C., Delgado, J. T., & Bryant, R. (2001). Screening for 2-acetyl-1-pyrroline in the headspace of rice using SPME/GC-MS. *Journal of Agricultural and Food Chemistry, 49*, 245–249.

Hashemi, F. S. G., Rafii, M. Y., Ismail, M. R., Mahmud, T. M. M., Rahim, H. A., Asfaliza, R., Malek, M. A., & Latif, M. A. (2013). Biochemical, Genetic and molecular advances of fragrance characteristics in rice. *Critical Reviews in Plant Sciences, 32*, 445–457.

Hien, N., & Yoshihashi, T. (2006). Evaluation of aroma in rice (*Oryza sativa* L.) using KOH method, molecular markers and measurement of 2-acetyl-1-pyrroline concentration. *Jpn J. of Trop. Agri., 50*, 190–198.

Izumi, Y., Tetsuya, Y., Mikio, N., Hidemasa, S., & Tsutomu, H. (1978). Volatile flavor components of rice cakes. *Agri. Boil. Chem., 42*, 1229–1233.

Jairin, J., Teangdeerith, S., Leelagud, P., Kothcharerk, J., Sansen, K., Yi, M., Vanavichit, A., & Toojinda, T. (2009). Development of rice introgression lines with brown planthopper resistance and KDML105 grain quality characteristics through marker-assisted selection. *Field Crop Res.*, 263–271.

Jantaboon, J., Siangliw, M., Im-mark, S., Jamboonsri, W., Vanavichit, A., & Toojinda, T. (2011). Ideotype breeding for submergence tolerance and cooking quality by marker-assisted selection in rice. *Field Crops Research, 123*, 206–213.

Jin, L., Lu, Y., Shao, Y., Zhang, G., Xiao, P., Shen, S., Corke, H., & Bao, J. (2010a). Molecular marker assisted selection for improvement of the eating, cooking and sensory quality of rice (*Oryza sativa* L.). *Journal of Cereal Science, 51*, 159–164.

Jin, Q., Waters, D., Cordeiro, G. M., Henry, R. J., & Reinke, R. F. (2003). A single nucleotide polymorphism (SNP) marker linked to the fragrance gene in rice (*Oryza sativa* L.). *Plant Science, 165*, 359–364.

Jin, T., Chang, Q., Li, WangFeng, Y., DongXu, L., Zhijian, W., Deli, L., Bao, & Liu, L. (2010b). Stress-inducible expression of GmDREB1 conferred salt tolerance in transgenic plant cell tissue and organ culture. 219–227.

Kawahara, Y., De la Bastide, M., Hamilton, J. P., Kanamori, H., McCombie, W. R., Ouyang, S., Schwartz, D. C., Tanaka, T., Wu, J., Zhou, S., Childs, K. L., Davidson, R. M., Lin, H., Quesada-Ocampo, L., Vaillancourt, B., Sakai, H., Lee, S. S., Kim, J., Numa, H., Itoh, T., Buell, C. R., & Matsumoto, T. (2013). Improvement of the *Oryza sativa* Nipponbare reference genome using next generation sequence and optical map data. *Rice, 6*, 4. URL: http://rice.plantbiology.msu.edu/index.shtml. Accessed on 15 July 2016.

Kovach, M. J., Calingacion, M. N., Fitzgerald, M. A., & McCouch, S. R. (2009). The origin and evolution of fragrance in rice (Oryza sativa L.). *Proceedings of the National Academy of Sciences of the United States of America, 106*, 14444–14449.

Londo, J. P., Chiang, Y. C., Hung, K. H., Chiang, T. Y., & Schaal, B. A. (2006). Phylogeography of Asian wild rice, Oryza rufipogon, reveals multiple independent domestications of cultivated rice, *Oryza sativa*. *Proceedings of the National Academy of Sciences of the United States of America, 103*, 9578–9583.

Lorieux, M., Petrov, M., Huang, N., Guiderdoni, E., & Ghesquiere, A. (1996). Aroma in rice: genetic analysis of a quantitative trait. *Theor. Appl. Genet., 93*, 1145–1151.

Maga, J. A. (1984). Rice product volatiles: A review. *Journal of Agricultural and Food Chemistry, 32*, 964–970.

Matsumoto, T., Wu, J., Itoh, T., Numa, H., Antonio, B., & Sasaki, T. (2016). The Nipponbare genome and the next-generation of rice genomics research in Japan. *Rice 9*, 33.

Myint, K., Courtois, B., Risterucci, A. M., Frouin, J., Soe, K., Thet, K., Vanavichit, A., & Glaszmann, J. C. (2012). Specific patterns of genetic diversity among aromatic rice varieties in Myanmar. *Rice, 5*, 20.

NBPR (2016). *Detail of Gene. Oryzabase*: Integrated rice science database sponsored by national bioresource project rice. URL: http://shigen.nig.ac.jp/rice/oryzabase/gene/detail/829. Accessed on 15 July 2016.

Niu, X., Tang, W., Huang, W., Ren, G., Wang, Q., Luo, D., Xiao, Y., Yang, S., Wang, F., Lu, B. R., Gao, F., Lu, T., & Liu, Y. (2008). RNAi-directed downregulation of OsBADH2 results in aroma (2-acetyl-1-pyrroline) production in rice (Oryza sativa L.). *BMC Plant Biology, 8*, 100.

Pandey, M. K., Shobha Rani, N., Sundaram, R. M., Laha, G. S., Madhav, M. S., Srinivasa Rao, K., Sudharshan, I., Hari, Y., Varaprasad, G. S., Subba Rao, L. V., Suneetha, K.,

Sivaranjani, A. K. P., & Viraktamath, B. C. (2013). Improvement of two traditional Basmati rice varieties for bacterial blight resistance and plant stature through morphological and marker-assisted selection. *Molecular Breeding, 31,* 239–246.

Paule, C. M., & Powers, J. J. (1989). Sensory and chemical examination of aromatic and nonaromatic rices. *Journal of Food Science, 54,* 343–346.

Pengkumsri, N., Chaiyasut, C., Saenjum, C., Sirilun, S., Peerajan, S., Suwannalert, P., Sirisattha, S., & Sivamaruthi, B. S. (2015). Physicochemical and antioxidative properties of black, brown and red rice varieties of northern *Thailand, 35,* 331–338.

Pengkumsri, N., Chaiyasut, C., Sivamaruthi, B. S., Saenjum, C., Sirilun, S., Peerajan, S., Suwannalert, P., Sirisattha, S., Chaiyasut, K., & Kesika, P. (2015). The influence of extraction methods on composition and antioxidant properties of rice bran oil. *Food Science and Technology (Campinas), 35,* 493–501.

Prathepha, P. (2009). The fragrance (fgr) gene in natural populations of wild rice (Oryza rufipogon Griff.). *Genetic Resources and Crop Evolution, 56,* 13–18.

Sakthivel, K., Rani, N. S., Pandey, M. K., Sivaranjani, A. K. P., Neeraja, C. N., Balachandran, S. M., Madhav, M. S., Viraktamath, B. C., Prasad, G. S. V., & Sundaram, R. M. (2009). Development of a simple functional marker for fragrance in rice and its validation in Indian Basmati and non-Basmati fragrant rice varieties. *Molecular Breeding, 24,* 185–190.

Sakthivel, K., Sundaram, R. M., Shobha Rani, N., Balachandran, S. M., & Neeraja, C. N. (2009). Genetic and molecular basis of fragrance in rice. *Biotechnology Advances, 27,* 468–473.

Salgotra, R. K., Gupta, B. B., Millwood, R. J., Balasubramaniam, M., & Stewart, C. N. (2012). Introgression of bacterial leaf blight resistance and aroma genes using functional marker-assisted selection in rice (Oryza sativa L.). *Euphytica, 187,* 313–323.

Schwab, W., Davidovich-Rikanati, R., & Lewinsohn, E. (2008). Biosynthesis of plant-derived flavor compounds. *Plant Journal, 54,* 712–732.

Second, B. Y. G. (1982). *Origin of the Genic Diversity.*

Sekhar, B. P. S., & Reddy, G. M. (1982). Amino acid profiles in some scented rice varieties, *37,* 35–37.

Shan, Q., Zhang, Y., Chen, K., Zhang, K., & Gao, C. (2015). Creation of fragrant rice by targeted knockout of the OsBADH2 gene using TALEN technology. *Plant Biotechnology Journal, 2,* 791–800.

Shi, W., Yang, Y., Chen, S., & Xu, M. (2008). Discovery of a new fragrance allele and the development of functional markers for the breeding of fragrant rice varieties. *Molecular Breeding, 22,* 185–192.

Singh, R., Singh, A. K., Sharma, T. R., Singh, A., & Singh, N. K. (2012). Fine mapping of grain length QTLs on chromosomes 1 and 7 in Basmati rice (Oryza sativa L.). *Journal of Plant Biochemistry and Biotechnology, 21,* 157–166.

Taylor, P., Kinnersley, A. M., & Turano, F. J. (2010). Critical reviews in plant sciences Gamma Aminobutyric Acid (GABA) and plant responses to stress Gamma Aminobutyric Acid (GABA) and plant responses to stress, *2689,* 37–41.

Toryama, K. (2005). Rice is life: scientific perspectives for the 21st century. *Proceedings of the World Rice Research Conference,* Tsukuba, Japan. International Rice Research Institute (IRRI), Los Banos, Laguna, Philippines; Japan International Research Center for Agricultural Sciences (JIRCAS), Tsukuba, Japan. CD-ROM: *590.*

Tsugita, T., Kurata, T., Kato, H., Kurita, T., & Kato, H. (1980). Volatile components after cooking rice milled to different degrees. *Agricultural Biological Chemistry, 44,* 835–840.

Turano, F. J., Thakkar, S. S., Fang, T., & Weisemann, J. M. (1997). Characterization and expression of NAD (H)-dependent glutamate dehydrogenase genes in Arabidopsis. *Plant Physiology, 113,* 1329–1341.

Vanavichit, A., & Yoshihashi, T. (2010). Molecular aspects of fragrance and Aroma in rice. *Advances in Botanical Research, 56,* 50–73.

Vanavichit, A., Kamolsukyurnyong, W., Wanchana, S., Wongpornchai, S., Ruengphayak, S., Toojinda, T., & Tragoonrung, S. (2004). Discovering genes for rice grain aroma. *1st International Conference on Rice for the Future.* Kasetsart University, Bangkok, Thailand, 71–80.

Vanavichit, A., Tragoonrung, S., Theerayut, T., Wanchana, S., & Kamolsukyunyong, W. (2015). Transgenic rice plants with reduced expression of Os2AP and elevated levels of 2-acetyl-1-pyrroline. US Patent (US 9,057,049 B2), pp. 1–27. URL: https://patentimages.storage.googleapis.com/9a/5b/3f/7d513881a556af/US9057049.pdf.

Verma, D. K., Mohan, M., & Asthir, B. (2013). Physicochemical and cooking characteristics of some promising basmati genotypes. *Asian Journal of Food Agro-Industry, 6*(2), 94–99.

Verma, D. K., Mohan, M., Prabhakar, P. K., & Srivastav, P. P. (2015). Physico-chemical and cooking characteristics of Azad basmati. *International Food Research Journal, 22*(4), 1380–1389.

Verma, D. K., Mohan, M., Yadav, V. K., Asthir, B., & Soni, S. K. (2012). Inquisition of some physico-chemical characteristics of newly evolved basmati rice. *Environment and Ecology, 30*(1), 114–117.

Wakte, K., Zanan, R., Hinge, V., Khandagale, K., Nadaf, A., & Henry, R. (2016). Thirty-three years of 2-acetyl-1-pyrroline, a principal basmati aroma compound in scented rice (*Oryza sativa* L.): a status review. *Journal of the Science of Food and Agriculture, 97*(2), 384–395.

Wanchana, S., Kamolsukyunyong, W., Ruengphayak, S., Toojinda, T., Tragoonrung, S., & Vanavichit, A. (2005). <Title/>. *Science Asia, 31,* 299.

Wang, C., Chen, S., & Yu, S. (2011). Functional markers developed from multiple loci in GS3 for fine marker-assisted selection of grain length in rice. *Theoretical and Applied Genetics, 122,* 905–913.

Wang, X. S., Zhu, H. B., Jin, G. L., Liu, H. L., Wu, W. R., & Zhu, J. (2007). Genome-scale identification and analysis of LEA genes in rice (*Oryza sativa* L.). *Plant Science, 172,* 414–420.

Win, K. M., Korinsak, S., Jantaboon, J., Siangliw, M., Lanceras-Siangliw, J., Sirithunya, P., Vanavichit, A., Pantuwan, G., Jongdee, B., Sidhiwong, N., & Toojinda, T. (2012). Breeding the Thai jasmine rice variety KDML105 for non-age-related broad-spectrum resistance to bacterial blight disease based on combined marker-assisted and phenotypic selection. *Field Crops Research, 137,* 186–194.

Yi, M., Nwe, K. T., Vanavichit, A., Chai-arree, W., & Toojinda, T. (2009). Marker assisted backcross breeding to improve cooking quality traits in Myanmar rice cultivar Manawthukha. *Field Crops Research, 113,* 178–186.

Yoshihashi, T., Nguyen, T. T. H., & Kabaki, N. (2004). Area dependency of 2-acetyl-1-pyr-roline content in an aromatic rice variety, Khao Dawk Mali 105. *Japan Agricultural Research Quarterly, 38*, 105–109.

Zhang, Y., Shan, Q., Wang, Y., Chen, K., Liang, Z., Li, J., Zhang, Y., Zhang, K., Liu, J., Voytas, D. F., Zheng, X., & Gao, C. (2013). Rapid and efficient gene modification in rice and brachypodium using TALENs. *Molecular Plant, 6*, 1365–1368.

Zhou, P. H., Tan, Y. F., He, Y. Q., Xu, C. G., & Zhang, Q. (2003). Simultaneous improvement for four quality traits of Zhenshan 97, an elite parent of hybrid rice, by molecular marker-assisted selection. TAG. Theoretical and applied genetics. *Theoretische Und Angewandte Genetik, 106*, 326–331.

CHAPTER 11

PROSPECTIVE FUTURE OF AROMA, FLAVOR AND FRAGRANCE IN THE RICE INDUSTRY

DEEPAK KUMAR VERMA[1] and PREM PRAKASH SRIVASTAV[2]

[1]*Research Scholar, Agricultural and Food Engineering Department, Indian Institute of Technology, Kharagpur–721302, West Bengal, India, Mobile: +91-7407170260, +91 9335993005, Tel.: +91-3222-281673, Fax: +91-3222-282224, E-mail: deepak.verma@agfe.iitkgp.ernet.in, rajadkv@rediffmail.com*

[2]*Associate Professor, Agricultural and Food Engineering Department, Indian Institute of Technology, Kharagpur–721302, West Bengal, India, Mobile: +91-9434043426, Tel.: +91-3222-283134, Fax: +91-3222-282224, E-mail: pps@agfe.iitkgp.ernet.in*

Aroma is considered by rice consumers in India and the world as the third highest desired trait after taste and elongation after cooking. This desired trait is one of the critical aspects of rice quality, which determines the acceptance or rejection of rice before it is tasted. It is also considered as an important property of rice that indicates the high quality and price in the international market. From an assessment of all known data, it is shown that there are around 500 volatile aroma chemical compounds, which have been documented so far in various aromatic and non-aromatic rice cultivars throughout the world. Does the rice industry have a future in volatile aroma compound analysis? It would be very surprising for the editors of the entitled book "Science and Technology of Aroma, Flavor, and Fragrance in Rice" and all the authors and contributors of the various chapters. Perhaps, we should look back on rice aroma compounds up to at least four decades, so that we cannot likely stop to say that the rice industries have broad and prospective future in volatile aroma chemical compound analysis. On the basis of the collected literature, it can be concluded that there are various extraction technologies for volatile aroma compound analysis from the rice sample, but presently, researchers and scientists who are working in this area are continuously looking for economic and environment-friendly feasible technology. There are various extraction technologies, but constant efforts are being made in the development of new analytical methods for rice volatile aroma compounds analysis. In traditional methods, large quantities of solvent and time are consumed. These methods cause additional environmental problems as well as increase operating costs due to use of a large amount of solvent. Besides these traditional and normalized methods, several modern novel technologies for extraction of rice volatile aroma compounds, such as solid phase micro extraction (SPME) and supercritical fluid extraction (SFE), appear as an alternative to conventional and traditional methods of extraction coupled with modern instruments, viz. gas chromatography–mass spectrometry (GC-MS), gas chromatography–olfactometry (GC-O), gas chromatography–flame ionization detector (GC-FID), gas chromatography–pulsed flame photometric detection (GC-PFPD), etc., which are partially or fully automated. These modern novel technologies have a great potential and offer advantages with respect to solvent consumption, extraction time and yield, and reproducibility. However, most of these modern novel extraction technologies are still conducted successfully for the determination of volatile aroma compounds from rice sample at the laboratory, although very few applications like SFE are available for rice aroma extraction at the industrial level. More research is needed to exploit applications of these modern novel extraction technologies, which can be characterized by low limits of detection and quantification, and high precision and accuracy of the obtained results. This chapter addresses the importance of rice aroma, exploration of its modern and novel extraction technology, and research opportunities in the development of innovative strategies that can be moved from hope to reality in the field of aromatic rice industries in the future.

GLOSSARY OF TECHNICAL TERMS

2-acetyl-1-pyrroline: 2-Acetyl-1-pyrroline (2AP) is one of the aroma volatile compounds and the key aroma compound in aromatic rice varieties responsible for the delicious popcorn-like fragrance.

Abiotic stress: Natural external stimuli other than living organisms which can limit or restrict plant growth and development, such as unfavorable soil conditions, drought, flooding, extreme temperatures, etc.

Adaptation: The dynamic evolutionary process of becoming suited to new or different environmental conditions to perform specific functions.

Agricultural productivity: The rate of production of crop per unit area per unit time.

Agriculture: Art and science of growing crops and the raising of animals for food, other human needs, or economic gain.

Agronomic efficiency: Units of crop produced per unit input (nutrient) added.

Agronomic traits: The traits relating to or promoting agronomy in a plant. This includes high yield, grain weight, soil condition, disease resistance etc.

Agronomy: Branch of agricultural science which deals with the management of field to provide favorable environment to the crop for higher productivity in terms of quantity and quality both.

Allele frequency: It refers to frequency of a given allele in a population.

Allele: Alternative form of the gene found on the same position of homologous chromosome of organism.

Antisense: Having a sequence of nucleotides complementary to (and hence capable of binding to) a coding (or sense) sequence, which may be either that of the strand of a DNA double helix which undergoes transcription, or that of a messenger RNA molecule.

Aroma: Aroma also called as fragrance, odorant. It is a chemical compound that has a smell.

Aromatic rice: Aromatic or scented or fragrant rices are the special rice varieties that emit delicious fragrance when cooked.

Association mapping: Association mapping is a method of mapping quantitative trait loci (QTLs) using historical meiotic recombination events performed over several generations to link phenotypes with genotypes in large germplasm populations.

Background selection: It describes the loss of genetic diversity at a non-deleterious locus due to negative selection against linked deleterious alleles.

Binary vector: These are shuttle vector that has the ability to replicate in multiple hosts. Used for *Agrobacterium* mediated gene transfer in plants.

Biochemistry: The branch of science concerned with the chemical and physico-chemical processes and substances which occur within living organisms.

Bioinformatics: The discipline encompassing the development and utilization of computational facilities to store, analyze, and interpret biological data.

Biotic Stress: Living organisms, which can colonize, weaken and restrict the growth and yield of plants and include viruses, fungi, bacteria, and other harmful insects, birds, etc.

Bombardment: A techniques for insertion of naked DNA into explants or plant tissue using a special machine.

Breeding: A process of genetic-crossing between two selected male and female parents for the production of offspring progeny.

Candidate-Gene Based Association Mapping: In this approach, only selected candidate genes are targeted for mapping which has specific role in controlling the desired trait of interest.

cDNA: The cDNA or complementary DNA is a double stranded DNA known to be synthesized, or manufactured from an mRNA or messenger RNA template, via a reaction catalyzed by the enzyme reverse transcriptase. cDNA is often used to clone eukaryotic genes in prokaryotes.

Cell proliferation: Cell proliferation is the process that results in an increase of the number of cells, and is defined by the balance between cell divisions and cell loss through cell death or differentiation.

Cell signaling: Cell signaling is part of a complex system of communication that governs basic activities of cells and coordinates cell actions.

Climate change: It may refer to a change in average weather conditions, statistical distribution of weather patterns or in the time variation of weather around longer-term average conditions.

Climate resilient farming: This includes development of strategies to boost up crop production under climatic variability.

Climatic variability: It refers to the sum-total of events which represents changes in the monsoonal pattern and prevalent occurrences of climatic extremes.

Comparative mapping: Once the two species have been mapped, the relative evolution of the two species can be compared.

Competent cell: Cell that have ability to take up a foreign gene or plasmid.

Degradation: The condition, process or phenomena of degrading (changing to a lower state) or being degraded, destroyed and/or spoiled.

Drought stress: Prolonged water deficit conditions in plants leads to cause the damage in photosystems.

Drought tolerance: Plant species that able to growth and proliferate in environment with shortage of water or water scarcity.

Drought: An environmental phenomenon associated with minimum rainfall and acute water shortage.

Eco-friendly technologies: Technological process that promote development in an Environmental compatible manner.

Encoding: Sequence of gene that represent or might be translate to the respective amino acid sequence or peptide.

Environment: The sum-total of all physical, chemical and biological components surrounding human.

Enzyme: A protein molecules produced by a living organism which acts as a catalyst to bring about a specific biochemical reaction.

Exogenous: Chemicals or substances that came from outside of the cells.

Expression: Protein or chemical that synthesis or released from a sequence of gene.

Flavonoids: A large class of plant pigments.

Food security: Physical and economic access to safe, healthy and nutritious food by the general public at all the times to fulfill its nutritional demands for a healthy life.

Gas chromatography: Gas chromatography is a common chromatography technique used in analytical chemistry for separating and analyzing the compounds that can be vaporized without decomposition.

Gene editing: It is also called as genome editing, is a type of genetic engineering in which DNA is inserted, deleted or replaced in the genome of an organism using engineered nucleases or molecular scissors.

Gene ontology analysis: It unifies the representation of gene and describes the gene function. Especially, maintain and develop its controlled vocabulary of gene and gene product attributes.

Gene silencing: Switching off of the expression of gene at transcriptional or at translational levels.

Gene: Small part of DNA that transcribe to produce RNA molecule and finally express their effect.

Genetic diversity: It is the total number of genetic characteristics in the genetic makeup of a species.

Genetic engineering: Genetic engineering can be defined as the deliberate modification of the characteristics of an organism by manipulating its genetic material.

Genetic inbreeding: Process of exchanging genetic material between individuals of same species.

Genetic loci: Specific location or position of a gene's DNA sequence, on a chromosome.

Genetic transformation: The process by which the foreign genetic material introduced into the plant nuclear genome. The two most common methods employed for plant genetic transformation are direct (micro-projectile bombardment) and indirect (*Agrobacterium*-mediated transformation) genetic transformation.

Genetic variability: It is the measure of the extent to which the individual genotypes in a population vary for a given trait.

Genome: A complete single set of the genetic material of a cell or organism; the complete set of genes in a gamete; the single DNA/RNA molecule of bacteria, phages, and most animal and plant viruses. In plants, it is composed of the nuclear genome, the mitochondrial genome, and the chloroplast genome.

Genomics: Genomics is a discipline in genetics that applies recombinant DNA, DNA sequencing methods, and bioinformatics to sequence, assemble, and analyze the function and structure of genomes.

Genotype: It is the genetic constitution of the germplasm of an organism.

Germpalsm: The living genetic resources like seeds or tissue that is maintained for breeding, preservation, and other research purposes for plants and animals.

Global warming: Global phenomena involving average increase in the temperature of the earth surface.

Glycophytes: This is a class of plants that will grow healthily only in a soil with low concentration of sodium chloride or other sodium salts.

Grain shape: The physical shape of the rice grain which is defined by its dimensions, i.e., round, slender, etc.

Grain weight: The weight of the grain often recorded as grams per 1,000 grains.

Grain width: The measurement of the grain across the broadest section.

Grain yield: Weight of harvested grain expressed as tons/ha with a moisture content of 14%.

Grain: Syn. rough rice, paddy, padi, caryopsis, seed). (i) A fruit in which the pericarp is fused with the seed. (ii) The ripened ovary and its associated structures such as the lemma, palea, rachilla, sterile lemmas, and the awn if present.

Gramineae (Poaceae): A large and nearly ubiquitous family of monocotyledonous flowering plants known as grasses. It includes the cereal grasses, bamboos and the grasses of natural grassland and cultivated lawns (turf) and pasture.

Headspace solid-phase microextraction (HS-SPME): Solvent-free sampling technique by which the volatile compounds absorbed to a fiber coated with polymer introduced into the headspace, subsequently desorbed into a gas chromatography for analysis.

Heat stress: Heat stress often is defined as where temperatures are hot enough for sufficient time that they cause irreversible damage to plant function or development. Plants can be damaged in different ways by either high day or high night temperatures and by either high air or high soil temperatures.

Heterosis breeding: A method of breeding to develop Fl hybrids which exploit the phenomenon of hybrid vigor to increase yield potential and yield stability.

High yielding varieties (HYV): Varieties of crop species having higher potential for crop production under various situation.

Histochemistry: The study of identification of chemical components within tissues using histological techniques such as staining, microscopic examinations etc.

Homologous gene: These are partly homologous genes possibly developed from a common ancestor.

Homeostasis: Is the ability to maintain a constant internal environment in response to environmental changes.

Hybrid rice: Hybrid rice is the commercial rice crop obtained from F_1 seeds of cross between two genetically dissimilar parents, produced by crossing two inbreeds – genetically fixed varieties of a particular crop.

Hybrid vigor: The increase in vigor of hybrids over their parental inbred types; also known as heterosis.

Immobilization: The conversion of an element from inorganic to organic combination in microbial or plant tissues. This has the effect of rendering unavailable (and usually not readily soluble) an element that previously was directly available to plants.

Indica rice: One of the two major eco-geographical races of *Oryza sativa* (see also japonica). This is major type of rice grown in the tropics and subtropics. It has broad to narrow, light green leaves and tall to intermediate plant stature (except for the semi-dwarf). Indica plants tiller profusely, grains are long to short, slender, somewhat flat, and awnless. Grains shatter easily and have 23–31% amylose content. They grow mostly in the Philippines, India, Pakistan, Java, Sri Lanka, Indonesia, central and southern China, African countries, and other tropical regions.

Indo-Gangetic plains: It is known as the Indus-Ganga and the North Indian River Plain encompassing most of northern and eastern parts of India, the eastern parts of Pakistan, and almost all of Bangladesh.

Integration: A condition when the inserted foreign gene is correctly ligate or link with the host genome.

Intrinsic markers: Internal signal that presence or derived from host family or itself.

Introns: It is a non-coding section of an RNA transcript.

In-vitro **callus:** A mass of undifferentiated plant cells induced by hormone treatment in tissue culture under controlled conditions.

In-vitro **plant regeneration:** A morphogenetic response to a stimulus that results in the entire plant generation from a single or group of cells under a controlled condition.

Isoelectric point: The isoelectric point (pI, pH(I), IEP), is the pH at which a particular molecule carries no net electrical charge in the statistical mean.

Japonica rice: Also known as temperate japonica, is one of the two major eco-geographical races of *O. sativa* (see also indica). A group of rice varieties from northern and eastern China and are grown extensively in some areas of the world. It has narrow, dark green leaves, medium-height tillers and short to intermediate plant stature. It is found in the cooler subtropical and temperate zones. Grains are short, roundish, spikelets awnless to long-awned, panicles low-shattering, and have 0–20% amylose content.

Javanica rice: Also known as tropical japonica, is a designation for the bulu and gundil varieties of Indonesia and many upland rices. They belong to the japonica race of *O. sativa*. Rice varieties are with broad, stiff, light green leaves. It is low-tillering and has a tall plant stature. Grains are long, broad, and thick, awned or awnless. Javanica grains are low-shattering and have 0–25% amylose content.

Landrace: Old cultivated form of a crop which are potentially adapted to local growing conditions and have wider genetic base but unimproved by modern plant breeding methods.

Linkage disequilibrium: Non-random association of alleles between genetic loci on the same or different chromosome.

Linkage: Inheritance of two or more genes together as a single haplotype without any substantial recombination frequency in a family or pedigree.

Lipid peroxidation: Oxidative degradation of lipids. It is the process in which free radicals causes release of electrons from the lipids in cell membranes, resulting in cell damage. This process proceeds by a free radical chain reaction mechanism.

Liquid chromatography: It is an analytical chromatographic technique used for separation of molecules dissolved in the solvent.

Mapping: Process used to identify the locus of a gene/QTL and the distances between genes/markers/QTLs in the genome.

Marker: Polymorphic sequence of nucleotide in the genome which can be easily identifiable and detectable.

Marker-assisted selection: It is an indirect selection process where a trait of interest is selected based on a marker linked to a trait of interest rather than on the trait itself.

Mass spectroscopy: It is an analytical spectroscopic technique to determine mass of the substance by measuring the mass to charge (m/z) of its ion.

Metabolic activity: It refers to the physiological process by means of which living dry weight of an organism is increased or decreased.

Metabolome: The total number of metabolites present within an organisms, cell or tissue.

Metabolomics: The total number of metabolites present within an organisms, cell or tissue.

Microarray: A set of DNA sequences representing the entire set of genes of an organism, arranged in a grid pattern for use in genetic testing.

microRNA: microRNAs (miRNAs) are small sized, non-coding, single stranded riboregulator RNAs that play critical role in post-transcriptional regulations of gene expression in response to various stress environments. They are 20 to 24 nt long molecules abundant in higher organisms.

Molecular biology: The branch of biology that deals with the structure and function of the macromolecules (e.g. proteins and nucleic acids) essential to life.

Molecular breeding: production of new clone using genetic engineering methods.

Molecular marker: It is refers to a stretch of DNA or RNA that exhibits sequence polymorphism when assayed across individuals.

Multigenic: Group of genes with similar nucleotide sequences and various functional responses.

Next generation sequencing: These are revolutionized non-Sanger based high throughput DNA sequencing technologies used to sequence DNA and RNA much more quickly and cheaply.

Non-coding RNA: It is a functional RNA molecule that is transcribed from DNA but not translated into proteins.

Non-reducing sugars: A carbohydrate that is not oxidized by a weak oxidizing agent (an oxidizing agent that oxidizes aldehydes but not alcohols, such as the Tollen's reagent) in basic aqueous solution, e.g., sucrose.

Nucleic acids: Complex organic substance present in living cells, especially DNA or RNA, whose molecules consist of many nucleotides linked in a long chain.

Omics: Omics is a generalized term for broad science avenues for analyzing the interaction of biological information in various 'omes' of the biological systems including genome, proteome, metabolome and transcriptome etc.

Orthologs: Genes in different species that evolved from a common ancestral gene by speciation. Normally, orthologs retain the same function in the course of evolution.

Oxidative stress: It is state of imbalance between the production of free radicals and the ability of the organism to detoxify their effects through neutralization by antioxidant production.

Peptide fingerprinting: It is an analytical technique for protein identification in which the unknown protein of interest is first cleaved into smaller peptides, whose absolute masses can be accurately measured with a mass spectrometer.

Photorespiration: A respiratory process in many higher plants by which they take up oxygen in the light and give out some carbon dioxide, contrary to the general pattern of photosynthesis.

Photosynthate: The amount of glucose produced by green plants during photosynthesis per unit area per unit time.

Photosynthesis: It is the biochemical process of production of glucose in plants with due presence of sunlight, water and chlorophyll.

Physiological pathways: A sequence of enzymatic or other reactions by which one biological material is converted to another.

Physiology: The branch of biology that deals with the study of mechanism and metabolism that are involved in growth and development of living organisms.

Plant type: A set of plant characters (e.g., tillering, leaf and panicle characteristics, plant height) that give a variety its peculiar architecture and geometry.

Pollution: Introduction of contaminants and or pollutants into the natural environment that may cause adverse effects/change.

Polyunsaturated fatty acid (PUFAs): Fatty acids that contain more than one double bond in their backbone. This class includes many important compounds, such as essential fatty acids and those that give drying oils their characteristic property.

Population structure: The level of genetic differentiation among groups within sampled population individuals.

Promoter: These are DNA sequences located near the transcription start sites in the upstream that initiates transcription of a particular *gene*.

Protein sorting: It is the biological mechanism by which proteins are transported to the appropriate destinations in the cell or outside of it.

Proteome: The entire complement of proteins that is or can be expressed by a cell, tissue, or organism.

Proteomics: The entire complement of proteins that is or can be expressed by a cell, tissue, or organism

QTL mapping: It is to determine the complexity of the genetic architecture underlying a phenotypic trait.

Qualitative character: A character which is simply inherited and exhibits discrete variation in a population. A discrete heritable character that is transmitted with well-defined limits and in a simple alternate manner.

Quantitative character: A heritable character that follows a continuous variation in a population and exhibits high Genotype x Environment interaction.

Quantitative trait locus (QTL): A genomic region which governs a quantitative trait.

Recessive gene: It is a gene that gets trumped by a dominant gene.

Regulatory genes: These are genes that regulate the expression of one or more structural genes by controlling the production of protein which regulates their rate of transcription.

Resistance: In plants, it is an attempt to prevent something by action.

Ripening period: The time after anthesis during which the fruits plant become soft textured, and accumulate soluble sugars, pigments and aroma volatiles.

RNA fingerprinting: A new method to screen for differences in plant litter degrading microbial communities.

Salinity stress: Increasing the saline concentration to soil or water. It adversely affect plant growth and development.

Salinity tolerance: Explants or plants that able to growth and reproduce under extreme or high salinity levels.

Sensory analysis: Organoleptic sensory analysis is one of the simple detection methods involve smelling of the sample after physical or chemical treatment.

Signal transduction: It converts a stimulus into a response in the cell. There are two stages in this process: A signaling molecule attaches to a receptor protein on the cell membrane. A second messenger transmits the signal into the cell, and a change takes place in the cell.

Signaling molecule: Molecules which interact with a target cell as a ligand to cell surface receptors, and/or by entering into the cell through its membrane or by endocytosis for signaling.

Single Sequence Repeat (SSR): Mono-, di-, tri-, tetra-, penta-, or hexa-nucleotide tandem repeats which are hypervariable and present throughout the genome in copies ranging from 100 to 1000.

Soil fertility: Soil fertility refers to inherent capacity of a soil to supply nutrients to plants in adequate amount and in suitable proportion.

Soil health: Capacity of soil to function continuously as a vital living eco-system that sustains plants, animals and humans.

Soil productivity: It is the present capacity of a soil to produce crop yield under a defined set of management practices. It is measured in terms of the yield in relation to the input of production factors.

Soil: Soil is the uppermost weathered layer of the soil earth's crust it con-sists of rocks that have been more or less changed chemically together with the remains of plants and animals that live on it and in it.

Spikelet sterility: It is measurement; deal with counts of well-developed fertile spikelets in proportion to total number of spikelets on five panicles at harvest.

Stress: it refers to the deviation from the normal living condition due to various forms of biotic and abiotic factors.

Sustainable rice production: This refers to the production of rice in an environmental compatible way.

Thylakoid: These are membrane-bound compartments inside of the chloroplasts. Thylakoids are the epicenter for photosynthetic light-reactions. They contain the chlorophyll for the plant, which is the light-collecting pigment.

Tolerance: Plant that have capability to survive under extreme conditions.

Trait: A distinct variant of a phenotypic characteristic of an organism; it may be either inherited or determined environmentally.

Transcriptional factors: It is a protein that binds to specific DNA sequence, thereby controlling the rate of transcription of genetic information from DNA to messenger RNA.

Transcriptome: The sum total of all the messenger RNA molecules expressed from the genes of an organism.

Transcriptomics: The field of transcriptomics allows for the examination of whole transcriptome changes across a variety of biological conditions.

Transformation: Modification of plant genome, where the inserted foreign gene is stably integrated with host genome and expressed.

Transgene: Gene or genetic sequence that transferred into plant tissue using genetic engineering protocols.

Transgenic: These are genetically modified organisms whose germplasm has been altered to meet the desired need through Recombinant DNA technology.

Two-dimensional electrophoresis: It is protein gel electrophoretic technique in which proteins are separated and characterized according to the isoelectric points and molecular weights.

Unsaturated fatty acids: They are liquid a room temperature. Unsaturated fats are derived from plants and some animals. They contain at least one double bond in their fatty acid chain.

Vegetative phase: The period of growth between germination and flowering is known as the vegetative phase of plant development.

β-oxidation of fatty acids: It is the catabolic process by which fatty acid molecules are broken down in the cytosol in prokaryotes and in the mitochondria in eukaryotes to generate acetyl-CoA, which enters the citric acid cycle, and NADH and FADH2, which are co-enzyme used in electron transport chain.

INDEX

Milton Keynes UK
Ingram Content Group UK Ltd.
UKHW022045141024
449569UK00022B/814